# A Century Passing

## *Carnegie, Steel and the Fate of Homestead*



**Richard Krooth**

**University Press of America,® Inc.**
**Dallas · Lanham · Boulder · New York · Oxford**

4501 Forbes Boulevard
Suite 200
Lanham, Maryland 20706
UPA Acquisitions Department (301) 459-3366

PO Box 317
Oxford
OX2 9RU, UK


Printed in the United States of America
British Library Cataloging in Publication Information Available

**Library of Congress Cataloging-in-Publication Data**

Krooth, Richard.
A century passing : Carnegie, steel and the fate of Homestead /
Richard Krooth.
p. cm
Includes bibliographical references and index.
1. Homestead (Pa.)—History. 2. Homestead (Pa.)—Economic
conditions. 3. Homestead (Pa.)—Social conditions.
4. Carnegie, Andrew, 1835-1919. 5. Steel industry and trade—
Social aspects—Pennsylvania—Homestead—History.
6. Steel-works—Pennsylvania—Homestead—History.
7. Working class—Pennsylvania—Homestead—History. I. Title.

F159.H7 K76 2002
974.8'85—dc21 2002032036 CIP

ISBN: 978-0-7618-2415-2

™The paper used in this publication meets the minimum
requirements of American National Standard for Information
Sciences—Permanence of Paper for Printed Library Materials,

*To know*

*the shape of the world*

*As it was,*

*as it could be*

– Barbara Kingsolver

**Andrew Carnegie**
**(Circa 1892)**

# Contents

**Henry Clay Frick**
**(Circa 1892)**

# LIST OF TABLES AND FIGURES

## Tables:



## Figures:

**Proposed Monument to the Heroes of Homestead**
**(Credit: A. G. Burgoyne, 1892)**

# FOREWORD
## [Winter 1985]

Homestead, dead or dying, lights fewer homes each night. It's haunted, too, by mocking ghosts and near-forgotten names, bundled down from yesteryears, conjuring up spectral airs, shadows scurrying around its humble mews and silent streets as a winter night comes on. Monongahela, once the river of corn and fish down there in a tangle of rusting railyards and ancient fencing; Amity Farm, the kernel which once held the fate of this tragedy in microcosm; Poor Farm Lane, the Carnegie Library, only one floor now lighted to save a paltry sum, six books on a local history shelf getting little circulation. Homestead. Snug-sounding name, with its summer-green hills and broad river that might once have suggested warmth, stability and prosperity . . . a hundred years ago.

A graveyard, Homestead today lies between two other graveyards. At the highest points of the hill above the city and Munhall rest the neat slabs of stone and wild baroque images of the sleeping dead, Slavs on one side, Anglo-Irish and Germans on the other, looking off towards the new security and wealth of Pittsburgh or down the unsteady hill streets, Ann and the others, to the graveyard below, the empty steel mill.

On this plateau of deadness above the town, facing winter's icy gusts putting out the vital flame, one may roam free in the bone-chilling cold. Near the river, a more insidious winter-cold probes the body as security guards at the old mill's mausoleum door hound cameras away, demand passes. Agents of a dying empire, they plan to go down with every button polished, their rubber stamps, inked ready, all lined up in martial order.

The library, halfway up the hill, is halfway, too, lying between the honest dead of mill-life past and the embalmed, ever-lasting grimace of Carnegie's portraiture at ease. No, the guards are not Pinkertons, nor will one find an Emma Goldman elementary school, or for that matter a Henry Clay Frick anything in Homestead; but they who once sculpt the city's decades, the mill and much of Homestead life today seem more unknown than ever alive.

The fateful tale of my last Christmas in Homestead, yes, Dickens could have done it; but, then, exactly how? Would Carnegie be the Scrooge haunted by Frick, his alibi, unrelenting factory manager and captain of the Monongahela's coke supply? Who would be Tiny Tim - Carnegie's ten thousand exploited and maimed laborers, or those who are abandoning mortgage payments, homes, and families now? Would Carnegie, whose shrine today lives in the second-hand Pittsburgh museum that scarcely recalls the name of Homestead, have that graceful empathy, healing change of heart, when the whole town, as once before in '92, lay prostrate and hungry before him? Then, too, will the Ghost of Christmas Future here be a chimera, as the mills are being dismantled for scrap, sold, as the common talk in Homestead goes, to the Japanese to make automobiles to sell back to us?

Alas, local papers tell us nothing now. For Homestead has none, no radio station either, as it blindly stumbles to an unknown fate, ethereally bruiting of race tracks, new lotteries, entertainment parks, any source of money to salve its fiscal crisis and care for its desperate poor.

Even from without the gates, before the whipping wind blurs vision, one can see the stark grandeur of the old Homestead mill. Wind whistles down the lonely chimney stacks and makes the loose-shingled eves and boarded windows chatter. Winter sparrows slice through frosty gusts and dive through broken skylights or up, up, to the control ramparts overlooking switching yards and inert hulks with no more ore, no more coal, to carry.

Only imagination can make all live once more: mammoth cranes rolling along the top of these great sheds, spidery laborers held aloft by machine-like fingers over the boiling jaws of the dragons that created steel. The hisses of these fire-breathing furnaces, the baritone roars of Bessemer converters, the seething vats of molten metal, tons of scrap, the air red with dust, lung black with smoke, armies of workers rushing madly about, screeching nickel-steel saws, rollers gnashing metal . . . ready to devour all as Satan did his children, orchestrating the Monongahela's fame.

A private undertaking, then and now, it is not the public's business, U.S. Steel's official lie evokes at the wake of the three River Valley: coal, coke, iron and steel works sitting dead along the river in Homestead, Duquesne, Clairton, McKeesport.

Its mortuary has now been departmentalized, too. Through this last winter of Homestead steel, the plant barely functions, the bent mill gates being secured by deadened bolts, locked firmly against an outside world of would-be trespassers. But few of these linger to confront the security of the abandoned mill.

**Where Once A Mill There Was**

Winter and I arrive in time for the steel shutdown taking the last few jobs away. Wages are no longer handed out every second Friday. And with little savings, after several months living on Social Security and welfare, many residents are barely scrapping by. Most steelworkers' families have delayed paying their monthly bills, some have defaulted on their mortgages, others have vanished, generous lenders having initiated proceedings to put the wayward debtors in the streets.

The sheriff has reviewed the matter with unexpected consideration, his deputies waiting for Spring to evict, saying it is the least that should be done for the men who are unemployed, especially for the veterans and people who have been working all their lives. With this decision the leading banks are displeased, loudly complaining that mortgage loan investors are being treated unfairly. "Borrowers who have lost their jobs need to protect the equity they have earned in their homes, while investors who have bought mortgage loans need to be repaid," offers the vice president of one mortgage company angling for foreclosure proceedings.

All along the Monongahela and Ohio Rivers near Pittsburgh, one-third of the steel workers are this frigid season without work, recalling the winter days of 1892. Yes, this Christmas, 1985, is *déja vue* . . . in the unheated Carnegie Library in Munhall, at the graveyards on the top of the hills behind, where minus five degrees in a whistling wind, one looks down on the Rainbow soup kitchen far below, images the Carnegie museum in Pittsburgh, or chances a glance at McKeesport, once a Pinkerton armory for attack on Homestead's workers, now center of the new growth industry of unemployment. Comparable offices in the valleys are called variously labor centers, job offices, or union halls, but all are fixed with apparently aimless, shuffling men, and a few women who have put schedules behind them.

Poverty, personal helplessness, seems to overwhelm nearly everyone in Homestead, making the social atmosphere gray, bereft of life's vitality. In Phil's Grill on Main street, a bar patron keeps counting his change, an easy task for one specializing with great sums less than a dollar. Ranking his pennies opposite a small row of silver, he stacks the coins by size, forms his copper troops into a semi-circle around the small glass of whiskey next to his beer, then moves them randomly with a lumpy hand when he reaches into the worn layers he is wearing.

Homestead beer nearly as cheap as it was in the 1890's, rings the cash register at 35 to 50 cents. A shot and a wash or a boiler maker takes less than a buck to ice one's pain; drunk for a dollar, dead drunk for two. But the bar crowds on Christmas Eve and the day after Christmas this dreary '85 are quietly soused, pickled, stewed . . . not boisterous, belligerent, or confident; everyone seems to play everything close to the chest.

Winds knife-edged outside still give the heavily bundled crowd that robust, secure look, doubly so as they peer down the bar praying to the TV rendering "The Money Show," featuring a Baptist-style stem-winder holding forth on the virtues of establishing credit without cash and creating a daisy-chain of letters of credit, personal banking, first and second mortgages, the intricate refinements of high finance. Relief in securing independent wealth flickers over their faces, the bartender noting confidently the re-run nature of the show, he himself having committed to indelible memory the esoterica of terms and procedures few viewers would ever try.

Fellowship of the unemployed, the old clothes holding up the rail in Phil's Grill, seems to keep the cold brotherhood in command. Young men, skin weathered like plastic cutting boards, relying for their scarred and lingering personal prestige on an animal-grip hand, weigh the odds of extracting commissions by lining up their fellows for government training and the like. On the streets, knotted men stand outside Zarkov's tavern down by the railroads, across the tracks from U.S. Steel in Homestead, as they did at the Munhall Station in December ninety-three years ago. As then, a few make their spaces official residences, bending the environment around them this icy winter day, projecting the feeling of someone coming to meet them soon, issuing an expectant air, tossing a challenging glance as if it is my fault that I am not the one waited for – when probably no one will ever appear.

Then there's Chirorcio's – like the Army and the craftsmen at Homestead in '92, the bars attract men of different ranks – with mill, drink and play souvenirs hanging from the ceiling. A muzzle-loading gun, football and baseball gloves, kerosene lanterns, someone's beer coaster collection, newer relics like a hard-hat. Younger guys here, a decent sandwich for a dollar, a would-be Wall street entrepreneur selling ornamental, steel-handled great jugs full of beer. You may have guessed, the jugs are German imports.

Here young husky guys can spend all afternoon plotting how to beat the video game – their unemployment benefits haven't run out, or maybe they're on welfare. They talk about moving out, going South, ask me about Texas. I tell them Austin and Dallas seem to be going along okay, Houston is having its troubles with oil. One of them says the job office offered him a course on becoming an Administrative Assistant; he hasn't decided to try it yet.

At the Rainbow soup kitchen, I talk to the assistant director, a pleasant fellow in his 30's, who says he served 300 meals on Christmas Eve and found the office broken into on Christmas day. The burglars took petty cash of less than a hundred dollars and quite a bit of food. The kitchen also runs a job program. On its walls are pictures of Jesse Jackson, who visited it when it opened two years ago. Other pictures show a "We Want Work" parade and picketing of the Mellon Bank. He says he was laid off three years ago and went to the Union, where they said that's the way it goes, you get laid off when things are slow and re-hired when things pick up. But now U.S. Steel is taking apart some of the plants; one in McKeesport is completely gone, the scrap metal, he reaffirms, sold to Japan, "so they can make cars to sell to us."

On the wall, too, is a chart of Rainbow clients showing sources of their income: 25 percent on welfare, 30 percent on unemployment, over 30 percent no income at all. "People are just leaving," he says. "The unemployment rate is supposed to be 25 percent, but no one's really counting." The union, he adds, now has two choices, to organize secretaries instead of steel workers or to become another social welfare agency. It didn't put up much of a fight, he offers, against the layoffs when the first wave hit three years ago and was too weak to do much about the present shutdowns. The workers bought one metal-workshop down towards McKeesport; it went bankrupt.

Carnegie's millions in philanthropy seem to have barely dented the city's consciousness or anything else. Three old ladies sit in the cold Carnegie Library, a giant step up the hill in Munhall. Two girls are doing research from a tattered set of Encyclopedia Americana ; one old man tells me I should read a book on parapsychology. I check the local history shelf, about eight books on Homestead, including a novel, a sociological survey of the population during the times of the great strike of the Nineties. I look at the cards on these books, *Out of This Furnace*, *Lockout*, *The Households of a Mill Town* – *Lockout* was checked out four times in 1985. On one wall of the library, near the heavy double doors at the entrance, hangs a map of Homestead in the '90s, a pleasant old etching. In it, as is true today, no smoke comes from the mills. Quite a few churches are shown on the map, and outside again I find some of them once dividing immigrants by wage scales and ethnic bonds – the Hungarian Reformed, the Polish Catholic, a couple of newer Protestant churches.

West Street has become semi-commercial as it crawls up the hill; past the funeral home there's a health foods store, a small bar, a gas station; around the corner is the old high school, now the city hall and police station. Higher on West is the Munhall Borough Hall, bigger than Homestead's city building. Three boys are inside the door, puffing, huddling from the cold.

Local color recalls Homestead's famous founder at the 150 anniversary of the Carnegie Museum in Pittsburgh, where his portrait is kept up much more lovingly and at a more constant temperature than in the Homestead Library. Carnegie there is beginning to crack and fade a little; one of the librarians told me that the money is about dried up. Even in Pittsburgh the permanent exhibit of Carnegie's great success story is little-visited, tucked back behind the dinosaurs in the Natural History wing. Most people go there for Renoirs and Dutch paintings, I suppose, or to the Christmas Tree exhibit.

I ask a few people in Homestead about Carnegie and draw blanks. It's like asking about Millard Fillmore. In the memorial brochure on him at the museum, selling for 96 cents, tax included, there's a reference to the "unfortunate incident" at Homestead, but more about the 300 million he gave away. You get much more reaction if you ask about the mills – small shop owners will tell you there are jobs but these men won't take them. They're used to working for $20 an hour, the word of the middle class goes, echoing the $30,000 fabled salaries of skilled workers in the '90s, and they won't take $7 an hour positions. They live off their wives, they don't save money, they were stubborn, they even bought Sonys and Datsuns, some of them, says a man who owns a Cadillac.

A few of the older people tell me how some relative at the plant earned a dollar a day with a subtraction of ten cents to maybe as little as one cent for ice – the smart ones got out of the mill and went to work for themselves in groceries and penny candies after the great strike. The McKeesport paper even has articles now about former steelworkers finding wealth running a pizza parlor – positive stories about a proposed racetrack for the city.

Murphy's Department Store is going out of business the day after Christmas, a sale offers everything for 85 percent off. The old train depot is for sale, too; and houses are going for $2,000, mansions for $60,000, everything's on the market, 'though some people don't even bother to put up a sign.

Well, it's a bare, barren little berg now. Goodwill Industries is the second largest store on the main street, but there are dozens of little second shops, trinket places, one bookstore too narrow for two people to get by each other. The town doesn't give up its secrets easily either; the few people I tell I'm trying to write something about Homestead tell me I have no right to – they know, they've been here umpty umpt years. It was the blacks who ruined the scene, they say, or the Japanese, something happened, what can you do.

At the foot of West, next to the bridge across the river is the only reminder, a small stone memorial to the steel workers who died in the war against the Pinkertons.

Things could be worse, some say – a few steel bars are still being trucked out of the plant, a few cars are still going in. The security guards tell me though that I can't pass the gate into the plant: "Everything comes out of Pittsburgh," they say, permission, plans, maybe something for Homestead too.

Yes, it's a cold winter here in Homestead and the town won't be heated by the mill-fires anymore. The iron heart still pumps from Pittsburgh, 'though now it's microtechnics and finance in the flow, and the extremities, the Homesteads, Clairtons and McKeesports up and down the three river valleys, have lost their circulation.

Morris Edelson
Homestead, Pennsylvania
2 January 1986

**Present Day Monument to Homestead Strikers**
**(Photo Credit: John C. Susko, Homestead Steel Workers E. M. P.)**

# Acknowledgements

This study describes how greed devastated a 19th century town and a great steel mill. The conflict foretold the death of a vibrant national industry, perhaps basic U.S. industry itself.

The city had grown up around and was specialized to accommodate mill production. The company became a magnet for the people who worked, lived, loved – and went into debt around smoke-belching furnaces of the mill. It was a classic company town, emerging from a culture centered on uninterrupted output at the time of the events described here.

In the language of the times, the owners and executives of the mill were captains, barons, behemoths. In a modern metaphor they could be termed manipulators of stock – of equipment, mind and muscle.

The mill realm was a complex, layered industrial protectorate, a few directing the many. Ranking at the top levels of mill operations were men of genteel greed and stark arrogance; at the lower reaches emerged the rough play of ethnic divisions and racism, ignorance and a Dantesque economic underworld.

Above the level of sweat-labor in the mill and the physical geography of the town, a *coterie* of well-paid, full-time managers held court. Down in the fire and heat of the furnaces labored their charges – union workers from Western Europe. These union men, trade-proud, as in an earlier age, guarded their crafts and ranks from migrants, so-called "black sheep" of Italian, Eastern European, and African ancestry. For they viewed these helots of low-wage and day labor as potential "scabs" that steel managers might permanently install as replacements.

As the factory and town on the Monongahela pulsed with vigor, other trades took life. The hillsides thrived with retailers and consumers, merchants of every class taking up license, appealing to human passions for goods and services of the latest variety. Already congested neighborhoods imploded, densely packing new generations destined to become mill hands and beer-hall patrons, or housewives trying to make do with leftovers of Friday's wages.

* * *

No doubt the reader will be familiar with the spatial contours and class-riven designs of other cities of the 19th century. But though this city was not Honoré de Balzac's Paris, Charles Dickens's London, or Fedor Dostoevski's St. Petersburg, it developed its own complicated and explosive relations of caste, class, and race.

The great cities get attention. Many people have at least notions of how New York, Chicago, and San Francisco grew after rounds of skullduggery or genocide justified by religion or patriotism, though at root in the service of moneyed classes. Gold Rush or Land Rush – some sought; others bought. Even in the Wild West gun-totting frontiersmen and urban freebooters can be recalled driving out Native Americans, styling themselves as militia, as land surveyors or developers – laying out small centers where one can still see traces of those original warlords and landlords, the results of land grabs and corruption, or how hostile forces were kept at bay with armed posses, walled streets, stockades, and restrictive ghetto boundaries.

In other towns the walls are less apparent, antagonists living and working nearly side-by-side or at different reaches of the same neighborhood, bayside or hillside community. Such divisions become clear after looking at the purveyors of wealth and employment; examining working hands and conditions, spatial quarters and attire of varied inhabitants; uncovering their culture or its absence; focusing on the population's habits, good, bad and horrible; depicting the natural, topographical design and planned or unplanned alterations of the space and place setting of different social classes.

To picture this town's portrait and recount its oral and bibliographic gyrations, as story teller, I want you to see, feel and smell the place, visit its mill and shops, walk its streets and squares, hear its sounds. My hope is to provide some details, describe the city's fugue, recount the battles here in this city – projecting its violent and degrading past upon a more placid order of deprivations here and now.

* * *

Also in order is a word about my adventure in trying to recreate this past, one of the best-known encounters in U.S. working class life.

My own intersection with this town of labor's toil and troubles began more than twenty years ago when I was an overzealous Ph.D. candidate in the Sociology Department of the University of California, Santa Barbara. I had begun research on the United States steel industry and its central role in the development of American political economy. The vast field that confronted me soon led me to concede the better part of blind ambition was some modesty, as I reluctantly cut the scope of my project to a single mill city, owned by a modern day Prometheus.

That mill town was Homestead; and its owner, seeking to transcend human limits, was Andrew Carnegie – a money mogul, well-known raconteur, and, in his own mind, seeking Utopia.

So, as a later day interloper in this tale, I made my way through a labyrinth of documents surrounding the life, travail and trauma of this mill town, its people, and its would-be Caesar.

Carnegie put in motion a drama of greed and poverty, life and death, gilded wealth and workers' deprivations, wage negotiations failed, and a military encounter. Carnegie and his hirelings could and would claim victory, but they had ripped the social fabric and Andrew Carnegie – who early on lived by lies on good men's lives – helped push the industry on a treadmill leading to its present weakened state.

In this adventure, I was not alone. Sociology professors Richard Flacks, Morris Friedell and Richard Appelbaum from the University of California at Santa Barbara offered encouragement and assistance. And as the story line and dynamics of Homestead's *imbroglio* of steelworkers and Carnegie unfolded, critical assessments were made by Professor Morris Edelson, then at the knowledge mill of the University of Houston.

Thanks also go to those who contributed their broad and expert knowledge in the period before this manuscript took final shape. Without their ever knowing, teachers from my past moved my pen, offering solid, if subliminal, support. Critical insight about the 19th century transformation of the U.S. political and economic landscape was tapped from the wells of University of Wisconsin history professors William Appleman Williams and Harvey Goldberg, as well as drawn from my brilliant mentor, friend and intellectual guide Hans

H. Gerth, who before his death was Professor Emeritus in the Sociology Department at the University of Wisconsin.

Also notable here are the background discussions with Dr. Robert Girling in the Management and Economics Department at Sonoma State University; Professor Sherry Keith of Interdisciplinary Studies at San Francisco State University; Dr. Hiroshi Fukurai at the Board of Studies at the University of California, Santa Cruz; and Professor William J. Chambliss in the Sociology Department at George Washington University.

At various stages of completion, parts of the manuscript were screened by Professor Edgar W. Butler, Chairman of the Department of Sociology at the University of California, Riverside; and by Professor Paul Stevenson at the University of Winnipeg.

Correspondence about Andrew Carnegie, Big Steel, the Amalgamated Association of Iron and Steel Workers, the United Steelworkers of America and their progeny was carried forward with dozens of experts, including Harold J. Ruttenberg, former research director for the United Steelworkers of America, William J. De Lancy, former chief executive and chairman of Republic Steel Corporation, and Mr. Paul Mellon, heir of a banking fortune supporting worthy social and environmental causes.

Professors Ann Baxandall Krooth and Morris Edelson helped sift and refine the story line and presentation of the study. Katherine and Lawrence Moore, editors of Ramparts Press, helped cut the manuscript by half. Historians without portfolio, Lester Radke and Miriam Paras Hall, also offered critical insights that helped shape my thinking about the U.S. development of production relations and international affairs impacting the steel industry. San Francisco environmental and human rights attorney Karl W. Krooth helped interpret the legal scenarios of the great strike of 1892.

Aiding me, too, to develop other social nuances of Homestead of the 1890s were James Danky of Special Collections at the State Historical Society of Wisconsin and Sara Cooper at the Southern California Library for Social Studies.

For the expedition to these and other distant libraries, travel grants were offered by the Sociology Department of the University of California, Santa Barbara, and the Faculty Senate at the University of California Riverside.

And as a Research Associate, I was furnished writing and research facilities at the University of California at Berkeley. Bell and Howell helped provide some among the large collection of newspapers cited in footnotes and listed in the bibliography; and a group of historians divulged other insights, including some of those appearing on PBS, "The American Experience. The Richest Man in the World: Andrew Carnegie."

For advice on graphic design thanks are due Lana and Michael D. Husser. Architect Allan Gatzke generously drew the map of Homestead's visage in 1892. Credits for help in locating images are due to *Harvest Quarterly*, Wm. C. Hollister & Bros., Homestead Steel Workers E. M. P., *The New York World*, The Library of Congress, *Pittsburgh Commercial Gazette*, State Historical Society of Wisconsin, Moe Foner founder of the "Bread and Roses" cultural program and longtime executive secretary of 1199/S.E.I.U. (New York's Health and Human Service Union) and the heirs of A.G. Burgoyne and M. R. Stowell or their printers. No permissions were secured for figures in the public domain.

Finally, this project could not have been completed without the encouragement and support of Diana Lavery, Beverly Baum, and office staff at the University Press of America.

I gratefully acknowledge the aid offered by these colleagues and friends, correspondents, experts and institutions, but the proverbial buck stops at my doorstep: I did the research, writing, and final editing.

Mine was the task of presenting in my own way "A Century Passing," a never-to-forget epic staring Andrew Carnegie and Henry Clay Frick, the working population of Homestead that took on the Pinkertons and the Militia in the strike of 1892, and a host of other actors in this bitter and deadly story.

I alone remain responsible for the facts presented and the conclusions drawn.

Richard Krooth
Berkeley, California
28 April 2002

Homestead Steel Works
(Photo Credit: I. Robbins & Son, 1892)

# Introduction

Coal and coke, iron ore and lime; shipping fleets, railroads and mills; workers and company towns – these were the basics for making steel during the 1890s.

Great sums of capital mobilized all of them; and in the Monongahela Valley none were richer and better positioned for such tasks than Messrs. Henry Clay Frick and Andrew Carnegie, overlords of an emergent empire centered on Homestead.

Steel had become the essential ingredient and fulcrum for the wonders of industrialization – machines to manufacture, equipment for petroleum extraction and refining, heavy rails and giant shipsides, structural beams and girders for bridges.

The magnet of steel and coke Works meanwhile drew an emergent workforce, seemingly without end, bringing mass deprivations, industrial mayhem and death for those laboring in the forges of the mills – also leaving to their own meager resources those laborers periodically laid-off, locked-out, maimed or, with their families, taken sick with disease and microbic epidemics.

What these coke and steel magnates were able to accomplish in the space of a few decades roils both the imagination and a sense of equity. And thus the story recounted here of financial resources devoted to employ an army of workers to extract coal and iron ore, to labor in the coking ovens and in the mills, to build rail tracks and run locomotives over what was once dale and farmland, to build whole communities and company towns on the back-flap of giant coking works and steel mills spewing smoke and lethal chemistries over country and city alike – ruining crops, enveloping mill towns, and degrading those who lived under their onslaught.

It is not a pretty portraiture of the building of the nation's steel industry. And thus a few snippets from the cavil of judicial affairs – microcosms of the rights of the propertied, the public, and labor – heralding an age to come and setting the stage of our story, eliciting some fundamental facts and describing assorted social pressures that, in turn, became locked into emergent class relations and turmoil residing at the heart of the mill system of the times.

## 1

On the October 28, 1891, appeals docket in the Pennsylvania Supreme Court seated in Pittsburgh, before the Honorable P. J. Hunter, defendants Carnegie Bros. & Co., Limited, are heatedly contesting a $4,798.10 judgment on a verdict rendered against them in the Court of Common Pleas. It seems a small sum for the Carnegies, but something deeper is at stake.

The plaintiff, Mr. Robb, contends that Carnegie coking operations have ruined his farmland, so little if anything can grow. But the Carnegies assert the land is second rate at best and, by comparison with the other comparable land in the agricultural district of Westmoreland county, is far below average, so by its own merits, despite Mr. Robb's efforts at husbandry, nothing could ever lushly grow.

Robb's land in North Huntingdon township, Westmoreland county, is indeed a modest parcel, comprised of 82 acres more or less, bought for $4,000 on March 24, 1866, and another 22 acres purchased for $2,000 on September 22, 1887. He believes such prior ownership rights to land place him in an enviable location between city, town and country, situated as it is about one-mile west of the village of Larimer, about twenty miles east of the city of Pittsburgh, and slightly north of the Pennsylvania Railroad.

But there is a worry. For his nearest property line falls about fifty rods from the railroad - Robb's parcel running along the top of a high bluff, more than two hundred feet above the grade of the road, and some three hundred feet above the location of Carnegie Brothers' coke ovens located in a deep valley.

It seems that part of the last tract purchased by Mr. Robb is fortuitously underlain with coal that he mines and sells to the Carnegie

Works and others in the neighborhood, bringing him considerable revenue. But the surface, Mr. Robb insists, is limestone soil, part black loam, and all good for farming.

As for the domain lying between the Pennsylvania Railroad and Robb's land, a marsh valley is an apt description. Unfit for cultivation, it is a low parcel of about twenty acres, the very one three hundred feet below Mr. Robb's domain, and a full nine hundred feet below the brow or top of the hill, where Robb's tillable land actually begins. Brush Creek runs through this lowland, a small stream emptying into the creek at the east. Robb calls it the frog pond – a swamp, bottom ground, grown up with papaw under-bush, willows, elms and sycamore trees.

But considering coal mining operations carried on at nearby Larimer, Irwin, South Side, Turtle Creek and other points in the neighborhood, this low-lying terrain is centrally located, and particularly advantageous for the establishment of a coke-making Works, able to utilize the large quantities of coal from the several mines.

Another advantage of the place is not so obvious: secluded, isolated from the thickly settled country about the mining districts, near the railroad for transport of coal and coke, with ample water for washing operations, it is a perfect place to burn coal to eliminate its carbon in forming coke. Knowing this, the Carnegies purchase this low-lying ground and here, in 1871, build their coke ovens.

A small enterprise at start, additions are made at various times until there are fully three hundred ovens that, together with the washing and hoisting apparatus and all the other machinery and appliances necessary to the coke plant, cost the Carnegies nearly $200,000.

This is no small undertaking, as coke operations require clockwork coordination. The Carnegies negotiate to buy the coal from the several mines in the neighborhood; hire out cars on the Pennsylvania Railroad to haul the coal to these ovens; there employ workers to wash it from the water of Brush Creek removing all possible sulfur and other impurities; use the most improved machinery to prepare it for coking; and then burn it into coke in the ovens for rail-shipment to their steel mills. One-hundred-thirty to two-hundred men are constantly in their employ, these workers either living nearby in their own homes or in the houses that theCarnegies built about their works.

Not that Mr. Robb gives the Carnegie Brothers notice not to erect the ovens or workers' houses. For he with his boys and horse teams at different times assist in building them. But Brush Creek is now flush with sulfurs and other impurities, the air reeks of smoke, gas, heat, dirt and carbons; the fallout of effluents and dioxides is unstoppable. And Mr. Robb is fuming that his crops are being destroyed.

No longer content to suffer such abuse, Robb sues the Carnegies, calls several witnesses at the trial to show that there has been a gradual falling off in crop yields six years before bringing suit, and deftly argues that he has lost some timber and fruit trees; that crops put in have grown to a certain height, then died away; that the stalks and blades of grass are covered with a black crust preventing maturing; that even manure and fertilizers have been useless to bring plants to fruition.

In legal terms, the case turns on damages sustained, not closing the Carnegie coke ovens. And in this regard, Robb's son testifies that the value of the loss can be gauged by the rental rate of the farm – worth $350 in 1881, $300 each year from 1882 up to 1886, and $275 in 1887. He adds that, as during two of the six years before bringing suit Robb has rented the farm for cash, any loss of such crops was suffered by the tenant.

Though these are tiny sums for the likes of the Carnegies, in an effort to ensure there is no diminution of their future coking operations, their attorneys produce upwards of fifteen witnesses of the neighborhood. These gentlemen testify that Robb did not sustain any substantial injury, but rather that he reaped reasonably fair crops – as good as he might expect for the varied attention he gives his farm.

So the court is put through the paces, its ruling critical not only for the moment, but for the future of the Valley's coal and steel industry as well.

Turning on the monetary measure of damages done Robb's farming by the Carnegie coking operations, the court decides that Carnegies' private enterprise can not be restrained from owning and operating their coking Works for profit. Nonetheless, the court adds, in such operations the Carnegies have no right to claim exemption from the natural consequences of their own acts; and Robb as plaintiff should be allowed compensatory money damages sustained. But exemplary damages are not awarded, as these would treat the Carnegies as wrongdoers in the establishment of their plant on well-selected and secluded land belonging to themselves.

So though they lose a paltry dollar sum, the Carnegie Brothers win the right to continue coking coal.

The ruling proves prescient, too.

In the Three River Valley, the future belongs to the owners of railroads and coal, coke and steel – to operate their various track-lines, mines, Works and mills as they decide, so that those with less financial resources and power must legally or otherwise protect themselves . . . if they are able and dare.

## 2

Death of his minor son in the employ of the Carnegies is the reason Mr. Thomas Tunney courageously takes the Carnegie Brothers to the Pennsylvania Court of Common Pleas, No. 2, of Allegheny county in 1891.

Working for the Carnegies at the Edgar Thompson Steel Works the night of the 24th of April, 1889, the recent immigrant, 17-year-old John Tunney, is moving bricks on a wheelbarrow to men repairing one of two large flues, each measuring twenty-feet in length, four-and-a-half feet in height, and four-and-a-half feet in width – placed about a foot beneath the surface of the ground and used for the purpose of conveying gas from a battery of boilers to its discharge stack.

The two flues are parallel, fifteen to twenty feet apart. The one being rebuilt is closed at the end next to the boilers; and throughout the night of April 24th, a party of more than twenty bricklayers and their helpers are at work along its length. John Tunney and at least two other persons are wheeling bricks to them, sometimes passing over the flue that is in use.

As the night passes, though, a time lapse is unaccounted for. Between half-past-ten and eleven o'clock, the workers have seen the boy warming himself at the boilers; but then he simply disappears. Of the twenty or more workers around at the time, no one has seen any mishap or accident, though about midnight some of the bricklayers notice a bright light coming from a hole eight or ten inches wide in the flue which is still in use. By seven o'clock in the morning, the hole has enlarged to twelve to fourteen inches.

At the same time, about 7 A.M., hearing the news, the boy's father comes to the mill to search for him, reports that the lad has not returned home, and hearing of the hole in the flue insists that he must have fallen in – the basis on which Mr. Tunney sues the Carnegies for damages for his son's death due to their negligence.

Knowing little more, Tunney simply assumes that his son has disappeared into a flue large enough to admit his body, that the flames within had consumed him – leaving no vestige of bones, enamel of teeth, or steel plates upon his boot heels – though the father later testifies that when he looked into the hole of the flue that morning he saw, at the foot of the stack, what he thought were bones and a metal heel-plate.

Were the Carnegies responsible for the missing lad?

At trial, no one suggested the Carnegies' general plan of construction of equipment was not proper. Competent witnesses testified that the flues were well-built, capable of sustaining great weight. Built a foot or more underground to withstand intense heat for eighteen-months to four-years, the arch over the flue in question had been constructed about six months before, and inspected three months previous to the night the hole had opened. But open it had.

And the trial judge rejected the Carnegies' argument that there was no evidence of their neglect for the jury to consider. The jury's sympathies were meanwhile touched by the sensational character of the action, enlisted by the happenstance theory of a missing boy, a faulty flue on which the case rested, and the horrible fate to which circumstances supposedly consigned John Tunney.

The legalities were not yet at rest, though.

Playing its assigned role, the Common Pleas jury was left without any means of ascertaining whether or not the Carnegies had omitted any safeguard which was ordinarily employed upon flues constructed like this one. Too, there was no satisfactory evidence that John Tunney had actually fallen into the fiery flue.

So was it or was it not a factual, logical or other matter for the jury to decide?

Could the jury infer purported negligence from the mere appearance of a hole in the top of the flue on the same night John Tunney disappeared? And before the scrutiny of the jury, did Thomas Tunney have the burden to prove with factual evidence the basis for Carnegie negligence?

This further begged the question: Were the Carnegies negligent in maintaining their equipment on their own mill grounds that endangered and led to the possible death of their employee?

On January 13, 1892, an appeals court thought not:

> When it is remembered that the foot of the stack was ten feet or more away from the hole, and that the flue was filled with the burning gas and vapor rushing with great velocity from the boilers to the stack, it is apparent that this story [related by Thomas Tunney] is an improbable one; but when to this is added the fact that the fire was turned off that [following] forenoon so that by one or two o'clock [p.m.] the flue was cool to admit a thorough examination of it, and its contents, and that nothing was found in the ashes and soot but a few small nails used several months before in rebuilding the arch, its value as evidence

> altogether disappears. There is a coincidence between the appearance of the hole in the flue and the disappearance of John Tunney that suggest[s] the possibility of their relation by each other as cause and effect. On the other hand, there is the circumstance that between twenty and thirty men were at work all night within a few feet where the hole appeared, and neither saw nor heard anything to suggest to them that any person had fallen into it. This is all there is in the case to prove that John Tunney is dead.

Thus the appeals court threw out the jury's decision and ruled for the Carnegies, arguing that:

> the fact that the judge in the Common Pleas submitted the question of the defendants' negligence to them was to their minds an assurance that there was evidence of negligence sufficient to justify them in finding the defendants guilty of having caused the death of John Tunney by the neglect of such precautions as it was their duty to their employees to take; and they lost no time in finding it. We think there is no evidence, or at least a mere scintilla, from which negligence could be found, and that the question should not have gone to the jury.
>
> The judgment is reversed.

The Carnegies had won again: An appeals court had overturned the jury decision of a lower court finding an employer guilty of negligence when an employee was killed or forever missing within the domain of the employer's factory property.

True, one could assume that young John Tunney simply ran away, never again to be seen. But, nonetheless, an onerous, if not near-impossible, burden to prove an employer's negligence in directing operations in their own mills had now been placed on the shoulders of future workers and their kin in the coke and steel industry of the Monongahela Valley.

## 3

If railroad, coke and steel enterprises could without restrictions extend their operations in both rural and urban zones – being limited only by court rulings for damages to crops or proven injuries their equipment caused on their right-of-way or on their mill grounds – would those injured be sufficiently protected in the future? And, to complicate matters, would contributory negligence of an injured party undermine a claim? What rights or remedies, if any, could the many thousands of pedestrians, rail, coke and mill hands then secure or demand?

Clearly, matters would be made more difficult when a person's own actions or inactions contributed to the cause of injury or death – requiring that a claimant show due care was exercised in order to support a case at law.

Its lack was often alleged by employers and others, as in the case of the Pennsylvania Railroad Company concerning a certain fatal accident in 1892, at the crossing of Dark Run Lane, on the outskirts of the town of Holmesburg, Pennsylvania, near the Wissinoming Station, where at grade level five tracks of the company were crossed by the lane.

It seems that a Mr. Childs and his son were returning from their work at the close of the day, reaching the crossing in a fog after dark. When they reached the crossing they stopped near the track and looked up and down the railroad and listened.

According to the son's testimonial account they neither saw nor heard anything to indicate the approach of a train and started to cross the tracks. But, the boy recalled, when they reached the last, most western rail, a train suddenly came out of the darkness upon them.

He successfully got out of its path, but his father, less agile, was struck – thrown through the air some sixty or seventy feet, over the heads of two witnesses. When reached, he was stone dead.

His wife, Susan A. Childs, sued in the Court of Common Pleas No. 1, of Philadelphia County, to recover damages for his death, allegedly caused by the negligence of the railroad, her attorney arguing that the train was clearly going too fast without sufficient warning of its approach.

An express, the train had been barreling from New York to Philadelphia at an awesome speed of forty-five to fifty miles an hour, the locomotive engineer testifying the average running time at fifty-three miles an hour – running precisely on time at its scheduled rate when the accident occurred.

The engineer also pled that the bell on the train was repeatedly rung on approaching the crossing from a little over nine-hundred feet away. But neither Childs, his son, nor the young ladies who saw the accident heard it; and even if the bell could be heard a quarter of a mile away, it afforded a trifle of a quarter of a minute for a traveler to determine what to do and to do it. Besides, there was no warning whistle on the train, which could have been heard twice or three times as far away, affording a traveler a comparable time to escape from danger.

But there was another important matter. As the crossing was located in a rural section of the city and the lane was an unpaved, country wagon road, the appeals court took its cue to distinguish the standard of responsibility for both a speeding train operator to protect the public and a pedestrian to secure himself.

As for the Pennsylvania Railroad, the court outlined that:

> In the crowded centers of business and population the public safety requires the speed [of a train] to be moderated, but in the open country the single traveler over the wagon road may, under all ordinary circumstances, provide for his safety by compliance with the rule of law and of common sense that requires him to stop, look each way along the track and to listen, for an approaching train, before attempting to cross the track. The movement of trains must be regulated by the railroad companies in the exercise of business discretion, and upon consideration of the competition they have to encounter and the necessities of modern business. We do not think a jury may fix a maximum rate of speed at which a train shall be moved in the open country, or that a high rate of speed is negligence *per se*. But while railroad companies may move their trains at such rate of speed as the character of their machinery and road-bed may make practicable, they must not forget that increased speed for the train means increased danger to those who must cross the tracks, and that increased care on their part to guard against accidents becomes a duty.

The court thus found that the jury in the lower court could justly determine if the railroad's duty was not met and that the dead man could not have avoided the accident.

"If they should find the fact to be," the appellate court ponderously began, "that a train moving at the rate of speed at which this train was running, would cover the distance between a point from which its bell could be heard at the crossing and the place of the crossing, in so short a time as to make the signal of little or no use to one in the act of crossing the track, then the failure to give notice by the [lack of a] whistle from a longer distance away would be negligence."

Susan A. Childs thus got to keep the $5,000 awarded by the Philadelphia Court of Common Pleas. But, as we will see, the case nonetheless opened the way for future courts to consider the contributory negligence of an injured or deceased party in denying any award from railroads or steel magnates alike.

## 4

Another matter was central to the industrial advance and turmoil that accompanied building the nation's steel industry: Whose rights would prevail in a contest of employers and workers . . . or even among employers, their hired guards or sheriffs?

Would the legally constituted authorities side with the large mill owners, throwing their political weight and police powers against the men who made coke and steel? And who would pay for the services of sheriffs protecting mill properties and "scab" workers employed by the Carnegies and the other steel masters?

In 1889, these questions were raised, and partly resolved, as workers were threatening to strike at several mills in the valley of steel. And in April, workmen put down their tools and walked out at the Allegheny Bessemer Steel Company located in Duquesne, Mifflin township, on the Monongahela River some nine miles south of the city of Pittsburgh.

There large crowds of striking workers were confronting the company's managers, threatening violence against the mill. And fearing the consequences, officials of Bessemer called on Allegheny Sheriff McCandless to secure their mill by employing guards and watchmen in apprehension of trouble.

Though Bessemer officials expressly agreed they would pay all the Sheriff expenses for the guards, following the required legal script Sheriff McCandless asked Allegheny Bessemer officials to prepare and serve notice on him under the Pennsylvania Legislature's Act of May 31, 1841, making the county liable for any loss of property in case of riot.

As both McCandless and Bessemer officials knew, the sequence had its logic in aiding employers secure their mill properties and other works; and for this legislation had long since been successfully lobbied by the mill owners, being fortuitously extended to all of Allegheny county by an Act of March 20, 1849, designated P. L. 184.

Moving under the color of legal authority and relying on Allegheny Bessemer's promise to repay expenses, the Sheriff immediately employed special deputies and sent them to Duquesne, paying each of them for their services at a rate of five dollars per day, providing both subsistence and arms. Along with them, he also sent some of his regular deputies.

"I told the [Allegheny Bessemer] attorney, Mr. Dickey, it would be necessary for him to serve me with a written notice in the case before I took any action in the premises, after which I would proceed to investigate it," McCandless later testified; "and he did furnish such notice, served a notice on me as sheriff that they had trouble at Duquesne; that they anticipated destruction of property, and called upon me as sheriff of the county to protect that property or they would hold the county liable."

McCandless also recalled that, after receiving the notice, he went up to Duquesne, where he found a large assemblage of men along the railroad track from the station to the steel Works; and that he had a proclamation printed, posted and read to the crowd. But that did little good, he said, as the crowd reassembled each time, threatening and forcing him to leave with his men.

Matters worsened once Allegheny Bessemer moved to put scab workers in the mill. McCandless added more and more special deputies to his regular deputies on the scene, but that did little good to quiet the strikers. For now crowds of union men congregated in the neighborhood of the Works, threatening acts of violence and even firing into buildings on the mill grounds.

"The disturbances went on from day to day," McCandless reported; "we had to arrest a few people and bring them into court. It got a little worse and worse, and I ordered the only saloon in town closed; that is Mr. Downey; they carried it down in jugs from McKeesport afterwards, and shots were fired, and the baker's wagons were discharged into the river, bringing bread into town, and all that."

As McCandless had run out of options and operating funds, and Allegheny Bessemer refused to make good on its promise to repay his expenses, McCandless sued for the costs of the special deputies he recruited – only to find himself haplessly cross-examined at trial by steel company attorney, John McCleave:

> Q. – And you, as you have said, was [sic] the sheriff of the county at the time this application was originally made to you?
>
> A. – Yes, sir.
>
> Q. – And the application came to you in your official capacity as high sheriff of the county?
>
> A. – So addressed.
>
> Q. – As I recollect, previous to the serving of the writing you have produced, there was a verbal application?
>
> A. – Yes, sir.
>
> Q. – By whom?
>
> A. – By the attorney, signing himself as attorney for the firm. Mr. Dickey in company with a manager, Mr. Clark.
>
> Q. – And they represented to you that here was a danger of a riot – threatened riot, and danger of destruction to their property in the

neighborhood?

A. – And said they couldn't run their Works; afraid of its being destroyed.

Q. – Then you required them to serve you with the written notice you have produced; did you?

A. – I requested that, and it was served.

Q. – And this notice, I suppose, you understand to be given to you under the law, the Act of 1841, making the county liable for any loss of property in case of riot?

A. – Yes, sir.

Q. – You so understood it?

A. – I understood it to call on me as the peace officer of the county.

But was the county obliged to pay the sheriff for the special deputies?

The lower trial court thought so, as McCandless had gone beyond the call of his official capacity as sheriff, appointing special deputies besides deputizing Allegheny Bessemer employees, to protect the mill. And without calling attention to the dual loyalties of Allegheny Bessemer employees in their role as deputies to secure their own employer's property, Judge J. Slagle of the Court of Common Pleas No. 1, ruled on Allegheny Bessemer's appeal, February 20, 1892:

> The plaintiff [Sheriff McCandless] alleged that the men had been employed and the expenses incurred under an express agreement by the defendants that they would pay all expenses so incurred. . . .
>
> [I]t is apparent that the matter could not be controlled by the peace officer of the township. It was therefore the duty of the sheriff to take action to preserve the peace and prevent the destruction of property. If this could be done by the ordinary force under his control, it was his duty to use them. If this force was not sufficient, it was his duty to employ all other legal means necessary to effect the purpose. The only legal means within his control was the *posse comitatus*. . . . This he did not use or attempt to use. On the contrary, he deputized men in the employ of the defendants [Allegheny Bessemer Steel Company], and then appointed special deputies, – at first a few and afterwards in larger numbers, – as in his judgment the emergency demanded. As he alleged, and the jury found, upon the express promise of the defendants, to pay the cost of so

> doing. . . .
>
> Though the circumstances demanded the action of the [sheriff] plaintiff in his official capacity, the means employed by him were not those which he was bound by law to employ. His duty would have been done when he called upon the *posse comitatus*, and however ineffectual this may have proved, his personal responsibility would have ended had he honestly invoked this aid. He was asked to do more, to employ special officers to quell the disturbance and protect men in going to and from defendants' Works, and kept them [the special officers] there much longer than they would have been able to retain unpaid citizen [in a *posse*]. The defendants having promised to pay the expenses thus incurred, we are of opinion that there is no law or public policy which forbids enforcement of the contract.

The judge thus ruled against Allegheny Bessemer's request for a new trial – affirming judgment for Sheriff McCandless.

Henceforth, the sheriffs of Allegheny county could deputize whomever they chose among the community of men and the steel companies' loyal employees – and be reimbursed by the steel employers to secure their mill property and nonunion workers. And as this study reveals, along with the Governor's mobilization of the Pennsylvania Guard, this is exactly what they did.

Richard Krooth
Berkeley, California
28 April 2002

## THE THREE RIVER VALLEY, 1892

Legend: Not to Scale

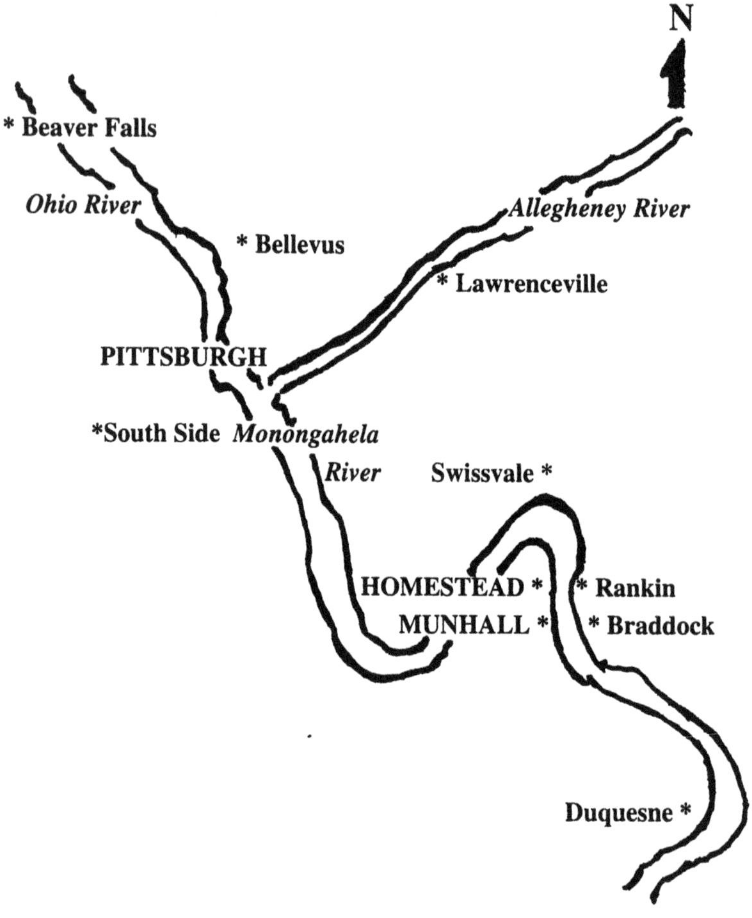

# Part I

# A Century Passing

The time passes without you even knowing it. When you start, they say you'll only stay a few years, but before you know it, it's been ten years on the job. Yeah, and ten years' seniority don't mean nothing anymore.

– Alice Alfonsi paraphrasing the warnings of steel working men in her family, during the hutdowns of whole mill towns in Pennsylvania during 1983 [1]

A century passing since Carnegie, Phipps & Company purchased the Homestead Works in 1883, an end to production was nearing in 1993. Steel workers throughout the Monongahela Valley were bitterly emoting, feeling an addled knot of anxiety as they were being stripped of seniority, security and jobs.

In the intervening 100 years, the complexities of steel production had turned on unrestrained nation-building, heavy labor and mayhem in the mills, and unimaginable degradation and strife in company towns. Among these, none emerged so central as Carnegie's Homestead, the daily affairs and fulminations of its exploited, determined population leading to the great strike of 1892.

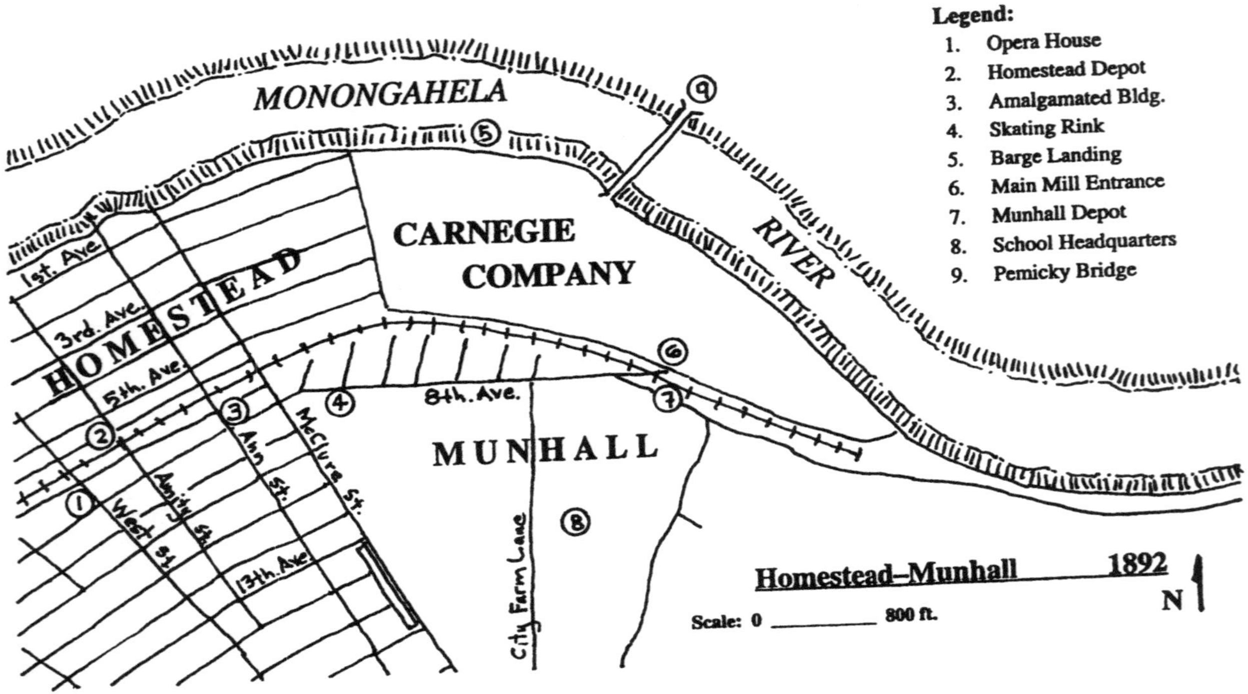


Map of Homestead, 1892
(Design Credit: Alan Gatzke)

# Chapter 1

# Walks Around Homestead

The steel-foundries belched huge flames that reflected the Allegheny hills blood-red and filled the air with soot and smoke. We made our way past the sheds where human beings, half men, half beast, were working like galley-slaves of an era long past. Their naked bodies, covered with only small trunks, shown like copper in the glare of the red-hot chunks of iron they were snatching from the mouths of the flaming monsters. From time to time the steam rising from the water thrown on the hot metal would completely envelop the men; then they would emerge again like shadows. . . . [T]hese slaves . . . had remained bli nd and continued in the hell of their own forging.

– Emma Goldman, reporting from the Monongahela Valley in 1892 [2]

I have always found it best to treat men well, and I find that my men are anxious to return my good will by working steadily and honestly, and instead of dodging are anxious to show me what a good day's work they have done. All haughty and disdainful treatment of men has a very decided and bad effect upon them. We must be careful what class of men we collect. . . . We must steer clear so far as we can of Englishmen who are great sticklers for high wages, small production and strikes. My experience has shown that Germans and Irish, Swedes and what I denominate "Buckwheats" – young American country boys, judiciously mixed, make the most effective and tractable force you can find. Scotsmen do very well, are honest and faithful. Welsh can be used in limited number. But mark me, Englishmen have been the worst class of men I have had anything to do with . . . .

– Advice to Andrew Carnegie from Captain Bill Jones,
manager of his Edgar Thomson Steel Works[3]

The birth of Homestead, Pennsylvania, attended as it was by bankruptcy and steel, industrial strife and the marketing of muscle, prefigured its later history.

Once the estate of squire John McClure, Amity Homestead had formed the original limits of the town, mapped out in 1872 by an insurance and banking firm that bought the farm, then went bankrupt in the Panic of '73. Building, production and populations came later, in 1880, when the Pittsburgh Bessemer Steel Company fueled city life by opening a modern steel plant covering 60 acres and bringing in 600 steel workers.[4] In 1883 Andrew Carnegie hastened the tempo for the 12,000 Homestead residents when he incorporated them into his growing complex of company towns. "Homestead," a contemporary wrote, "existed for the steel works; so, too," he concluded, "if the Carnegie firm were to move its works tomorrow, would Homestead be blotted off the map, or, at best, reduced to the rank of an insignificant village." [5]

## Authority and Wealth

Carnegie's iron and steel empire was unique in both size and power, combining the capital of three earlier empires of Carnegie, his initial partners and Henry Clay Frick in one powerful enterprise. Andrew Carnegie and his brother Thomas had, prior to Homestead, accumulated a fantastic fortune from their rolling mills. This had laid the foundation for the partnerships represented by the Carnegie Bros. and Company, Carnegie, Phipps & Co., and other "independent" firms, and was enlarged by Frick's dynasty of coal and coke.

Frick independently had bought out and consolidated his competitors in building a monopoly of coking ovens. By 1882, he controlled some 3,000 acres of coal land and 1,026 coke ovens. In 1890 his control had grown to 35,000 acres of coal land and 42 coke plants in the region, containing some 10,046 ovens. The company employed 11,000 men to oversee these ovens, 23 locomotives, 72 pairs of stationary engines, 172 steam boilers, and 816 horses and mules.

A million and a half dollars bought Carnegie a majority of shares and half interest of Frick's empire in 1882; and six years later, in 1888, Henry Clay Frick entered the Carnegie Company as a partner.[6]

By leaps and bounds Carnegie's holdings grew. Dateline, July 1, 1892, the Carnegie Steel Company had synchronized production in the Braddock Steel Works, the Duquesne Steel Works, the Thirty-Third Street Mill and Twenty-Ninth Street Mill of Pittsburgh, the Keystone

Bridge Company of Pittsburgh, the wire mill of Beaver Falls, an assortment of lesser works, and the all important Homestead plant.[7] Meticulously calculating, Carnegie personally scrutinized every operation in an effort to control labor and marketing, to reap high returns.

Homestead would receive his special attention. Its plant covered 600 acres along the Monongahela River, held sophisticated machinery capable of a wide variety of output, and represented an enormous investment for a single mill in those days – $6-to-$8 million. Its mills produced armor and standard plate, and held a Bessemer department, 16 open hearth furnaces of 20-40 ton capacity, and adjoined three other large mills and four structural plants.[8] The complex operated its own railroad system, with 18 locomotives and four switching engines, loading output on the Pittsburgh, McKeesport & Youghiogheny Railroad running directly through the Works. An observer put production at "25,000 tons a month, or 50 percent more in the rough," adding that: " . . . steel ingots so large that they weighed several tons or more, have been cast in this mill, which had in addition to other appliances of a similar nature twelve mammoth traveling cranes for handling objects of great weight."[9]

No other American steel mills matched Homestead's size or sophistication. The Works supplied huge slabs of heavy armor plate for the United States Navy. Hydraulic cranes operated by one man moved weights of 200 tons, and machines cut the steel like butter and variously shaped it.[10] The Works possessed the nation's largest hydraulic press, a vast machine shop, its own waterworks, and facilities to produce its own gas products. Yet, despite its vastness and variety, Homestead was rather specialized for the manufacture of structural iron and steel for fireproof buildings and for bridges, as well as for Navy armor plate.[11]

Central direction also offered an industrial advantage. In June 1892, Carnegie took a new offensive by retiring the mill's former patriarchal chairman, Abbott. He put Frick in as manager, while maintaining the controlling financial interest and making other moves to fit Homestead more neatly into his industrial empire. Frick became chairman empowered with supreme authority to deal with labor. And Carnegie Bros. and Company; Carnegie, Phipps & Co.; and other related firms were simultaneously merged into a limited partnership.[12]

The other partners were more or less "yes" men in the organization's design: Henry Phipps, Jr., George Launder, H. M. Curry, W. L. Abbott, John G. A. Leishman, F. T. F..Lovejoy, Otis H. Childs, and a group whose stock represented Carnegie promotions in the firm, rather than investment. Several of these took roles in the drama at Homestead. They fronted for Carnegie and his Homestead Works, where by 1892, lived 13,000 workers, operating more than $25 million in combined capital equipment under the nation's leading managers experienced in dealing with labor in the coke, iron and steel industry.[13]

**Walks Around Homestead**

As the Carnegie dynasty matured, Homestead came under Frick's unbending control, with the Works at the center of city prosperity.[14] Yet industrial expansion meant life itself lost all semblance of tranquillity as the filth, noises and blasts from the mill made Homestead a difficult, if not unpleasant, place to live.

The city fit tightly within the frame of Pittsburgh's steel-making, where growth had progressed far enough by 1849 to fill the sky with smoke and the river with slime.[15] And once begun there was no stopping the onslaught. "Six months residence here would justify suicide," concluded British social scientist Herbert Spencer, speaking of the smoke and grime when arriving for a Carnegie-inspired lecture tour to Pittsburgh more than three decades later.[16]

Even Carnegie managers and a *corps* of clerks, draughtsmen and engineers inescapably worked and lived near the mills. The grounds of eight handsome residences of assistant managers and a clubhouse built for the use of company officers and their guests, only a stone's throw from the mill yards[17] were tracked through from all directions by the mill hands seeking shortcuts to and from work, until Chairman Frick built a wall surrounding the mill.[18]

True, occasionally the best paid of Frick's staff might escape to the Amity Hotel, the finest restaurant in Homestead, with its elegant scroll partitions, bar fixtures and the dinning room put in by Messrs. Feather and Kerr, lumbermen and interior decorators.[19] But even the Amity's luxuries were barred to the managers, once conflict between the workers and the mill owners began.[20]

Homestead's mill property was foul at best, everything caked with soot and grime, particles of metal, minute shavings and debris. Piles of iron and other materials grew everywhere, sulfurous acid choked the air. The river along the mill was clotted with human and animal feces, chemicals, rotting foods and garbage, other discarded refuse, the excrement of the modern industrial world.[21]

Homestead shared the river not only with mill pollution, but also with the sewage of upstream towns that built tax-financed sewage lines directed to the Monongahela's flow. Duquesne, Braddock, Rankin, Munhall and other smaller municipalities multiplied the contamination downstream at Homestead, and the Carnegie Company simply added its share of sulfuric acid and other chemicals.[22] So well known was the death of the river that when two adults tried their hand at fishing the Monongahela in June '92, the *Local News* mused that they "attracted nearly as much attention as two delegates on their way to Minneapolis."[23]

Water for home use was always in short supply. The rare inside faucets of Homestead produced vile, viscous liquids, while the rain brought down tons of soot suspended in the atmosphere. The usual method for getting water was a trek over the refuse, cinder and scrap metal piles that lined the lifeless river bank. The water thus fetched was medically unsafe for drinking, though some claimed its pollution was antiseptic: "No respectable microbe would live in it," one Homestead resident allegedly asserted.[24] Homestead actually experienced a smaller incidence of typhoid than nearby Pittsburgh, possibly because, as chemist/engineer I. P. Roberts said in August of '92, the oxide released into the Monongahela acted like alum to clarify the water. The bottom could be seen at a depth of six feet, he deftly pointed out, though the pleasant dark blue color was formed only after the oxide destroyed the fish, whose carcasses clearly filled the river. Roberts concluded that the water was thus perfectly safe for household use.[25] But its acidity likely made it closer to vinegar than a life-sustaining element.

Pollution meanwhile contributed to a variety of communicable diseases beyond the means of city doctors to control, creating an ever-present danger of plague. "If there is not something done to improve the sanitary condition in and about Hays Row, fronting on the pike," the local media warned, "an epidemic might ensue at any time," adding: "The residents of the place would do very well to have this matter attended to at once."[26] Nor was this condition limited to the poorer neighborhoods near the river; even the better streets like Dickenson had sewage running upon them, acting as a breeding ground of fatal typhoid fever.[27] A Mrs. McWhinney, the first wife of the highly-paid Republican Postmaster, thus succumbed to typhoid in June of '92, just as surely as did people she might never have seen socially in the desolate Third Ward.[28]

The *Local News* issued repeated, yet tardy, warnings against the "filth diseases":

> Street sweepings, sewage and all filth should be removed from the possibility of again entering into any food or drink. Among the most

> common of germ diseases are typhoid, cholera, infantum and yellow fever. . . . it is plain that all contamination of water and food supplies should be carefully guarded against.[29]

The Homestead Council's austere *Ordinance Code* specified the dreaded communicable diseases and set down unerring obligations - with threatened penalties for householders, doctors and others for failure to immediately report such cases to the City Health Officer.

The *Local News* took a narrow view of the cumulative origins of milltown diseases, blaming everyone except the large polluters for deaths arising from unsanitary conditions. It excoriated the poor for not using proper medical care, although people of the tenement districts and in shanties along the river clearly could not pay for medical care, and doctors preferred clients of good standing in the city. The *News* also criticized physicians in Homestead who "utterly disregard the borough ordinances relating to sanitary matters," adding there were also "many instances where such disregard had resulted in communication of disease, and the death of children, all of which would certainly have been prevented, had the simple requirements of the health ordinances been observed."[30] Homestead doctors turned a deaf ear, perhaps fearing infection, many living in the better parts of the city, far removed from the tenement districts and the hillside pollution.[31]

Responsibility for illness, its cost, the loss of family members to plague were considered to be personal, not social, problems calling for public health solutions. Each family was on its own, and affected families were offered elaborate rituals by the *Local News* for disinfecting clothing, bedding, and rooms:

> In disinfecting a room, after the bedding has been put in the solution given [four ounces of sulphate of zinc with two ounces of common salt to every gallon of water] and well washed, and the tinware and crockery thoroughly cleansed, the doors and windows are thoroughly closed, the cracks are filled with cotton and sulphur is burned.
>
> When the sulphur begins to throw off its fumes go out of the room and fill up any crevice about the door with cotton. In twenty-four hours an outside window may be opened. The floors and walls are then to be washed with a disinfectant, and the furniture cleaned and polished.[32]

Odes to the continuing charms of the Monongahela made bitter reading, if one could, for those who drank its water. The *Local News* seriously suggested that: "Nothing could be more delightful than an hour's boat riding on the beautiful Monongahela during the pleasant summer afternoons and evenings . . . ."[33] Such were the opportunities for recreation and, if one fell in or grew thirsty, hygiene.

Rains churned together the pollution of the mills, river and hillside city, too. The yellow mud of the streets mixed with the soot of the mills and the refuse of the latrines and garbage heaps, forming a sludge that flowed into a mass of rivulets and puddles in the residential districts on Carnegie Hill. They then joined to form powerful torrents moving downward from the triangular apex of the hill towards First and Eighth Avenues, and straight down the city's streets of Hays, West, Amity, Ann and McClure, running northwest to the already polluted Monongahela.

**Clean-Up, Debris and Spatial Dispersion**

The only large-scale effort to clean up sewage in the Homestead area was made by none other than the Carnegie Company itself on its City Farm property in early '92.

The firm explained developments on what had been a farm for paupers by saying it would offer the lots for sale during or after July of '92.[34] But this talk of land sale turned out to be a diversionary tactic, since the tract later became the site of a barracks for non-union workers for the mills.

To city residents, though, the purposes remained unclear for the moment.

Meanwhile in June Carnegie graded the land and installed sewers. Carnegie managers had already signed a contract in May with Homestead brick manufacturer, H. T. Healy for 300,000 bricks to be used for sewers servicing the planned buildings.[35] To facilitate the project, the town of Homestead planned to finance the grading and uniform widening (to 80 feet) of the entire length of Eighth Avenue, a street connecting to the old poor farm property bordering the Carnegie mills.[36]

The City also helped clear Carnegie's City Farm property of rail lines by moving the rail station away, and turned 500 Homestead paupers off the land.[37] Though the plant and farm were outside the Borough, municipal expenses for Carnegie development, his managers suggested, should come from selling Homestead bonds repayable from tax revenues and/or by partial payment by the firm itself. This proposal, said the *Local News*, was an "easy solution of the sewage question so far as the Carnegie property is concerned." But the paper noted that the

City itself needed no new drainage and was "under no obligation to take care of the sewage from the steel works addition. If they wish to share in the benefits of public improvements in the borough, let them make application to be taken into the borough limits and pay their share of the tax." Besides, the proposed sewer line would endanger the supply of pure water: "If for no other reason than that the outlet of the proposed sewer is above the water works, that alone is sufficient reason why the proposition should not be entertained for one moment."[38]

Carnegie's scheme to have the Borough taxpayers pay for sewers to benefit mill additions failed, but the firm continued building its own sewage system. It was now clear to Homesteaders that Carnegie had other plans. And, in fact, once negotiations between the skilled workers' Amalgamated Association and the managers broke down, Frick accelerated preparation of the properties as a safe and sanitary domain to house the non-union workers he planned to employ in the mill.

Outside this narrow slice of land, however, the workers and inhabitants of the city had to fend for themselves against the rains, the refuse, the sewage and the air which invaded the streets and alleys. Homestead's sewage system remained clearly inadequate, the investigating Braddock Councilmen and Postmaster reported.[39] Pollution continued to fill the sky, while in the mills old machines, parts, piles of scrap and rails awaiting reconversion in the furnaces rusted red and were themselves covered with fine particles of air-borne soot, steel and heavy dirt.[40] Hundreds of tall mill furnace stacks belched without remission a dense, pitchy smoke, so that trees on the hills showed the same gray hue of soot as those near the Carnegie properties.

With the mill following the contour of the Monongahela for a full mile and the Munhall mills not far away, the air reeked of smoke, smog and acid fallout. A short distance further was the plant at Braddock, with its own dozens of stacks emitting more dust and gases, while other plants in the nearby vicinity added to the profusion of dirt and haze that hung between the ground and the clouds. Sunshine was usually blocked by this smog or filtered weakly through its yellow-hue. So Homestead was always enclosed in this atmosphere of grime, subverting the quest for hygiene, impacting all classes of people, their clothes, the interiors of their homes, the facilities of every commercial establishment. Few washed bed sheets in the city stayed clean long enough to dry in the air, and only a heavy rain briefly revealed the natural color of the grass and trees.[41]

There was no escape. People depended on the mills for their wages, their keep; and as in so many company towns in other parts of Pennsylvania, they located their homes near the factory gate, as Carnegie had planned from the start.*[42] They walked to work, met their needs, did their business and shopping in the stores near their homes.

The hilltops offered some relief from the overarching shower of soot and filth. Here a yellow haze of grass could be seen alongside soot-gray trees; even sunlight occasionally peeped through the smog cover far above the belching stacks. And here lived the wealthier residents: the old land and mine owners, some merchants, a handful of preachers and teachers, some town officials and a few better-paid millmen. Here, too, land was for sale, but prices were as steep as the unkempt roads spiraling upward.[43]

The less expensive residences appeared on the lower reaches, leaning on one another, presenting a drab uniformity – two stories, two rooms upstairs, two down. Crowded together with only narrow passageways between them, the houses afforded little light on the first floor, so that kitchens, dinning rooms and parlors usually required candles or gas lamps on dark days. The finest of these homes had outdoor privies, a single spigot of cold running water to service the kitchen, and a garden or plot of grass facing the narrow streets.

The "Saturday Night Bath" was a regular rite in this section of Homestead: Kettles of hot water came off the coal or gas stove to fill a wood or tin tub. The adventurous experimented with contraptions made of rubber such as advertised by retailer J. N. Lawson: "The Goodrich Folding Bath Tub and all other kinds of bath tubs, wash stands, etc."[44] The soldiers who were later to occupy Homestead sneered at the latrines and filth of the city and said the population was comprised of some of the dirtiest people they had ever seen.[45] But if dirty they were, the fault was the condition of the city, not their own.

Near the bottom of Homestead's hill, close quarters and zigzag paths connected a maze of unkempt buildings, privies, sheds and shanties. The memory of paint hid beneath caked layers of dirt and soot - wooden walls, either smog-gray or mud-yellow, blending with the soil under foot and the rarely-seen rays of sunlight.

Following the descending walkways, potholes, depressions and catwalks picked a way around natural ravines and those cut into the hill by torrents of sewage and rain. On this uneven terrain lay a crazy quilt of foundations, each building perched on the immediate elevation of the hill beneath. Two or three structures were often huddled in a clump, sharing yards, privy space, and the stench of garbage mounds growing at random. Streets and alleys variously appeared, so one house faced the alley of another without shame, and the second had a front yard that was the back yard of a third residence.

All this made Homestead a mass of obscure places and personages few could locate. A city *Directory* had been started in 1890, but by its third edition in 1892, only one man, James M. Campbell, was honored by the *Local News* for his prodigious knowledge and revolutionary census-taking in the clapboard casbah. "He is the only person in town who knows the names of the alleys. . . . Hundreds of people live on the alleys and sometimes they are hard to locate."[46]

Elusiveness of Homestead residents was offset by the occasional congruity of four or more buildings enclosing a square of land and forming an accidental playground or a gathering point for nearby families. These plots were alive with unwashed youngsters, gossiping and pipe-smoking adults, or would-be entrepreneurs, without portfolio, gaming over a hand of cards.[47] They formed tableaux out of Dickens, minus the neighborhood gin mill; the skein of town saloons lay next to the mill.[48]

These then were the scraps of pleasure on the flood of the harsh reality one visitor depicted:

> The streets were horrible; the buildings were poor; the sidewalks were sunken and full of holes; and the crossings were formed of sharp-edged stones like rocks in a river bed. Everywhere the yellow mud of the streets lay kneaded into sticky masses, through which groups of pale, lean men slouched in faded garments, grimy with soot and the dirt of the mills. The town was as squalid as could well be imagined, and the people were mainly of the discouraged and sullen type to be found everywhere where labor passes into the brutalizing stage of severity.[49]

## City Improvements

In 1892, however, the residents of Homestead began to take arms against a sea of civic problems. Their petitions and complaints prompted the City Council to put an arc light on the corner of Eighth Avenue and City Farm Lane, making 42 lights in all in use in Homestead. Petitions were considered to grade, curb and pave the major thoroughfares. And to aid the growth of commerce in the city, the Council also began to clear the clogged avenues with an ordinance to prohibit the parking of wagons on streets and alleys when not in use.[50]

The plans for improving the streets, providing an electric light plant extension, expanding public buildings, and extending fire protection for the growing number of buildings in Homestead had to be approved by the taxpayers, though. On May 28, Homestead voters thus balloted on the advisability of increasing the Borough debt by $105,OOO, with street improvements to receive the lion's share of $55,000. These measures would have more than doubled the city debt from $101,500 to some $206,500. And the voters proved cautious about imposing new future taxes, only approving $87,000 additional bonds for three key projects – streets, waterworks and the fire department. Yet, this vote marked a significant stage in the economic development of the mill city.[51]

Residents were becoming aware that Homestead was lurching from the gloom of a company-town to a full-scale city. Wages were high, a building boom was on the city's doorstep, and, so confident were homeowners that the future bode security, they fervently petitioned for more street improvements, especially to accommodate what most Homesteaders only dreamed of – an electric street railway. Though the contractor, the Birmingham Electric Street Railway Company, believed Homestead was too small to pay for the capital outlay, the people of the city went on wishing, planning and voting the required revenues from their own pockets in the May 24th election.[52]

Almost everyone in town wanted street improvements: "When we have free delivery electric cars, and at least two streets paved up the hill, and a city charter," a taxpayer pointed out, "then we can appropriate twenty or fifty thousand dollars to build a town hall."[53] The *Local News* agreed: street improvements worth $55,000 were "of special importance . . . for the good and welfare of the borough."[54] With hoped-for permanent street improvements, a boom might follow, said another resident, because, if "home labor was employed, the money would be kept and spent at home among our merchants, [who] have largely contributed towards the building up of the town." [55]

But the $55,000 in Borough debt allotted for street improvements soon became entangled in the relation between improving streets and grades and the hesitancy of the Birmingham Company to begin Homestead's electric railway. The city contracted with the Birmingham Electric, but the company refused to build until the streets were first paved and the area was cleared of the tracks of two existing railroad companies. And there was a Catch-22: street improvements waited for the prior installation of the electric rail line.[56] These barriers created delays until well after November 1892 – a date when, we will see, the people had no heart and lacked the funds to go forward with the exciting plans of May.

Street improvements were also blocked by the varied elevations of the hillside road terraces and the temperament of those who had built homes upon them. These old elevations were so varied that for new road building no one was sure whether to revert to the old grades of the hill or to tear out the old roads, pave the new streets on a uniform grade, and allow existing homes, doors and windows to rise above or fall below the grade depending on the original architectural design. The City Engineer, Hough, reported to the City Council that it was impossible to restore the original grade on McClure Street for which the residents had petitioned, because the elevation of the newer buildings prevented it. Taking the old and new structures, a compromise between their elevations could be made by the restoration of the street to within three feet of the original grade. And the Council ordered it done, so that now half a window might show above the street level and a quarter of a door might swing below it.

Controversy about road improvement soon broke out. The need for a level path for the electric train and the city's plan to meet that need would make permanently imprinted in cement curbs and asphalt streets unequal access and light, as well as the spill from storms, for homes along the road.[57] "We don't understand why it is necessary to change the established grade, when only curb-lines are required," the *Local News* editorialized on discovering that the engineer hired by the city had taken the prerogative to alter the old street levels.[58] Resident John F. Hill was hot under the collar because the city engineers had originally advised him to build on the old grade and now would cut off half his house from access to the new street level. By contrast, his neighbors would receive a windfall, having accompanied the city engineer on his expedition when the new grade was set, a situation Mr. Hill believed smacked of favoritism at his expense.[59] The streets might be improved, then, but there was no way all the buildings of Homestead could be given equal access to use them.[60]

Every resident and business owner on the street-ways seemed to have his own opinions. Some objected to using the $55,000 in Borough bonds to grade, curb and pave the way for the electric road. They granted that the original estimates of the costs were for streets on which the electric tramway was to move, but argued that the funds were not limited for use on these streets. Thus the Council should authorize the improvement of the streets that needed it most. "The condition of our streets is a reproach to any civilized community," wrote Homestead ex-Burgess, M. M. Wilson. He argued the councilmen should "select such streets for improvement as in the judgment of the Council will do the general public most good . . . . Business men, teamsters and others suffer great inconvenience and expense in delivering goods, building material, fuel, etc., on the hill in the Third Ward."[61] The most undeveloped and poorest section of Carnegie Hill, its streets reflected its dismal poverty.[62]

In all the bickering over street improvements, it quickly became clear that $55,000 was insufficient to pave Homestead. "The $55,000 would not go very far," the *Local News* informed its readers. "It would not improve the half of Eighth avenue," let alone all the other streets the contestants wanted paved.[63]

Every special interest and section of Homestead had competing priorities as well. The business men, commercial enterprises and home owners stressed the importance of the main thoroughfares in the more prosperous First and Second Wards, where the greatest property values were found.[64] Yet even the First and Second Ward residents remained at loggerheads over specific priorities, the residents of each preferring their streets be paved first, and the residents of every avenue wanting the asphalt roads begun opposite their own homes or businesses. Naturally, the real estate brokers and mortgage lenders preferred the paving to start on the tracts they were trying to sell and finance. Each special interest lobbied for its own benefit, though there was too little money to meet the needs of all.

Mobilizing the unpaid time and labor of Homesteaders was a proposed solution to the fiscal dilemma. The local papers stressed the virtues of cleanliness, encouraging each resident to build and maintain public ways on the street frontage through maintenance of individual holdings:

> Probably you will need to begin by forbidding the abuse of the public way before your house. Do not allow the road to be used as a dumping ground for cans and ashes, or as a rooting ground for pigs.
>
> If a sidewalk is necessary, make it of good and lasting materials. Lay down between it and the travelled way a grass strip of as great a width as possible. This travelled way, the central part of the highway, should be no wider than is absolutely necessary, for the care of a wide way is expensive, and the glare and dust from it are disagreeable.
>
> If the ground is uneven, the road grade may need to be cased by some cutting and fitting, but the foot-path and the grass strip should rise above the road or fall below it, if the natural surface of the ground can be more closely followed by so doing.[65]

If Homesteaders had followed this advice, though, the elevation of the sidewalks, grassways and streets would have become as irregular as the natural depressions and open sewer lanes of Carnegie Hill. And, whether or not changes were ever made, Homestead would remain a complicated patchwork of vistas no sane engineer or architect could ever completely restructure.

**Pressures and Politics in Urban Development**

Streets and public services nonetheless grew as important issues in 1892 Homestead because of the mill town's needs for expansion, roadway traffic, and security of homes and businesses.

When property owners voted to tax themselves to improve public facilities, they effectively spread to others and thus socialized the cost of building and maintenance, not only for themselves, but to benefit commercial and manufacturing enterprises as well. As the only large manufacturer in Homestead, the Carnegie mills were the greatest beneficiary. But small business owners, too, understood the connection between civic improvements, a prosperous mill and full employment. C. M. Stauft thus linked the opening of his new jewelry store, May 27, 1892, with the expectation of universal progress and prosperity: "he enters the business circle here with the belief that with the adjustment of the wage question for another term of years, the town will enter upon a period of greater prosperity than we have yet experienced," the local media paraphrased his thoughts.[66]

Despite the general optimism and civic pride, improvement benefits and prosperity would be awarded more or less based on the relationship between outlays and the position of each beneficiary in the business and residential community. New roads would facilitate labor punctuality at Carnegie's mills and efficient transport of raw materials and finished goods. The public water supply and fire protection provided business enterprises would lower operating and production costs. Residents also would gain by improvements in the streets, the water works and the fire department, though their taxes went up to pay off municipal bonds. If streets were improved, they could travel to work and elsewhere more rapidly and easily. If water was connected to their homes, they expended less effort fetching it from the river or the corner spigot. The faster a family could wash, prepare food and clean up the home, the more promptly could a workman be sent off to the mill. Workers in residences protected by a fire department would demand fewer raises to meet losses, rising home insurance rates and, gaining security, be likely to permanently remain in Homestead. Overall the labor force would be enabled to work with both a smaller exertion and need for money. The resulting lowering of the effort and cost to maintain the working class family was calculated by Carnegie -

then matched by him with a proposed lower wage scale.

Despite these underlying forces, the resulting battle over street grades, improvements of Homestead's alleys, waterworks and fire stations was presented more narrowly in terms of the development of the city. For these improvements, residents had actually voted to impose both present and future property taxes on themselves, accepting taxes, like work, as an individual responsibility to improve the conditions that would aid commerce, Carnegie's mill and workers alike.

Thus the *Local News* made no distinction between prospective beneficiaries, as it argued that the new water works plant worth $20,000, when added to the existing $70,000 in equipment, would make the Homestead facility "the largest and best plant in the Monongahela Valley." Since the mill needed a trolley system, the editors assumed everyone needed a trolley, first and foremost. The $55,000 for street improvements should thus pave, grade and curb the streets in order to prepare them for the electric street railway. "No part of the money should be used on any of the streets until after those are provided for which the street railway are to occupy. . . . The money cannot be used for sewers," the *News* concluded, though that paper had itself complained for years that sewers were essential to make the city safe from accumulated refuse and communicable diseases.[67]

For such one-dimensional growth, the only limitations were seen in terms of the uneven advance in technology and the undue use of the arteries of commerce for pleasure or pride. Invoking the morality of safety, the *Local News* condemned people who proceeded at unbusiness-like paces on the improved roads: "The temptation for fast driving is very great when some of our goers get on the asphalten pavement. On account of the narrowness of the streets, and it being a busy one, an ordinance has been passed prohibiting fast driving thereon. Of late the ordinance has been daily violated and numerous accidents and narrow escapes are reported. The officers of the borough are instructed to enforce the ordinance, and those who suffer in consequence are themselves to blame"[68] And so it was done, those who "were airing their steeds in a lively manner on Sixth Avenue, in violation of the borough ordinance," being sued by the enforcing policemen and compelled to pay fines upwards of $10.05 each.[69]

## Labor, Rents, and Real Estate

It was not reality but possibility that released the imagination of Homestead merchants, real estate developers, and skilled workers with secure jobs. Even the mechanics, the unskilled, and the day laborers sensed a new direction of the winds of economic revival. Electricity, like nuclear power a century later, claimed it was the genie that would transform the city and each home. The same power that drove the city's rail coaches would hum into kitchens, as advertisers claimed revolutionary advantages for electric refrigerators foretelling that the old ice box would go out with the horse and buggy.[70]

Dreams of home ownership also danced through the air, forwarded by realtors like Coen and Gillilend at No. 610 Ann Street, using the idea of possession to make a typical pitch:

> Do you want a house and [to] stop paying rent - Money that is gone never to return? We have a plan whereby you can pay for property in monthly installments, that will not amount to more than the monthly you are now paying, but instead of it being gone, never to return to you, you have it secure in your own property, which, in a few years, will enable you to own your own home. Come and see us, and if you are interested in owning your own property, and want the money you are now paying rent for to be placed where it will pay for your own home, we will assist you. We have some choice properties for sale in all parts of Homestead.[71]

All accepted the premise that as city improvements benefited the mill, mill profits and thus higher wages would enable labor to home ownership and other amenities, thereby benefiting the town. "It is altogether probable," the *Local News* sounded a familiar theme: "that Homestead will experience a building boom immediately after the settlement of the wage question at the mill. The Hays estate property and the Carnegie addition will come into the market about the same time."[72]

The expectation that Carnegie would sell his workers the new, modern buildings was cause for optimism. For years already, the skilled workers had purchased homes in the legal sense that they lived in them and paid off mortgages.[73] The money to buy them had been borrowed from either the banks or Carnegie; and Carnegie in turn had borrowed it from other workers whose deposits in his banks exceeded the loans.[74]

To those who could not or cared not to borrow, the Carnegie Company also rented houses in Homestead and nearby Munhall. With an eye on wage demands, Carnegie was a model landlord, generously lowering the cost of labor's maintenance by renting out these houses at 30 to 40 percent less than other landlords, and furnishing upkeep, water and gas at a price less than that charged by the gas company. Carnegie units were models in another sense: Supply never came close to demand, so there was a long waiting list of workers who wanted to rent company homes.[75] Only after the 1892 battle at Homestead and the blacklisting of the striking work force did the Carnegie rental space exceed the demand.

Meanwhile, more poorly paid workers – mainly male Hungarians and Bohemians, unmarried or with small families – made do in rooming houses, crowding into $2-a-week, single rooms, exhausting most of their mill wages. In these tenements and shanties along the river west of the mill, immigrants came with their sacks of worldly goods. The jobless erected shanties or shacks on the flat grounds east of the mills, until Carnegie bought the land and the authorities herded them away, or sent them to the City Poor Farm to be subsidized by taxes on Homestead's property owners.

Yet, by 1892, even the meanest day laborer was being encouraged to become a homeowner by Homestead realtors. "Why do you persist in paying big rents and living in crowded tenement houses? Munhall Terrace offers fresh air, good water, plenty of room, no rent," realtors Schuchman and West excoriated Homestead's poor: "Good lots are selling at $200 and $300 on payments of $1.00 and $1.50 only per week. These lots are sure to enhance in value soon, and this is the time to buy."[76]

Those who had accumulated savings were encouraged to place them in banks for safekeeping; those without were invited to borrow. The Security Building and Loan Association of Homestead took pains to explain this dualism, asking prospective clients: "Do you want to save a little money each week? Do you want to put it where it will make 8 percent interest ? Do you want to belong to an association that will help you to build a house?"[77]

New financial institutions and banks opened in Homestead every month and expanded, helping finance real estate agents who bought land cheap and divided it into lots for sale at a high price. In May of '92, one new company was formed to purchase the 40-acre Whittaker farm for $12,000, adjoining the already-divided Munhall Terrace, to be transformed from the growing of crops into large-size residential lots for sale.[78] Every month some new real estate scheme or another was announced by the city's leading brokers, enticing potential homeowners and commercial enterprises to buy, appealing to the workers' sense of a need for personal improvement and security.[79]

Skilled workers, such as those of the Amalgamated Association of iron and Steel Workers, believed their jobs were secure at the mills and remained confident they would be able to pay off mortgages on their homes. They took a short view of the past relationship between Carnegie and his workers, however, for jobs were secure only in the sense that labor had agreed to a pay cut every time Carnegie demanded that production costs be reduced in exchange for a reduction in the cost of living. The Homestead workers had forgotten that in 1883 Carnegie forced a pay reduction on his employees in the Edgar Thomson Works at nearby Braddock and, as a pacifier, given them Carnegie coal at one or two cents per bushel below the market price and a 25 percent reduction on one installment of the mortgages they owed him.[80]

Carnegie also realized his financial hold through real estate loans over his employees: "The labor force is more stable and there is less likelihood of a strike, the employees not wishing to jeopardize their positions after a house has been acquired, lest they have to move," wrote one investigator. "Loans up to two-thirds of the value of the property are made to employees to aid them in buying houses. Interest is charged at the rate of 5 percent per annum plus the state tax. A considerable length of time is given in which to repay the loan, and payments are taken out of the pay in such amounts as the employee may direct."[81] In the interim, if labor refused to comply with Carnegie's wage reductions and struck, Carnegie would halt wages, forcing mortgage payments into default, and become entitled to repossess the property. In this sense, workers were vulnerable to Carnegie's wishes. Yet, there were other, more significant, ways in which the Homestead population was obliged to accept the conditions, linkages and relations surrounding them.

## Fitting IntoThe System

By the early 1890's the population of Homestead was more an instrument of the wage, market and two-party political system than a force shaping it. Homestead workers and their families depended upon wages to enable them to buy essentials in the marketplace. Almost everything they needed was less and less made or grown by themselves at home and more and more produced by others in Homestead and elsewhere. They had succumbed to the process of selling their labor-time for wages, buying the products they needed in the market and, accepting the basic tenets of the economic system, seeking security by supporting promised policies of the Democratic or Republican parties at the polls.

Many people of Homestead had once grown their own food in garden plots or on rural farms. Some were forced off nearby land by

Carnegie's expanding mines, coking plants and mills that aerated smoke, gas, heat and dirt, or dumped acids, ashes, slate, slack, and other debris – polluting farm land, air and water sources, undermining crop and animal production.[82] And as Homestead grew and came to depend on the mills, the workers had neither sufficient land nor time to travel back and forth to nearby farms to tend crops or animals. A life in the mill simply did not square with devoting labor to raising one's own food. As well, the tiny parcels of land and absence of sun in the Homestead hillside districts made it next to impossible to grow enough to meet the needs of a family. Not only were there small and few remaining yards, there was also public sentiment against fences to protect crops and animals.

"The fence in town is a relic of days past and gone, when cows and pigs wandered at will about our streets," a paper noted. "As for mischievous dogs no fence could keep them away. Modern opinion says there is no positive use of fences in a city, except as boundaries between neighbors and in many cases two or three small stones serve this purpose, whereas by abolishing fences, a continual source of expense is removed and the general public gets many of the benefits of a number of small parks by being allowed to get an unobstructed view of the well-kept lawns."[83]

During the '80s and '90s, fences that had existed in Homestead's borough limits were torn down. A few farms remained within the town with a handful of urban animal-tenders, but they were besieged by laws and complaints. Homeowner V. S. Bullock caught the anti-husbandry sentiment of townsfolk in a letter to the *Local News*:

> Is there not an ordinance prohibiting the running of large cattle and horses? If so I wish to enter a complaint against the authorities, for I think I have cause, as they (not the authorities, but the animals) have destroyed property in the shape of flower vases, to the extent of about $20, and for me I must confess I am getting about tired of it, as I believe I am not compelled to fence cattle out or off my premises, and if I am compelled to, I certainly shall have to take action to protect my property.[84]

The cows must be retired, argued former Homestead Councilman Hoover to the Borough Council. The beasts damaged lawns and slowed traffic. "The matter was referred to the police committee," the local paper reported, "and the cows will have to go," to which the editor wryly added, "There is an ordinance against cows running at large, but that doesn't stop them."[85]

The original farms on the outskirts of Homestead were also being liquidated and turned into real estate divisions of the city. One old homestead dissolved after another, the last four absorbed in 1892 being Munhall Terrace, the City Farm properties, and the Cox and Whittaker Farms. Some farms were transformed into stables and horse clubs for the wealthy in the surrounding towns, others prospective sites for coal, oil and gas, still others real estate subdivisions.[86]

The Whittaker Farm, lying some two miles from Homestead and a mile and a half from the Steel Works, had been laid out into large lots, each comprising a quarter an acre for sale at $300 on easy terms. As these lots were so far from the Works, sale to Homestead workers came to depend on a means of rapid transport; and, as the *Local News* explained, "When the electric road is built, the distance may be covered in ten minutes."[87]

In all these ways, the old farms were taken out of food production, and fewer people could farm. Most of the people of Homestead now depended on the market and were compelled to earn an income sufficient to purchase these essentials. If Homesteaders ever had control over the land, the tools of farm and hand labor and workshop processes of production, they now lost them and worked under the direction of employers at the Carnegie Works and other substantial and lesser businesses.

There were changes in personal relationships related to the workplaces as well. Some business owners had once been moved by a sprit of *noblesse oblige* towards their workers in Homestead. But that attitude quickly dissipated when the workers and employers formalized the ways in which labor-time was bought and sold.

The changes made at Hays Coal Works in 1892 were symptomatic. Hays, had always operated as a family business based on benevolent paternalism towards its workers. But when its founder died and the mines were temporarily closed to divide his estate, that relationship was jeopardized. The *Local News* favored continuing the old personal ties and obligations in the mines, noting that "if they fall into the hands of someone as reasonable and as much in sympathy with labor as Mr. H. C. Bughman, the working people need have no fear, especially if under the control of the veteran Superintendent Watson."[88]

Elsewhere in Homestead, there was little hope that local businesses could fend off the "outsiders," or that these "foreigners" would act reasonably toward labor. "Recently," the *Local News* warned on May 28, 1892, several new Building and Loan Associations have been started in Homestead. They are what we call 'foreign' companies because they belong to Chicago or some other distant place.

The Security & Loan Association is not one of them. It is a Homestead institution and was organized five years ago."[89]

The largest "foreign" company in Homestead was obviously the Carnegie Works, where paternalism ended in mid-1892, when Frick began a centralized, cost-cutting campaign. Hitherto, Homestead and all other mills under Carnegie's control each had its own superintendent and kept its own books, so as to function fiscally apart from the others. Even the Carnegie auditors found it impossible to determine the firm's composite costs of operations, capitalization, net returns on investments and interest charges. Frick, recognizing the inefficiency of such a system and, trying to bring together the uncoordinated methods of management, sought to turn the firm into a centralized body under his authority. He finally convinced Carnegie to combine all subsidiaries under the aegis of the new Carnegie Steel Company, Limited, ending the old structure of local management at Homestead.[90]

Frick then took complete charge. Already infamous for his brutally repressive control over the laborers at his own coke fields in western Pennsylvania, Frick replaced the methods of measured consultation of his predecessor with his own "arm's length" approach to labor.[91] This, and his drive to centralize mill operations, destroyed the old relations between workers and employers, and also put the workers on guard against new methods that were being used to regulate them and undermine their control of the details of production.

Frick's management reflected part of a growing nationwide trend of viewing people in terms of labor-power to be valued and sold like any other commodity. Duties and hours were to be clearly defined, without consideration of personal matters. Even yearly terms of employment were proposed, so as to establish a going relationship for selling all sorts of wage services. And as wage-labor became dominant, workers everywhere began to realize that they had to bargain to sell their potential strength for, and quality of, work.[92]

## The Buyer's Market

Once "foreign" companies entered Homestead and employers adopted impersonal, bureaucratic means for dealing with their labor force, the workers adapted to the necessity of selling their labor-hours, settling into their various roles as wage earners. Workers' efforts produced, not goods for themselves, but money wages. And money in sufficient quantity seemed able to supply every need or desire. Not only goods such as clothes or furniture which families might have once improvised or made themselves, money also bought intangibles like real estate insurance and skilled services the wage workers no longer had personal time for – from wallpaper hangers to plumbers, gas fitters, roofers and carpenters.[93]

Services were being brought into the Homestead market by professional people as well. The city had its barbers who had learned their trade as apprentices. There were doctors who had gone to medical school, and dentists who pasted austere looking degrees on the walls of their shops. Dr. A. Troth, announced the local media, invited the public to his enlarged and refitted dental offices in the respectable Hough Building.[94] And for a price one could have faulty molars excised or Troth fill a tooth.

Not alone in providing services, there were others who could alter buildings, wagons, sidewalks for a fee. The old homily that "a good workman is known by his tools" was also extended. "This trite old saying does not necessarily apply alone to the mechanic," the *Popular Educator* journal pointed out the dimensions: "The man of letters, the teacher and the students are in the strictest sense workmen," requiring "their employment of the best tools to accomplish the best results."[95] The services of teachers and mental workers were thus entitled to compensation, and wage workers, either directly through the market or by way of taxes, would have to pay for these.

The paradox of consumer goods thus took hold in the city: spend your time for money, spend this money for conveniences saving time and money. "Every person that keeps house," J. Henry McClure of 321 Eighth Avenue advertised, "knows what a great source of convenience and saving a good refrigerator is during the summer season. . . . Those on the market this year are much improved and cheaper than heretofore."[96]

Certainly, the wage earners of Homestead could not always work sufficient hours to afford to purchase everything essential for a comfortable life. But, if one lacked sufficient wages, one could borrow or buy on credit. Pickering's Store at 227 Sixth Avenue specialized in financing short-term accounts for potential customers, calling to their attention it was "the only furniture and carpet dealer in Homestead that will sell you anything to furnish your house complete for a *small amount down* and let you *make your own* terms to pay the balance." The store pioneered in Eastern-style credit and mass marketing in the mill city: "We handle nothing but First-Class Goods and can sell cheaper than any other cash house in Homestead will sell you the same goods for spot cash. Ten dollars' worth of goods, one dollar down fifty cents per week."[97] Even the most poorly paid could hardly resist.

Earning wages to buy Homestead's expanding cornucopia of household goods and services created new dependencies on retailers and the manufacturers of consumer goods. Once low-cost commercial fuel reached Homestead, one could not cook without a range. And rather than take time and energy to construct one's own fireplace or stove for heat or cooking, ordinarily one *purchased* a mass-produced stove on the market from wages earned.[98] But to operate the range, coal and coke would no longer suffice in Homestead and gas had to be connected and paid for. Homestead households had the "choice" of buying gas at metered rates or at a contract rate for a certain period.

One could only ponder *what*, not *whether*, to buy, and the news and advertisers directed attention always to the beauty of the smart purchases – so consumers now had to ponder which was cheapest. "One gentleman's bill for a cook stove and grate for the month of April is reported to be $11.00, while another runs a cook stove and grate for $2.50," the *Local News* pointed out the dilemma. "One lost heavily while the other saved considerably."[99]

"The Low Pressure Gas Company!" J. Henry McClure advertised "has required you all to put in meters if you wish to continue the use of gas, and in the event you do, we are prepared to show you Stoves and Ranges that are Perfect Gas Savers. They cook with the least amount of lost heat and offer every other advantage." Of course, the householder could rip out the costly stoves and ranges; but merchandiser McClure provided the disenchanted an alternative: "If you wish to discontinue the use of natural gas, we will arrange for that, too, by furnishing Oil Stoves of all sizes and patterns."[100]

Such momentous "choices" did not lead most householders to wish a plague on both alternatives and go without a stove and grate fire in the dead of winter. The die was cast and, from here on, everyone would live with it.

Water and fire protection were also included in the new cash relationship. The only decent water was available from the town utility, and as Homestead improved its waterworks in '92, it set yearly rates by the number of spigots and rooms. If people did not pay the set rates, they simply did not receive water.

"In case the rent is not paid on or before the first day of said months" of January, April, July and October, the June 6, 1892, Ordinance of Water Schedules and Rates proclaimed, "the water shall be shut off from such premises, until said rent is paid in full, and if turned on again, a fee of 50 cents must be paid."[1101] And so few residents could afford not pay the following quarterly rates:[102]

| **Rates by Number of Rooms** | |
|---|---|
| 1-2 rooms | $3.00 |
| 3 | 3.50 |
| 4 | 4.00 |
| 5 | 5.00 |
| 6 | 6.00 |
| 7 | 7.00 |
| 8 | 8.00 |
| 9 | 9.00 |
| 10 | 10.00 |
| Add'l rooms | 0.50 each |

| **Additional Rates for Attachments in Private Homes** | |
|---|---|
| Sink | $1.00 |
| Basin | 1.00 |
| Wash Tub | 2.00 |
| Closet | 3.00 |
| Bath Room | 3.00 |

Rates for all kinds of "public houses" were about double those of private homes, and the owners of these establishments had to pay them to remain in business,[103] though without rent ceilings, they might shift the cost to their clientele – who, many being millmen, might demand higher wages.

To remain clean, to have heat, fire protection, drinking water, passable streets, and the like, now required each adult to make an outlay of taxes and money. It was a municipal duty to earn wages, to join the market as buyer and seller.

## The Tax Hook

Homestead taxpayers also had the duty to support the school system and its budget. Homestead might offer only the most rudimentary education to its young, and upward mobility for its students might only get them to the mill gates; but the school system became expensive for taxpayers, as outlays for teachers, equipment and other services expanded.

In 1892, the Homestead School Board had finally settled up the year's business by ordering all outstanding bills paid, including teachers' salaries, janitors, wages, etc. An agent of McGinnis, Smith & Co., of Pittsburgh was present and submitted plans and estimates for heating the First and Second Ward buildings in preparation for the winter season.[2104] As part of the city budget, School Board agreements thereby shifted and imposed new expenses on the taxpayers.

Other local taxes on Homestead residents were also increased from $16,026.73 in 1891 to $26,999 in 1892, but there were more expenses than tax revenues, calling on accountants to crunch numbers: floating debt to be retired of $10,000, running expenses of $20,000, repair costs of $4,000, and bonds and interest due of $4,000. Between expenses and revenues was a deficit of $11,000, of which state revenues covered $8,000 and licensing and fines counted for the rest.

But there were items of expense yet to be determined: Interest on bonds might go as high as $5,000, bonds falling due, $2,000, and water work salaries and street materials, $5,000. The police would be paid $2,880; the cost of electricity for lighting might rise as high as $3,400; and street work might require an outlay of $3,720. In all there might still be a shortfall of $10,000 or so.[105]

Such were Homestead finances as the city expanded its services, borrowed more money to finance improvements, and levied a growing tax burden on its residents. These taxes had to be paid, moreover, a "Fair Warning" being given by the Collector in Homestead, Mr. F. E. Baldwin, on June 10, 1892: "Persons knowing themselves indebted for State, County and Poor tax or 1891, will save additional costs if paid on or before July Ist, 1892."[106]

**The System of Exchange Value**

One no longer had the choice of paying taxes or not, or buying in the market or not. These were outlays that had to be made, and to make them one had to be part of the wage-earning family or have another source of income. As few people in Homestead were professionals or merchants, the typical wage-earning family packed its grown men and boys off to the mill, remitted its taxes to Mr. Baldwin the Collector, and bought all its essentials in the market from the McClures and the Munhalls.

Transportation was also an item on the market. Horses were being made obsolete by Homestead's progress, and railroads prospered greatly by connecting people with their jobs and markets. Anyone who could dredge up 25 cents could buy a round trip ticket from Homestead to Pittsburgh on the P. McK. & Y. Railroad.[107] Without the quarter, one could hike the rocky hillsides. The Pennsylvania Railroad enforced this spatial logic on a vast scale, spent $1 million a week to build facilities in the State in 1891, and $40 million the same year outside Pennsylvania, the figures indicating the line's heavy dependence for rails and bridge steel on Homestead's mills, as well as the dual need for revenues and "intimate relations" with Carnegie and "all matters pertaining to business in a most striking way."[108]

To send their wares to Pittsburgh and pick up commodities for transport back to Homestead in the same day, merchants also needed to employ L. T. Wall's Pittsburgh Daily Express by wagon or messenger.[109] At Wall's location the Pennsylvania Railroad had also set up a transfer yard, offering at stated rates the special services of lunch, reading and bathrooms. Expected to employ fifteen hundred men in Pittsburgh, the yards would clearly affect Homestead; and the *Local News* foresaw that "a fresh impetus will be given to the town, and perhaps a regular boom."[110]

The timing was evident. Transportation and fast delivery of all commodities, labor-power among the rest, had become essential to the profit and loss accounts of the steel mills, the railroads, the merchants and those who thrived by selling services. Mill towns had become centers of efficiency. Streets had to be improved to move workers and goods quickly. For such ends, city governments, police departments, fire and power stations, had to be streamlined.[111]

Homestead fell into the clutches of these demands, as did Braddock and the other mill towns. Thus, as at Homestead, Braddock voters imposed an indebtedness on themselves (some $100,000) for general street improvements in 1892.[112] Monongahela planned a similar vote a short while later.[113] Latrole sent a special committee appointed by its city council to examine the paved streets of Pittsburgh, Wilkinsburgh and Braddock.[114] Homestead examined the Braddock fire department alarm system, before spending $1,500 for eight miles of wire and fifteen alarm boxes to protect the city's population, real estate and business interests.[115]

The Homestead population was meanwhile being regularized and secured for mill production day and night. Amidst a sea of privation for some,[116] the streets had to be lighted to ensure labor's efficient, safe passage from the home to the mills. "When poverty erupts it brings violence on a new scale: attacks, robberies, . . . . loathings," a reported truism ringing as clear today as among Homestead's underclasses in '92.[117] Verse of the period speaks of beggars thinly clad and crying in the streets.[118] Young, hapless people also roamed at night in Homestead,[119] the growing need for after-hours security leading to the installation of a new electric light boiler at the end of April in '92. "It was a big job hauling it from the cars to the power house," the *Local News* informed its readers. "When the new boiler gets into operation, the arc lights will be kept running steadier than in the past."[120]

**Ideology and Politics in the Mill Town**

With jobs steady, labor regularized, and the market full, Homestead had become a town where nationalist and patriotic feelings held strong. From the grammar schools to the mill men's union halls and clubs, there was a powerful, fervent belief in the virtues of America and its revolutionary spirit. This belief was reaffirmed by the rituals of elections and patriotic holidays. For the fairness of the system of voting for candidates seemed ensured by the Pennsylvania "Baker Ballot Law," which provided for secret ballots, and police officers, constables and deputy constables to watch over the polls during the elections and vote counts. They preserved order and made sure there was no electioneering or showing of ballots to others to wield influence at the polls.[121]

Yet the Baker Ballot provisions also tied voters to the two-party systems, as only Republican or Democratic Party nominations were printed on the ballot. True, independent and citizen candidates could still be placed on the ticket by a certain number of voters signing nomination papers, or taking other trying steps. "But," the *Local News* insisted: "still the best way will doubtless be to make regular party nominations."[122] With the workers of Homestead in the Republican fold promising tariff protection of the market for American steel and their jobs – and having the party alternative of the free-trade Democrats – political choice was clearly narrow.

Upward mobility in the party machines also meant loyalty to these two political alternatives, and the ambitions of Homestead Councilmen and Burgesses kept them within these parties. There was a regular progression from Burgess to the state legislature. Ex-Burgess Captain M. M. Wilson had hoped to follow Ex-Burgess John F. Cox and D. R. Jones in serving Homestead and the Sixth District of Pennsylvania in the state legislature. Perhaps, too, the role of John McLuckie in the strike of '92 can be explained in part by his plan to advance along the well-worn path to the state capitol. "It would not be too much to expect that our present incumbent Hon. Jon McCluckie [sic], should some day aspire to the same distinction accorded the gentlemen referred to," the *Local News* editorialized.[123]

Homestead mill workers had voted the Republican ticket because that party promised to maintain tariffs to keep out cheaper foreign steel that might result in unemployment. Believing their vote was a powerful lever, when the workers' higher wage demands were presented in 1892, most felt the employers had to accept them because the mill owners also supported the Republican Party and its protectionist policy. "I don't think we have any reason for serious alarm this year," Ed Keil, the Treasurer of the Amalgamated Association of skilled steel workers, said at the wage negotiations with Carnegie in mid-June of 192. "In this presidential year, when indications are for a hard fight to reelect Harrison, it would not be good policy for the manufacturers to insist on their scale."[124]

Shortsighted or not, some sixteen years later, an investigator was to reaffirm labor's reliance on power-politics to win bread-and-butter demands: "Most of the Pittsburgh steel workers vote the Republican ticket, because they see no immediate hope of success through a workingmen's party; but they are ready to accept any political theory that promises something worth while to labor. If the workmen in the mills were once convinced that in an approaching election there existed the possibility of the election of socialist candidates, there would follow what could not adequately be termed a landslide; it would be an avalanche."[125] Yet, all information coming down to us today indicates the workers at Homestead were convinced the Republican Party, not the Socialist, was the political force promising something worth while to labor. Though they were to discover otherwise later in '92, what resulted was their demoralization, criticism of the Republicans, and a retreat from all politics – not an open turn to the politics of Socialism.[126]

## The Politics of Socialism and Reform

In the '90s, there were certainly socialists in the labor movement in other areas, but their influence seems not to have reached Homestead in a significant way. Classical Marxism, bred in places like Milwaukee, maintained that the workers did not harmoniously fit into the system of capitalism and wage labor, and that all of the commodities labor created could **not** be purchased by labor simply because part of their value was appropriated as a surplus and not paid to the workers as wages.

"There is a greater demand for the products of our farm, our factories, our mines and forests today than for many years," the *Milwaukee Advance* informed its readers, noting there was a difference between output and labor's purchasing power. "The improved power of production has only reduced the masses to greater extremities, to more abject dependency, to deeper misery and despondency. The masses have the demand, the few have the supply, but the demand goes unappeased because of the lack of . . . purchasing power." The *Advance* theorized a worsening state with increasing immiseration of the work force:

> The purchasing power is unattainable to the masses save through their labor, but their engagement depends on these same few who control the employment, who own the homes, and even the avenues of escape, of these same needy ones. The few are the barons of industry, the many are the helpless subjects; and yet we boast of our independence and our glorious civilization. The weak and helpless are jeered at in their leprosy of want and despair, their comforters telling them there is no hope of relief and nothing to do but to curse God and die. A thousandth part of this cruelty and wrong if perpetrated openly would set the world aflame with revolution.[127]

Yet Homestead workers who felt this cruelty and wrong did not make the revolution. Nor did they try to undermine the ownership rights Carnegie had in the city's mills, though labor was certainly exploited there, and may well have followed the national average - awarding capital from 20 to 25 percent of the commodity values labor turned out.[128]

The Socialists of the'90s might rail that eight individuals in the United States controlled assets of $720 million and that the richest man in Europe, the Duke of Westminster, was poorer than any one of the big four in America.[129] But on a practical daily basis, the workers had to, and became accustomed to, march off to the factory to earn sufficient wages to allow them to live. And in the steel industry the Amalgamated Association of Iron and Steel Workers stressed wage levels, not political independence or socialist revolution against the possessors of wealth and means of production.

Socialist beliefs and the Socialist Party were barely visible in Homestead during the strike year of '92. The only significant instances of radical activity were the arrival of the Anarchist Socialists from outside the city, advocacy of Henry George Reform-Socialism by Homestead preacher Galloway, and Alexander Berkman's vain attempt to kill manager Frick. Voting at election times remained the workers' "politics," though the real politics of Homestead workers were carried out at the union lodges and in the clubs, where the economic questions of the day were discussed and the workers' job security and wages were the principal issues. Outside these secretive and exclusive organizations, Homestead workers were Republicans and as "patriotic" as everyone else in America.

In the Carnegie mills, the methods of Captain Billy Jones also became a way of keeping down labor's protest. Jones had mixed the ethnic background of his labor force in an effort to stop all-English or all-German or all-Scottish crews from controlling the production process and presenting high wage demands. His policy of a "judiciously mixed" group of young American "Buckwheats" was designed to keep the workers from separating themselves into antagonistic ethnic groups that might lead to disputes over job jurisdiction, promotions and hiring.

Such differences had previously fomented constant turmoil in the Anthracite coal mines around Jones's district and at Homestead under the original owners. There the Welsh controlled the rail mill, the Irish the converting works. The little fiefdoms carved out by these and other nationalities often struck sparks when they overlapped, resulting in interrupted schedules and lost production.

Jones nipped such troubles in the bud, ethnically mixing crews traditionally pitted against one another, successfully maximizing production without adding a cent to the wage bill. Thus the blast furnace crews of "Buckwheats" labored mightily to achieve the highest output for the week to win the company-awarded symbol of supremacy – a steel broom which, displayed high on the furnace stack, proclaimed to Pittsburgh the mightiest men in town.[130]

Eventually many workers in the steel industry saw through these schemes and became convinced the only way to protect labor was a workers' party, which either took charge of the wage system and the instruments of labor or abolished the wage system altogether. Only later, after the Homestead strike, did some workers in Homestead and Munhall realized that Socialism and the Socialist Party were essential to their goals.[131]

**Pomp, Patriotism and Political Consciousness**

In the intervening years, the waving of flags and uttering of revolutionary slogans were the stuff of Memorial Day, Columbus Day, and other national holidays. In the crucial year of Homestead in 1892, the General Griffen Post 507, whose cannon was later trained on the Pinkerton barges, became the center of the Memorial Day Committee. The G. A. R., as it was called, respectfully requested "the hearty cooperation of all societies and schools in the borough of Homestead and vicinity," inviting them to send representatives to the great events for the end of May.[132]

Likewise, the schools prepared for Columbus Day, October 12, a day later to be recalled in bitterness as the children of the locked-out workers faced the approaching winter with cold hands and empty stomachs. But on May 7th, the national fervor of the *Local* News was still heard: "Love of Country, knowledge of her resources, and glorification of the public schools – are these not objectives worthy to justify the National Columbian Public School Celebration?" The national refrain it offered was to mean something quite different in October than in the heady days of May:

> Wake in our breasts the living fires,
> The Holy Faith that warned our sires.
> Thy hand hath made our nation free;
> To die for her in serving thee.[133]

Even nationalism and patriotism were to prepare the working people of Homestead to organize in the coming months. On Labor Day in 1892, the mills had closed and the eight union lodges of Homestead workers had planned to join others in a labor day demonstration. The future outline of the organization of the Homestead commune was thereby drawn. Hugh O'Donnell, a leading figure in the Amalgamated Association of Steel Workers in Homestead, had assumed command of the demonstration. He had appointed a "chief of staff" from the Munhall Lodge, and an "adjutant" from the Armor Lodge. He had directed the President of each of the eight lodges to appoint five "aides" from each, with each President in personal command of his respective lodge to act as the "line aide" in the parade. The lodges and bands had also been assigned places in the parade. Every move was under the control of Staff Headquarters in the union's McClure Building, where all "aides" had to report to the "chief of staff."[134]

What this military line-organization would mean in the contest of workers and mill owners was yet unknown, and the employers went on their merry way assuming they could buy the workers' allegiances. Thus in a grandstand gesture, Superintendent Schwab of the Carnegie Braddock mills gave $100 of the town's $300 in collections for the revolutionary Fourth of July celebration. In Homestead, Carnegie was more circumspect, donating a large, costly pipe organ to the planned Catholic Church.[135] "The price to be paid is not known," the *Local News* confessed, "but as Mr. Carnegie does nothing by halves, it will no doubt be a very fine organ."[136] Such gestures won Andrew Carnegie a great deal of good will, and made the working class towns like Homestead more habitable.

If there was consciousness of social classes in Homestead, it tended to revolve around the rights of the workers to freely organize to better their condition. The realities of social class were diminished by the assumption that class was chosen or at least a temporary state, that the possibility existed for one to move up in social station, that many were content to remain where they lived, with how they worked, and the life they had. Thus the *Puck* humorous comparison of the lifestyle of the rich that of the working stiff:

> Miss Van Million (on the east sidewalk of Ninth Avenue) – "Don't you feel sorry, Mr. Dudekin, for those poor overworked street laborers?"
>
> Dudekin – "Ya-as; and think of wearing those course and dwedfully cut blue trousers?'
>
> Phalen (in the trench) – 'McCarty phwat wud yez do av the boss ast yez to wear clo's loike thim on the thing younder?"
>
> McCarty – "Be Gorry! Oi'd stroike."[137]

It was a fine joke on clothes making the man or not, but such assumptions trivialized the deeper antagonisms of class.

Certainly everyone in the Homestead work force knew Carnegie owned the mills and was the principal employer of the workers on the payroll.[138] They knew there were hierarchical jobs in the mills, with superintendents in each department – the 110-inch mill, the armor plate department and mill, the machine and press shops, etc.[139] They also knew their interests were different than and antagonistic to those of Carnegie. But they viewed themselves as a force of production that Carnegie had to have to make steel; this, they thought, created a relationship with Carnegie that would ultimately compel the firm to accept the wage scale they demanded.

They had reason for optimism as well. Carnegie had published his guidelines for labor relations in two 1886 *Forum* magazine articles, calling resolutely for recognition of labor unions representing the right of working men to organize and to receive a sliding scale of wages linked to the ups and downs of steel prices the employers received. In Carnegie's view, the choice was either such recognition or, if the employers brought in strikebreakers, potential labor violence. Clearly, he pontificated, "to expect that one dependent on his daily wage for the necessities of life will stand peaceably and see a new man employed in his stead is to expect too much."[140]

At first, labor at Homestead believed this line. For Carnegie had insisted that, except for railroads and public service agencies, industries struck by labor should not hire scabs, but simply shut down and wait out the discord. He had also argued that scab labor performed poor work, were neither the best men nor the best workers. He understood, he said, that skilled laborers would present themselves in common solidarity, as "There is an unwritten law among the best workmen: Thou shall not take thy neighbor's job."[141]

The Amalgamated lost some heart in 1889 when Carnegie himself brought scabs to Homestead. But that only made the union more calculating, looking to the advantage of Carnegie's absolute need for their skills during heavy market demand. They also recognized that they would have to wait out periodic moments of overproduction to gain the benefits they demanded.

"The market is glutted with stock in nearly every manufacture," Ed Keil, the Treasurer of the Amalgamated Association of skilled steel workers thus explained in mid-June of '92, adding that "a shutdown for a month or two would give the manufacturers a chance to dispose of the surplus. Then, it is believed they [the manufacturers] will sign our scale and go to work again."[142]

Here was the limit of the consciousness of the Amalgamated rank-and-file in Homestead: They accepted the wage system, bargained over higher wages with the employers, and believed themselves essential as skilled workers for the production of any steel at all.

**Leisure, the Family, Battery and the Lock-up**

To some in Homestead, Andrew Carnegie appeared as a savior and a philanthropist. Others saw him as a skinflint who cheated his workers of their due. But until late June of '92, almost everyone in the skilled work force agreed that wages were high and sufficient to sustain the needs of the family for necessities and recreation. Thus, in moments of bravado, excitement and meditation, the young were seen escaping to and from the schools, the men to the lodges and saloons, and the women to the market, their clubs, friends and the churches. Just as there was a division of labor in the family with the women isolated from the lives of their men, so too there was one in recreation, which had also become an item of expense subject to the pressures of the market.

The 3,800 workers in the mills were all men and boys, the age of 12 or 14 being the break-off point between school attendance and mill employment. The young girls stayed in school longer, in preparation for their future role as housekeepers - preparers of meals, dishwashers, shoppers for family needs, wives, mothers, and, for those teen-age daughters who could stand the humdrum of Homestead life no longer, elopement or escape to the supposed excitements of Pittsburgh. Even Homestead's Catholic Church had been embarrassed by one or its priests, the Reverend Father Flickiger, eloping with 16-year-old Norma Friskie, who warned her father not to look for her and took off with her Godsend to spend the priest's several thousand dollars of savings in Europe.[143]

This was but an alternative from the great social drama of women being kept in their place in Homestead by the system of employment of males only in the mill and family dependence on the wages earned by fathers, husbands and sons.

**Mill Men's Pleasure, Mill Wives' Pain**

Heavy work in the mills turned the men into hard drinkers and, for many, rugged, two-fisted brawlers.

"It is essential," Fitch wrote, "that they should meet together somehow and somewhere in fellowship and in relaxation from their work. The saloon and the lodge remain as the social centers for the steel workers," adding that:

> There are other reasons, to be sure, than the desire to mingle with one's fellows, for the popularity of the saloon; drinking is traditional among steel workers. But there is no doubt that the craving for companionship is one of the strongest reasons for its hold upon a community of working men. The nature of mill work is such as to make the saloon habit one of the most natural ones in the world. Practically every man is affected by the heat even if he does not have a "hot job." The whole atmosphere is such as to induce perspiration and enhance thirst. All the workers drink water in great quantities as long as they are in the mill. Sometimes a man drinks too much, so that he leaves at the end of a day's work feeling half nauseated. Such a man steps into a saloon for a glass of something to set his stomach right. Or if a man does not overdrink during the day, he is still chronically thirsty, and it is to satisfy a real longing for drink that he stops for his beer. The dust of the mills, too, that the men have been breathing for twelve hours, sends another quota to their beer or whiskey to clear out their throats. Then comes the largest contingent of all, the men wearied with the heat and work, some almost overcome and dragging their feet. These feel the necessity of a stimulant, and they get it day after day, regardless of the waste of physical and nervous energy involved in keeping themselves keyed up to their work by artificial aid. I do not think I am far wrong when I say that a large majority of steel workers sincerely believe that the regular use of alcoholic drinks is essential to keep them from breaking down.[144]

During the work week most men were careful to be fit, not intoxicated, for the next work day, so conviviality did not emerge till Saturday nights and after paydays. "Then," said Fitch, "is the time the men relax; then the treating is done. The saloon becomes a social center and the men find the fellowship that they crave. It is a following of the line of least resistance that makes the saloon supplant all better forms of social life. A man does not need to change his clothing and get a shave before he is made welcome here. He may come covered with the grime of the mill and not feel out of place."[145]

The women were meanwhile isolated from the life of the men on the job and in the saloons. Women never entered the mills; nor were women allowed in the union halls. Even in slack times when the mills were not running, the men headed for the saloons as a regular meeting place, primarily for companionship and only secondarily for drinking.[146] In good times, after weeks and months of work, twelve hours a day, and no holidays, Fitch noted, "A man gets far behind in his accumulation of pleasure that he feels to be his due. When a holiday comes it is all too short to collect the overdue bill; pleasure of a concerted sort must be sought in order to make up for lost time. As a result of these cumulative promptings, the saloons take more of the steel workers' money than do many of the legitimate business interests of the mill towns."[147] At the saloons, the men frequently spent $20 in a single night after paydays.[148]

Work hours may indeed have been long and hard, but saloon hours meant a tremendous drain on the family budget, as well as hours of family separation. Such divisions of money and time increased pressure on the family, the *Local News* itemizing that a marriage would fail: "When money that should go for a book goes for what only one side of the house knows anything about. When the lord of creation pays more for cigars than his better half does for hosiery, boots and bonnets. When 'father' takes one half of the pie and leaves the other half for the one that made it and her eight children."[149] Time spent at the local saloon also meant time away from the home and household chores, though this lack was seen as a personal failing, not a reflection of working class releases: "When the husband runs away from a little bit of house cleaning and then thinks it isn't much of a job to clean house," concluded the *News* in smoke-shrouded Homestead, then the marriage would soon suffer.[150]

**Reproduction of the Family And Low Wage Work**

With the men caught up in the tidal rhythm of mill work and escape from it, family maintenance was women's work: "The wife of the steel worker, too, has a hard day, and even a longer one than her husband's," Fitch reported. "To prepare a breakfast by six in the morning she must rise not later than half past five. The family cannot sit down at the supper table until seven or later, and after that the dishes must be washed."[151] The women of Homestead knew the burden only too well though they were encouraged by the woman editor of "Our Ladies" column of the *Local News* to bear it - making the helping of others the reason and excuse for their existence.[152]

Women were largely responsible for the sustenance and reproduction of the family in the working class towns. It was especially clear during the transition from rural production using self-labor to the system of wage-work. Mothers of working class families were paid with something less than respect, not money, just when money was becoming all important. For women's work in the family was not viewed as productive labor, a child responding to the *American Farmer* with society's blinders:

> "My mother gets me up, builds the fire, gets my breakfast, and then sends me off," said a bright youth.
>
> "What then?" said the reporter.
>
> "Then she gets my father up and gets his breakfast, and sends him off; then gets the other children and their breakfast and sends them to school, and then she and the baby have their breakfast."
>
> "How old is the baby?"
>
> "Oh, she is 'most two, but she can walk and talk as well as any of us."
>
> "Are you well paid?"
>
> "I get two dollars a week, father gets two dollars a day."
>
> "How much does-your mother get?"
>
> With a bewildered look the boy said: "Mother, Why, she don't work for anybody."
>
> "I thought you said she worked for all of you?"
>
> "0, yes, for all of us, she does; but there ain't no money in it."[153]

Too little pocket money reflected the unrewarded situation of women: "I never have five cents even for postage stamps without asking for it," a young Homestead wife lamented in "Our Ladies" column of the *Local News*. In her girlhood, she said, she had earned regular wages as a seamstress but, when married, found her financial situation changed. Her husband now "held the purse strings and made plenty of money." But, she added, he was in business for himself and the pressure of competition weighed heavily on his resources: new machinery was often needed; improvements had to be made; hired hands cost a great deal; consequently, no allowance was provided for her. She functioned as his "nurse, seamstress, housemaid, cook," with the added duties of motherhood. But because she was also his wife, he felt no obligation to provide her with a financial incentive. Quite the contrary.

"I always have a lump in my throat when I ask for a dollar," she continued, "and I used to go to his pocketbook for spare change, for at the marriage service he said, 'With all my worldly goods I thee endow.' But when little Tom [their child] began to steal pennies because he wanted something and could not get it, I began to wonder if I had done wrong and the sin was visited on him."[154]

Even when women entered the wage market, they were paid far less than men in Pennsylvania. Female public school teachers, for example, for the year ending June 6, 1892, outnumbered male teachers more than two to one (17,177 against 8,162), but earned one-fourth less for the same work ($42.15 per month for males, $31.41 for females).[155]

Whether as cheap labor in the market[156] or confined to household chores, taking care of children and doing the shopping in the market, the woman was also deprived of the man's companionship. "There is little time for husband and wife to have each other's company," Fitch concluded. "It is only by an extra exertion that they can spend an evening out together, and the evening at home is robbed of much of its charm by the projection of domestic duties beyond the time that would be required if the meal were served earlier. The father, too, has little time with his children. If they are quite small, he may go for weeks without seeing them except in their cribs. What the ultimate social effect will be of this stunting of family life may only be conjectured."[157]

The women of Homestead were not so doubtful of the effects. They understood that their husbands cheated them of money and time by belonging to ethnic clubs as well as fraternal and union lodges. They also protested against the splintering of families, the creation of male-only activities and clubs, saloon society on a higher plane at best.

"This is a great town for secret organizations," wrote one woman, L. A. Lews, of the Amalgamated locals and fraternal lodges of Homestead. "Now is it right for our husbands and brothers (mine belong to five or six) to spend their evenings in lodges, or other places, where ladies are not admitted? Should husbands have secrets that their wives may not share?" Meanwhile, she thought the man's membership in a lodge and one other good "order" was an uplifting place for men to be, far better than in a private club. But when the man became a member of more than two orders, she continued, the woman "does most decidedly object, for she knows full well he must neglect something else. No businessman can give proper attention to his family and church, and social duties, and attend more than two lodges. No man has a right to belong to five or six orders and spend every evening in some lodge or club room."

The logic that time and money spent in the saloon or lodge would take away from that spent in the home was irrefutable, the "Our Ladies" columnist thought:

> If it is not a man's duty to spend a few evenings of the week with his family, after being away all day, we should like to know in what it does consist. No true woman is selfish enough to exact all the leisure hours of her husband, but she has a perfect right to insist on a portion of that time. Again to belong to these societies is no small item of expense to a man. The plea and excuse for membership generally is that the money paid as dues, will eventually come back to the wife and children in the shape of a benefit . . . .
>
> It can not be doubted but that many families in Homestead have been greatly benefited from the lodges and organizations. It is certainly right for a man to provide as well as he is able for the future of his family; no right-minded woman will object to this, but in the meantime the household should not be impoverished in order to make this provision.
>
> No membership in such orders should be encouraged, when the money used each month to meet the dues has to be withdrawn from the household fund, and that which is actually needed for the present necessities of the home.[158]

**Battered Wives**

The women of Homestead continually tried to deal with their men's lodges, clubs and saloon holidays. Since they did the housework and the shopping, and had to make do with their men's take-home pay from the mills, every other Friday -payday - led to a battle in many families to divide the wage between cigars and drink and necessities for the home. Not only did the women have to pick up the pieces after their men's drinking sprees, but from a multiple of causes the battle in the home led to frequent instances of wife beatings.

"Officer Gingher had quite an exciting chase on Yesterday morning," reads an account of May 7, 1892, "after Walter Lewis, who had been arrested on charges of assault and battery on his wife."[159] Some accounts were more detailed, a Mr. Rayburn and his wife having a tumultuous ordeal leading to his suicide: " . . . he quarreled with her over some trifling affair, and a fisticuffs . . . took place. He struck his wife several times and knocked her down. She screamed lustily and her cries were heard by neighbors, who thinking they signaled nothing made no move to investigate the cause. Mrs. Rayburn then sent for Policeman Gessner and asked him to arrest her husband. Having no warrant the officer went before Squire Kuhn and made information against her husband, charging him with assault and battery."[160]

Wife-beating was frequent enough that neighbors accepted the cries of women as nothing out of the ordinary. To charge a husband with assault and battery was also a lengthy procedure that often allowed beatings to continue. A wife or policeman had to go before the judge and make allegations. The Homestead Squire then had to issue a warrant of arrest to have the man taken in custody and brought in front of him. Next the offender had to be caught by a police officer in possession of the warrant. And once the man finally appeared in court, the Squire could release him on bail in return for the man's promise to keep the peace.

It was a fine system for ensuring a man his freedom, but a poor one for protecting women from retaliation by angry husbands. Hardly a week passed when some man was not compelled, hat-in-hand, to appear in front of a Squire on charges of "surety of the peace," brought by his wife, only to be released on bail. In a typical case, a Mr. Mark Beatty of Munhall appeared during the second week of June in '92 in front of Squire Kuhn for a hearing on charges of breach of the peace which were preferred by Beatty's wife. He was "held for court" in the sum of $300 bail – which meant he was released from jail once he raised the bond.[161]

Though liquor was not the sole cause of the battery in Homestead, women were advised to win their men from beer by substituting root beer, lemonade and ice tea in the hot summer months, when the pollution of water was at its height - though there was never any pure H2O variety any time of the year. "Men and boys, especially those who are engaged in very laborious work have formed the habit of drinking a great deal," Homestead's "Our Ladies" columnist offered Temperance Society advice. "They seem to be afflicted with a continual thirst that is not easily quenched. And there are times during the heated period when sufficient draughts of water to satisfy their thirst cannot be indulged in without discomfort and injury. And again, the water they are obliged to use during the summer is not always cold and sparkling, [and] is often, at this time when most needed, warm and muddy and instead allaying, only serves to aggravate thirst. For these reasons, many of our men (who would otherwise never taste such a beverage) are driven to quench this great demand of nature by drinking lager beer; and thus is often formed the vile habit of using the intoxicating drink, a habit which may be prevented by a little care and forethought on the part of the wife or housekeeper to have on hand at this season some simple summer drink."[162]

Heavy and hot labor, impure water, as well human thirst made the turn to beer and other intoxicating liquors logical. Few men in Homestead would succumb to lecturing wives echoing the temperance movement's Utopian shrillness – "It's here we pledge perpetual hate to all that can intoxicate."[163] Battered wives and violence were not infrequently the consequences.

## Women's Leisure and Class Culture

With the work and leisure time of the male wage-earners dominant in the family, the women who sustained them had some time but little money to enjoy Homestead's offerings. Many pleasures were free or inexpensive, and women crowded the skating rink at the corner of Eighth Avenue and Ammon Street. They filled the Opera House at Eighth Avenue and West, when it hosted occasional performers who happened through Homestead on their Pittsburgh tours. As the most cultured group in Homestead, the women often went alone, the men being too exhausted from the 12-hour stint to tell the difference between a baritone and a falsetto. "The eight-hour men get a lot more pleasure out of life than the twelve-hour men do," a steelworker told investigator Fiche. "We can go to entertainments and social affairs as we couldn't if we had to get up next morning and go to work at six o'clock."[164]

"Opportunity for mental culture would seem to be ample in the mill towns," Fitch wrote in 1908. "Each has its Carnegie Library. Big, splendidly equipped affairs are the libraries in Homestead, Braddock and Duquesne. Each has its auditorium and music hall with a fine pipe organ, where lectures and concerts of high grade are held. There is a club in connection with the library in each of these towns, and membership in the club is open to all. The dues are small, and the privileges of the baths, bowling alleys and billiard tables are open to members. . . . In the libraries themselves there is a good stock of books and periodicals, the supply on metallurgy and mechanical arts being especially ample. But the steel workers seldom make use of these privileges . . . . The trouble is the same as that which has already spoiled half of the home life, and bids fair to blight the rest. There is not enough energy left at the end of a twelve-hour day to enable the average man to read anything of a very serious nature, and the reading done by even the most intelligent does not extend much beyond the limits of the daily paper. As for lectures and concerts – to attend them would necessitate a change of clothing and a preparation for which a weary twelve-hour man has little heart."[165]

In his verbatim notes, Fitch recorded the words of a steel worker: "The library and the lecture course are fine things for businessmen, women and children, but they are absolutely useless, so far as the mill men are concerned."[166]

Nonetheless, those with ample resources continued to dispense cultural charity. Gaining good reports in the media as foresighted philanthropists, they carried out their cultural pacification programs with a personal sense of *noblesse oblige*. The manager of the Braddock Mills, Schwab, thus recruited troops of actors and musicians to tour the Carnegie mill towns. The Munhalls and the Potters, famous names in Homestead for their role in business, land ownership and mill management, were also high on the register supporting highbrow culture. In May of '92, for instance, a half dozen shows were put on by the local talents in Homestead, giving great cause for concern lest the traveling artists be ignored. "Had the little Vernon Brothers come here at any other time than so late in the season. . . ," penned the *Local News*, "they would not have been greeted by such a very slim audience here . . . . All who were there felt that they were enjoying the best entertainment of the season. Mrs. John Munhall and Mrs. J. A. Potter were so well pleased that they want them to come to Munhall later in the month and give a performance in the new chapel."[167]

There were lowbrow entertainments as well, the Opera House hosting the Children's Carnivals, with a purpose beyond the mere frivolity of youth. "Aside from the enjoyment the little folks get out of it," the *Local News* pointed out the regimentation involved, "the training does them good, and the pride of the parents in witnessing the bright acts of their promising offspring is ample reward for their efforts in getting up the entertainment."[168]

One interesting event after another came to Homestead for the enjoyment of the locals, all of them performed under the city's blanket of smog,[169] most having a Potter or Munhall or Schwab or some other famous name, like that of McLaughlin of commerce or Mr. Carnegie of steel, among the list of supporters. In these ways, the people of Homestead had their cultural events planned for them in advance. And even in sports, Schwab and Carnegie provided subsidies and facilities.

For themselves, the wealthy had exclusive clubs and clubhouses, their entertainment consisting of simple, semi-idyllic occasions which emphasized exclusivity and graceful granting of recognitions. "The May Carnival held by the Cuckoo Rod and Gun Club," the *News* reported a typical upper crust event, "was a successful affair. The ballot was taken, the lady receiving the greatest number of votes to be Queen of May. The result was in favor of Miss Lina McLaughlin, daughter of the Captain of the Club. . . . The Cuckoo Club consists of about twenty gentlemen members, and in a few weeks will encamp near Granville on the upper Monongahela River, and the members will be accompanied by their wives and families."[170] Readers would know that McLaughlin was a famous name in industry of the time.

The Carnegies rusticated in more distant climes, Andrew at his Scottish castles and Thomas at his Valhalla, a resurrected pre-Civil war plantation called Dungeness on Cumberland island off the coast of Georgia. It was a world apart from that Homestead residents would ever know. "Men like the Carnegies, who took the millions that were suddenly bursting from their coffers and built castles and mansions, also bought whole islands; instead of going to Newport or Charleston for the season, they imported society to them. Guests visited Dungeness for a month at a time to enjoy life with steam-heated towel racks and wooden elevators, to wander around the dairy farm and sawmill, tennis courts, golf courses, artesian-fed indoor swimming pools . . . . [G]enerally the place was, as [Thomas's] granddaughter Nancy Carnegie wrote in 1917, 'nature's sanatorium, without malaria, nauseous odors, pestiferous insects . . . and with a salt ocean atmosphere that wonderfully dissipates the lingering effects of *la grippe* . . . .'"[171]

Homestead frivolity was a cut far below that of the chosen few. As experiments in recreation for some restless souls, women took up the "manly" sport of bowling, which the *News* applauded as "a fashionable feminine accomplishment." But the male editors drew the line at baseball, which they reserved as a sport for men. When the "young ladies' team" played at Homestead and were entertained afterwards at one of the hotels, "Some of Homestead's swells," the *News* editorialized, "got into a fight over them and succeeded in having their optics decorated and their clothes spoiled. As base ballists, the ladies are not a success," the editors concluded of the men's brawl.[172]

**Women Without Power**

Girls were prepared in Homestead from infancy for a life of housework and compliance to the male members of the family. One twelve year old girl showed an ironic understanding of female socialization in an 1892 essay "on boys":

> The boy is not an animal, but can be heard to a considerable distance. When a boy hollers he opens his big mouth like frogs, but girls hold their tongue till they are spoken to, and then they answer respectable and tell how it was. A boy thinks himself clever because he can wade where it is deep, but God made the dry land for every living thing, and rested on the seventh day. When the boy grows up he is called a husband, and then he stops wading and stays out at nights, but the grew up girl is a widow and keeps house.[173]

The rights of men to "holler," express their thoughts freely and to join organizations requiring their evening hours was certainly in sharp contrast to the structures and limits on women to hold their tongue, only speak when spoken to, and live within the secure realm of "dry land" under the control of fathers, husbands, and, in the twilight of widowhood, community mores.

Women were to be passive all their lives, follow the rules, speak only when spoken to, and then "answer respectable and tell how it was."

Women had to earn, not deserve, a right to their passive existence even in marriage, according to the female editor of "Our Homes":

> Girls who are anticipating matrimony, and of course they all are, we would advise very emphatically to learn the art [of cooking] as the way to keep a man's love is through his stomach. When "love flies out the window," it is generally the dining room window.[174]

The women of Homestead were to show devotion and cook as well. The man's love could be withdrawn at the sight of a poor meal!

Women were also expected to submit in other ways to domination in Homestead, the political sphere and voting being viewed as the exclusive domain of the male of the species. Justification offered in the local media was the supposed triviality of feminine thought:

> "If woman should be granted the right to vote, how would you vote?"
> "I don't know. Are the ballots printed in colors?"
> "I don't know. I guess so."
> "I should vote the color that harmonizes with my complexion."[175]

The going assumption was that the woman would remain a delicate churchgoer and discerning buyer of the latest fashions, not a social and political being. She used feelings, intuition, not brain power, and was not concerned with the economic and political life of Homestead. The newspapers promulgated the supposed limitations of the interests of the "fair" or "weaker" sex: "Considerable space in the local paper is always occupied by the proceedings of the borough council, and articles pertaining to the streets, the taxes, public officials, etc., very important in themselves but not of much interest to ladies," the male editors of the *Local News* wrote: ". . . we know that they also want their local paper to contain more that is of special interest and value to them. To make a local paper useful to housekeepers by having a department devoted to home matters, giving helpful hints and suggestions and the latest and best fashion notes, requires a careful reading of the magazines by a person competent to select, condense and prepare the matter for that department. A lady has consented to edit such a department for *The Local News*, and we believe it will prove of interest and value in the many homes into which it finds its way."[176]

In her early columns, the new editor encouraged the Homestead woman to follow her man from rags to riches and *vice versa*,[177] also calling attention to upward mobility, which she thought was inevitable in organized society. "People in all walks of life are constantly on the watch for an opportunity to break into the charmed circle of the select set just above them," she wrote. "No matter how exalted their social standing, there seems to be a goal just beyond for the attainment of which they are always striving."[178] Like other professionals, she viewed society as a process of continually moving up; but actually there was no ladder to climb for most Homestead women.

Rather, women faced a life of external self-denial on behalf of their men and families; and they were not even supposed to be interested in men's affairs such as politics or the mill.[179] Yet, they were usually better informed than the men, having gone to school for more years and being part of social clubs in which important information about politics, the community and Pittsburgh was discussed. So too the women understood the meaning of Carnegie's company town at Homestead. They may have understood their "place" in the family structure, but they were also aware of the pressure being put on their men in the mills. When the contact between the mill men and the employers ended and a new one had to be signed or a strike was in the offing, the women understood the basic issues as well as, or better than, anyone. As the mothers, wives and sisters of mill men whose pay they had to have to maintain the family structure itself, they would share with their men any struggle against the managers of the mill.[180]

**The Milieu of Family Strife**

Despite understanding and sharing in Homestead, the family

absorbed the pressures of scarcity and toil. Conflicts emerged from its market-dominated *milieu*, especially pay scales insufficient to release many families from want and crowded living quarters. Violence in the working-class neighborhoods became routine, a joke for newspapers that repeated a familiar theme: "A couple of Watsontown ladies had a set-to the other day, one wielding a broomstick and the other a clothes prop; no bloodshed or black eyes ensued."[181]

Unstable families amidst the ghetto environment also put large numbers of Homestead children on the street. "The street," complained the *Local News*,

> is a poor school for young girls and boys and it is a sad thing that so many are learning the lessons which are likely to bring them a dishonored and wretched life. One who is obliged to be out late at night sees strange sights, sights which would startle and grieve many an indulgent and careless parent, sights which indicate but too surely that accessions will soon be made to the ranks of the vile men and women who are to be found most everywhere. The girls of our households should be the object of particular care. Their companions should be known. Their hours should be regulated by loving but watchful parental discipline. A little firmness in their formative years may spare them a life of misery . . . . of course there are exceptions, but the rule that the child is the product of the home and its influences, and if the home be good, cheerful, pure, under a wise, kindly but firm discipline, the children are likely to grow upright and [become] useful men and women.[182]

This advice was difficult if not impossible for many Homestead families to follow in the rough and tumble hillside community.

Turmoil was also built into family life in Homestead because of the implicit belief in separate but unequal rights for men and women. Strikes and wage negotiations were viewed as the business of the men, their lodges, clubs and saloon discussion. The "little women" were assigned to the sphere of taking care of the children, the home, the shopping for essentials. The domineering spirit even colored the language of the workers' struggle for rights, the strike for higher wages, secure jobs and then spending money with saloon friends being seen as a "good blow" for "manhood, for honor and fame . . . ."[183]

Few workers connected equality in the home with the measure of economic egalitarianism they struggled for. There was a chasm between what woman was held up to be in the marriage relationship and her real role in the family. Idolized in the abstract for their beauty and "lovingness," women were the object of both repression and violence. Tributes to enduring, physical perfection and undiminished joy, such as Mrs. M. S. Kidder's printed the *Local News*, might well have appeared as mockery to many a Homestead wife beset with insoluble family problems:

She is shapely
She is fair –
Eyes like stars,
And golden hair;
Yet not for these
Alone, you know,
Do I love
My darling so.

She has lips
Of cherry red,
Teeth like pearls
In river's bed;
Little hands
As white as snow,
That's not why
I love her so.

But her eyes,
Like stars above,
Are so kindly
Full of love,
Bringing joy
Though there be woe;
That is why
I love her so.[184]

## John Law in Homestead

With increased work pressures and hardening economic and social divisions, conflict became systemic in Homestead's ghettoized working-class communities. In times past conflicts had been resolved by families, neighbors and the balance of power between employers and workers; but now a small police force was created by the City Council for the purpose.

With the police dependent on tax revenues, the prison was poorly kept in 1892. "When a prisoner is under arrest," a taxpayer complained, "our police have to take him down Ann Street or McClure Street to the alley. The proper entrance to a lockup is on Sixth Avenue. It is at present closed up, and used as a tool house. . . . [O]ur lockup . . . has the poorest accommodations of any town of this size in the state."

There were also dangers inherent in the procedures for making large numbers of arrests to fill a small jails, this taxpayer added. "Our police [would be] . . . as safe in regard to riot as . . . any other citizen of our town were proper accommodations made for placing prisoners in our lockup. . . ."[185] Short of riot, precautions were also seen necessary in the light of frequent jailings of the unruly, even of the residents of the more exclusive streets and avenues. "John Tonkle of Dickenson Street," runs one account, "was arrested last Saturday night for being drunk and disorderly. He was taken before Squire Kuhn, who, upon giving him a hearing, fined him five dollars and costs."[186]

The city had its share of promiscuity, toughs, gamblers and a red light district. So many were arrested and jailed, in fact, that the increasing activities of the lockup had to be placed under the central control of a Police Committee. The Committee authorized new locks to be fitted to the cells, and sole possession of keys by the town's three policemen, the ordinance officer and Burgess McLuckie. The other judicial officers were turned into second-grade administrators under these untried procedures. "When the squires and constables wish to use the lockup," wrote the *Local News*, tongue-in-cheek, "they will have to apply to those who hold the keys before they can enter. They will also be held responsible for cleaning the cells when used by their prisoners."[187] Judges without keys and obliged to scrub the cells of convicts would surely became rather lenient in sentencing offenders. As well, prisoners were secretly released, keys were miraculously distributed, and the privately-hired "coal and iron police" under Frick's command gained keys to keep those they wanted in the lockup. These keys, a local resident complained, were handed out "without any authority whatever." And though the *Local News* wanted names named, the "coal and iron police" were names enough to understand the complicity of the authorities and what was to happen in 1892.[188]

Moderates called attention to State Statute Section 891 that provided the Justices were entitled to use the lockup; but without keys to the cells, they were clearly being deprived of this power. They were, the *News* took pain to emphasize, "responsible officials," the clear implication being there were others with keys who were irresponsible.[189]

## Homestead's Growth Process

Such were the pangs of Homestead as it grew into a working class town. Wage labor and the buying and selling of goods had become the dominant ways of survival for the population. Impersonal relations in employment and the market had become essential to the functioning of production and commerce. The pressures of working class life had led towards a division of labor and recreation in the family, resulting in family antagonisms and instability, increased oppression of women, ghettoized neighborhoods, widespread brutality, and the growth of a regulatory system of policemen, judges and jails.

By 1892 the wage labor system had developed so far in Homestead that most working men were completely used up by their jobs. They functioned in the mill, relaxing outside, preparing to re-enter the endurance contest for wages. Not always fully aware of the crushing importance of the men's respite from work and regimentation, women, religious people and occasional cultural up-lifters railed against Saturday night saloon-going and Sunday work. The Prohibitionist Party and the churches also joined in demanding that men control their drinking and do God's work to stack up heavenly wages. Yet the men still drank, continued under orders to work most Sundays, and left it to the women and children to people Homestead churches. Sundays, the priests orated, the women prayed, the men worked and the children played.

**Opposing Sunday Work**

The *Local News* continually cried for Sundays to become an ethical holiday of sorts for workers, despite the fact that the mills worked round-the-clock and could never allow a furnace to go out without a substantial financial loss. Nor in the view of the *News* was it to be a day of recreation. Rather, after many days of being told what to do and performing conscientiously according to orders, the worker was to profit by a Sunday of the same regimen, that paper quoting from the moralistic *Youth Companion*: "It is certainly a mistake on the part of the 'friends of the working man' to advise a seventh day of mere pleasure-seeking. It can be better spent. When we use Sundays for religious worship and quiet home associations, they [the working men] make it emphatically a day of rest. For the change from their work-day routines could not be greater, and besides refreshment, they carry back to their daily work a strength of moral purpose which needs no definition, but has been known since Sabbaths were first kept."[190]

The mill workers saw through the exhortations, as the churches had usually missed the point of the social context of their existence.[191] "There are in the churches of Allegheny County," said Fitch, "ministers who are doing heroic service, but there are too many preaching every

Sunday to steel workers' families, who have never stepped inside a steel mill and who do not know a Bessemer converter from a puddling furnace. It is because of this lack of knowledge of some of the real problems of life that the ministers sometimes deliver their heaviest blows against secondary evils while the primary wrongs, the ones that dry up the roots of community life, may escape their wrath."[192]

Men working seven days a week, including Sunday, were run down physically, intellectually and morally - to say nothing of their inability to attend God's services when they were at work. Ministers more often directed their energies against small, retail businesses open on Sundays than against Sunday work in the mills; for the steel companies were considered too powerful for a successful attack. Naturally, the steel workers had an instinctive feeling that the church was "on the other side."[193] A skilled steel worker, expressing what others also told Fitch a number of times, harshly summed up the feeling:

> There are a good many churches in this borough, supported generally by women. The preachers don't have any influence in securing better conditions for the men and don't try to have. They never visit the mills, and they don't know anything about the conditions the men have to face. They think the men ought to go to church after working twelve hours Saturday night. They could accomplish a lot if they would try to use their influence in the right direction. Let them quit temperance reform until they get better conditions for the men. It is no time to preach to a man when he is hungry. Feed him first, then preach to him: so with the workingman. Get a decent working day with decent conditions for a man, then ask him to stop drinking. Let the preacher find out how the men work; go and see them in the heat and smoke and heavy toil; let them notice the crushed hands, broken arms and amputated limbs and find out what the working man is up against, and then try to better his condition.[194]

**The "Blind Spot" and Dimensions of Danger**

Before the Homestead struggle of 1892, most Homestead preachers were ignorant of the harsh conditions under which the mill men worked. Later, in the heat of battle with the Carnegie firm, almost every preacher in Homestead took a stand for the workers' cause, a leading minister stylizing Carnegie Chairman Frick as a man lacking the "sensibilities of a toad." But by this time Carnegie and his managers had secured the whip hand.[195] The ministers of Homestead came forward too little and too late. They had avoided preaching against the terrible dangers in the mill too long. They had brought God's message to the women and the children alone. While they preached from the pulpit, lives and limbs were being crushed – not only on Sunday, but other days as well.

The combination of long work hours, unprotected machinery, unremitting tension and speed, and maintenance of old equipment beyond its term despite dangers to the workers, made accidents almost unavoidable. [196] Many accidents were attributed to the "green hands" who were put to work before they had an adequate knowledge of English to understand commands, or knowledge of the processes in which they took part. "If the foreigner is a menace to the skilled worker's pocketbook" by earning lower wages, said Fitch, "he is probably a greater menace to both their lives. Unacquainted with the machinery, and with an exasperated boss shouting unintelligible orders, the Slav workman is as likely to run into danger as out of it, and may throw the lives of all his fellows into jeopardy."[197]

Tension during long hours of work also caused accidents: "It is a rare man who can keep his mental faculties keenly alert and centered on one object for twelve consecutive hours," Fitch reported. "In blaming their employees for lapses of attention which have resulted in accidents, manufacturers may often be demanding a self-control and mental alertness such as few men can sustain throughout practically their entire waking hours."[198]

Accidents at the Homestead plant were frequent, and almost every issue of the local papers reported one kind of dangerous situation or another. Technological advances brought lethal machinery into Homestead's mill, but Carnegie sought to keep production costs down and refused to lay out the money to protect his workers by investing in guard rails and other safety mechanisms. Without protective shields, workers were killed or maimed, a few examples illustrating the dimensions of the dangers.

May 10th, 1892: worker Edward F. Follansbee was talking to a friend, Henry Brown, who was an operator of the beam straightening machine in the Homestead mill. The machine was composed of a set of rolls which removed the curves on beams once they were cooled. Inadequate room in the plant caused the machine to be located in a tight place. Where the beam came out of the machine, the space was, the *Local News* wrote, "exceedingly dangerous for anyone to attempt to pass through as a beam might shoot out and catch them."

The *News* said, "Follansbee having concluded his conversation with his friend, stepped away and . . . endeavored to pass through the space referred to. Brown not seeing or dreaming that he had gone that way, pulled a lever and passed the beam through. It caught Follansbee about the middle of the body and crushed him in a frightful manner." About an hour later he was dead. [199]

A typical month included reports such as: "Mr. Rosser Davis, of 415 Fourth Avenue, had the large toe on his right foot painfully crushed . . . at the mill by a heavy channel bar falling upon it."[200] Charles Barrett, managing the hook in the armor plate mill, had a piece of scrap he was handling fall on his left leg, smashing the bones.[201] On June 15th mill worker George Boggart was badly injured by a chain striking him on the head.[202]

Homestead workers were not compensated for such injuries and, if they could, paid their own medical bills. Judicial practice followed the English Common Law that held a master free of any wrongdoing if his "servant" or that servant's "fellow worker" contributed to, or was responsible for, the injury sustained by the servant. The employer might install dangerous machinery, but if the employee and his co-workers did not escape the dangers by exercising reasonable care, that was their own fault and they had to pay the awful price.

No court would have held Carnegie liable unless it could be shown the machinery installed was defective, there was no contributory negligence by the worker or his fellow workers, and that the injury or death was a proximate cause of the employer's sole negligence.[203] And that was not the situation in the Follansbee case, the press reporting: "His friend Brown, hearing his cry, stopped the machine instantly, and leaped to see what had happened. 'My God, what did you go in there for?' You are not to blame, Brown, it was all my fault.' "[204]

After the Carnegie lockout at Homestead, the green hands also became a tremendous danger to the skilled workers. One William Lebeater, in a case of fellow-worker negligence, was crushed between two molds when one of the inexperienced workers suddenly swung a crane around. "The unfortunate man was not looking and was caught between the two hot molds," the *News* reported. "He was badly crushed and burned about the hips. He was removed to the hospital where he died a few hours later. It is said that the crane is operated by one of the new hands and has been the means of several accidents. . . . Mr. Ledbeater . . . leaves a wife and two small children. He was about 28 years old."[205] Having waited for a job from June 29th to January 9th and in financial need, Ledbeater had applied for assistance from the Strike Relief Committee the day before he was killed. And now without funds and no chance of collecting anything from Carnegie, the Ledbeater family was put on its own to survive.

Fault was so narrowly defined that workers on a break for lunch, sitting on an employer's powder keg, could be held responsible for injury resulting from an explosion. And the Homestead Works was in danger of exploding at any moment, as in the week of May 21st, when a portion of the plant was burned after chemicals exploded.[206] This had happened often in the past, and the injured and their survivors could only sue the employers and try to change the law itself.

An example was the case of John Shields. Shields had been burned to death in an agonizing accident at Homestead's steel works, and his widow, Kate Shields, sued Carnegie, Phipps & Co. for $30,000 damages, alleging that the death of her husband was the directly caused by the negligence of the defendants. "This suit," wrote the Homestead *News*, "is one of great interest here and will be stubbornly contested by both parties."[207] If Carnegie was found guilty it would have changed the old master/servant law on negligence, imposing a new cost of production. Almost every Homestead mill worker and family sided with Kate Shields in the suit. She, they knew, was suing for her security, as well as their own.

Both theoretically and economically the issue behind the legal cases and juridical forms was whether or not Carnegie would maintain low production costs by shifting the financial burden of injuries to the work force.[208] Eventually when the Carnegie mill was pressured to accept this burden, it recouped the loss by installing new machinery, which enforced a work speedup - increasing output, displacing some high-price skilled workers with low-price unskilled hands, and again lowering the per ton cost of making steel.

**Schools, Skills and Mill Hands**

Only the dangerous, hard, dull life of the mill awaited the young boys of Homestead. The excitement of their youth, the grammar schools, and the games of the field, were soon to give way to the necessity to earn one's keep and provide for the family. Yet, as Homestead expanded, it was not so obvious to the young that their futures would lead in this direction. Even in Homestead, youth had the momentary privilege of dreaming for a future that went beyond the sullen presence engulfing their parents' lives.

The backdrop for such hope in the late '80s and early '90s was that the Homestead grammar schools were being consolidated with the nation-wide educational system. Despite economic pressures that motivated Carnegie's development of an unskilled class of "steel serfs," technical education was the rage of the day, paralleling the rapid spread of technology that had taken place in the post-Civil War years.

"We have not yet decided whether manual training shall take the form of trade schools, and fit our boys and girls for some special work by which they can earn an honest living; or shall be educational and disciplinary without any direct reference to special departments of labor," wrote the supervisor of the Boston school system in 1891. Yet, he advised, the foundation stones of a well-rounded education already included drawing, sewing, and penmanship. "Cooking," he added, "is now claiming a larger share of attention, and, surely if we consider its direct bearing upon the comforts of a well-ordered home, we must acknowledge its claim to . . . an honorable place in the school curriculum." Both boys and girls should be able to choose this form of manual training, he thought, with the use of kitchens established in the school buildings to be coordinated with the teaching of other subjects. Carpentry and all other branches of manual training should have the same advantage, and the teaching *corps* would thus have to give its "hearty support" to make the transition possible.[209]

Others had also called for a transition to the departmentalization of the school system along the lines of industry. But the process had lagged. The *Popular Educator* complained that: "The manual work begun in the Kindergarten is rarely pushed forward into the primary grades. There is an interregnum here between the manual work of the very little folks and the sewing of the girls and the tools of work of the boys of the grammar grades." The removal of all barriers to unifying technical education and teaching manual tasks was thus a goal to be pursued in the primary schools.[210]

In 1892 this scenario for departmentalization of subjects and skills was pursued in the grammar schools in both Pennsylvania and Homestead, and the city was abuzz with rumors about expanding upper level education. Homestead teachers were active in the State Teachers' Association, which had been instrumental in unifying the curriculum and promoting local fund-raising to expand library facilities with essential technical texts. And, with support from the local newspapers, Professor Kendall, the energetic principal of Homestead's public schools, was raising $600 to finance libraries in each of the Homestead ward school buildings.[211]

The Division Headquarters for the school district was near the City Farm Lane where there were few houses or structures, making all appear orderly and progressive, though near enough to hear the blast of the mills and the grinding of machinery, suggesting the fate of graduates. Here, where all academic records were kept, the Homestead School Board met periodically to discuss curriculum, examinations, the costs and revenues for operating the schools, the yearly hiring and assignment of teachers, and other matters relating to the administration of the school system.[212] It was empowered to provide an educational system appropriate to the mill town, its commercial enterprises, and the needs of the population - all seen as one goal.

**Technology and Public Education**

The purpose of the technical education sought and sired by the School Board was to initiate the young into the technological world they were to become part of.

Technology had already changed the nature of work from its traditional crafts and skills to simple functions that were parts of the old skills or were entirely new operations. Tools had been fixed to machines and steam power substituted for human and animal power, transforming the worker from laborer to machine tender.

In one sense, labor was being simplified. In another, new techniques called for more training and effort in school and on the job. Students had more practical crafts and machine-related skills to learn. A contemporary wrote that "a higher intelligence and higher knowledge is required to direct the mechanical work of the machine than is required in direct labor, and thus the steam engine, while increasing the power of man a hundred- and a thousand-fold, make education a necessary requirement of his industrial usefulness."[213] Such education, mechanical training, was designed to make potential workers more efficient machine tenders.[214]

Educational efficiency fitted to industrial demand in Homestead included, first of all, a division of boys and girls along the lines of shop training for the young men who would enter the mill, and sewing, knitting and traditional household crafts for the young women who, as potential wives and mothers, would enable this future work force to enter "industry" and stay there. Both boys and girls received training in the basic skills of reading, arithmetic, and, to a lesser degree, writing. But by and large only the girls stayed in school after age twelve.

The wonderland of technology was not for all, as even the educational system made clear. The unskilled Eastern European workers could not and did not learn English. Most of them lived, ate, drank, and associated together in one-room shanties; spoke only to their peers; and avoided relationships with the Yankee workers and other Europeans.[215]

The foreign workers usually had no free time to go to school or church, either. And when they did, "They can't keep awake," said Father Kazincy, a Polish priest in Pennsylvania, complaining of the long hours and Sunday work that prevented his people's practice of religion. The community sponsored Americanization schools, but the response was lukewarm, Father Kazincy explained: "They are not a very great success for the simple, reason that the men are overworked and they do not feel like going to the schools and depriving their families of their company after these long hours. Sundays, they have none, for most of them go to work."[216] A Serbian steel worker said it better: "They accuse us of not becoming Americans. When do we get time? Can a man working in a blast furnace – and anybody knows that ain't no boy's job - twelve hours a day, or even ten hours, get time to learn English – or learn anything else? What in hell do they expect of us?"[217]

For the sons of skilled workers the situation was different. Their fathers were generally of Anglo-Saxon, German or Celtic origin, and the largest proportion were American born, not educated so far as school and university training were concerned, but "graduates in the school of experience."[218] Their sons remained in the schools of Homestead until they were able to take skilled jobs in the mill - crafts and their unions being closed to foreigners and their children who lacked American schooling.[219]

In addition to accustoming children to the division of labor and the industrial caste system, the schools emphasized selected employee virtues such promptness. The Homestead population moved through a routine punctuated by mill whistles calling out the turns of work, the penalties for tardiness and accidents; fire and church warning bells; and school sirens. Local papers encouraged punctuality, publishing a *Roll of Honor* for those pupils who "were neither absent nor tardy during the entire term."[220]

In such an environment, the young people learned the art of being on time and being dependable - factors that permitted teachers to designate who would be responsible in the workplaces or the households of the humdrum town.

## Limits of Upward Mobility

Could the boys or girls of Homestead ever escape mill or household labor through education? At first sight it seemed so, for open to them was the developing academic sequence of grade school, one year of preparatory high school, two years of senior high school, graduation, and possibly entrance into the university or a skilled trade school thereafter.

In fact, few advanced beyond preparatory high school, only established in 1888. Homestead's high school offered a one-year course of preparation, after which an examination had to be successfully hurdled for entrance into the next level, senior high school class. After two more years of studies, another rigid examination had to be passed to graduate and receive a diploma. Out of the many hundred of children in the high school, only three graduated in the initial class of 1890, eleven in 1891 and twelve in 1892, together making up only a handful of Homestead's eligible population.[221] A few of these graduates went on to college and others went into the technical trades.

The twelve high school graduates of 1892 compared with 1,669 students enrolled in the grammar and high schools. Besides these, the *Local News* noted, "there must be two or three hundred young people of school age, many of them working, who were not in the schools at all."[222] It was no small accomplishment to graduate, either; and only a single graduate, the valedictorian, received a university scholarship.[223]

College was a privilege of the well-to-do that no working class family could afford. Displaying diamond rings and silver bracelets as graduation gifts from their families, the six women in the class of '92 were backed by sufficient funds to send them beyond the city limits for upper level studies. Appearing in delicate pinks and pastels at the graduation ceremonies[224] in smog-covered Homestead, these few young ladies were described lovingly by the *Local News*:

> The gown in which the "sweet girl graduate" reads her wonderful essay over which she has toiled so long and anxiously, is always a thing of interest and comment, as those of the class of "'92" were unusually pretty, and will bear the following description: Miss Bertha Lloyd was dressed in cream colored silk, chiffon lace and pink roses. The slippers and gloves were white kid. . . . Miss Bertha received several. . . pretty tokens of reward, one of which was a gold hair ornament from friends . . . and a very handsome fan. In the fall Miss Lloyd will attend the California State Normal to complete her studies and prepare herself for a teacher. Miss Mamie Cronin's costume was of cream India silk, and yellow roses. Her fan was of white feathers, slippers and gloves white kid. This interesting epoch in her school life was remembered by her father in the gift of a diamond ring, and a ruby ring from a relative in Uniontown, Pa. Miss Cronin expects to take a course in music, and for that purpose will attend a college in Pittsburgh. Miss Hilda Kuhn wore a very becoming gown of pink silk, trimmed in chiffon lace and pink satin ribbon. The flowers, fan and slippers were of a corresponding color. The delicate tint of Miss Kuhn's costume was quite suggestive of her recitation, "Briar Rose." The diamond ring worn that evening was a gift from her father, a well-earned reward for her perseverance in completing the high school course. Miss Hilda, with the others of her class, will leave home shortly to attend some school or college to continue their studies.[225]

The account of the finery and gifts of the female graduates went on to convince readers that Homestead's upper classes even exceeded the limits of women graduates elsewhere -- having gone beyond the recommended standard, that "good taste in the manner of making is more essential than expensive fabrics."[226] With a knowledge of such extravagances,[227] readers might agree with the *News*' prediction that a bright and promising future for these ladies would in many ways be "a repetition of the past."[228]

Few Homestead workers could afford to dress their daughters thus, give their children gifts of this kind or to send them away to school. The names Squire Kuhn, a Homestead judge, and Hilda Kuhn, his graduating daughter, may indicate the social caste of the other graduates. The commencement speeches given by the graduates were pretty, too, more concerned with the delicacies of life and the vague, abstract virtues of being an "American" than with the mill town they emerged from and the mechanical technologies of the age.[229] Amidst the finery of chiffon and crepes, John Schaffer, Jr., who was to become one of Homestead's important Republicans, explored the abstract emotionalism of "Patriotism," the *Elizabeth Herald* remarking that the graduate "pulled vigorously at the tail feathers of the American eagle and made the bird of freedom scream lustily."[230]

In Homestead few questioned the scope, nature or class differences in education. The rich insisted they had gained money by virtue and wisdom, advising that others could be taught such qualities. The *Local News* also sang the advances of the age, the benefits to labor that came with the development of capitalist industry, the wage system and the public schools. The school system was doing a good job it thought, and "To the High, School Graduates," the *News* offered a "Shake."[231]

Yet the most successful graduates had a few more resources than the public education offered to those of limited means.[232] The professional people in Homestead carefully guided and catered to their own children's advanced education, emulating what the upper classes were doing in other places.[233] Thus, for example, in June of '92, Dr. A. M. Barton took the time to accompany his son to the State Normal School to view the instructors and the methods they used, finding himself well pleased that this was the proper place for his son to study.[234] Few Homestead workers could afford such time off the job or money from the family budget to personally look into their own children's education.

Though post-high school education was touted as a way to improve one's future earning capacity, most schools of advanced education were private and costly. Many of them taught what would now be called basic skills, as, for example, writing, which was becoming increasingly important to hold a skilled job in a shop or the office of a mill, or to function as a commercial agent.[235] This art was neglected in the public schools and was taught for a fee to a handful at a private summer school,[236] signifying but another limit to upward mobility for Homestead's population.

**Industrial Education in the Mill Town**

The educational system of Homestead was both rudimentary and inadequate. Nor were the educational tax benefits and burdens equal in the borough. Most young men had to leave school after the age twelve to work in the mills, though the working class families with modest homes had to pay taxes to finance further schooling for both girls and boys who did not immediately have to support their families.

With some 1,669 pupils enrolled in 1892, and many already working part time, the average daily school attendance of 1,131 meant that over one-fourth were out of the classroom each day. Yet to support the city's twenty-five school classes and twenty-nine teachers at all average cost of 96 cents per pupil each month, the homeowners of the city paid out $16,359.50 in taxes to cover the 1891 school budget, and the Pennsylvania taxpayers provided another $2,434.25.[237]

There were still insufficient revenues, however, and, whether or not their children were in school, the Homestead taxpayers had their tax assessments increased the following year. True, many homeowners did not meet their tax obligations. Nonetheless, Homestead taxes assessed for local school purposes were recorded as $20,033.54 in 1891 and $26,999 in 1892. Payments on school bonds and interest charges amounted to $4,335, and were almost a third of the teachers' salaries of $13,605. "Taxation," wrote the *Local News*, "is a luxury this year that comes high, but it seems we must have it."[238] Without a consideration of equability, the technical public schools thereby placed a growing burden on Homestead's taxpayers, required the borough to borrow money and pay interest for buildings, supplies and teaching staff.

Costs, the quality of education and teaching standards created sharp controversy in mid-'92. The School Board had met in conclave without Principal Kendall in the second week of June. And so secret was the discussion over technical versus general education, the *Local News* editorialized that "now a Philadelphia lawyer would hardly be able to tell how certain matters stand concerning the plan of school government for the coming year." The issue was whether or not professor Kendall would continue to raise the general educational standards by directing the schools in all three city wards or would be replaced in the commercially advanced First and Second Wards and be demoted to the position of principal of the impoverished Third Ward. This fit the pattern of what was taking place elsewhere, the old fashioned social studies approach being confronted by more practical pedagogues wanting the rapid regimentation of technical class work in preparation for the world of labor. The media opined that the good professor had been deposed and two new principals put in control, a situation of divided authority that would prove unsatisfactory and lowered the standards of the schools.[239]

By contrast, rather than lower the standards, some citizens wanted to raise them, so that graduates could go to institutions of a higher order without having to "tutor" in some branches or enter college with conditions to be made up. Moreover, a higher education for students could thereby be had in Homestead itself, under "the roof of their natural guardians," "without incurring the unnecessary expense of traveling, tuition, board, etc., in order to receive the same education in other institutions."[240]

These were utopian suggestions for Homestead, since there were few places for those with a higher education in the work force. There were also insufficient funds to upgrade teaching standards and create new opportunities for higher education. Too little money not only squeezed the school budget and its teaching staff, but rudimentary training continued to prepare young Homestead boys for manual jobs.

* * *

During this period, the main form of education was still mechanical. Though mechanical skills were gradually being wiped out by the advance of industrial technology in Homestead, many were still thought essential and were acquired by long physical training. The *Manufacturers' Gazette* also stressed that nationwide industry required a whole repertoire of physical skills, attained by training the eye, the muscles, the voice, and the sense of hearing. The mechanic, for example, had to perfect "sight" to detect imperfections in measurements and color, and the blacksmith had to learn muscular control to "hammer square" as an essential to that occupation.[241]

A good common school education was consequently viewed as most suitable for future craft workers. And the *Local News* encouraged boys to enter the schools of trade to learn plumbing, brick laying and other skills: "their native smartness will enable them to master their trades, and by avoiding all unclean and unsteady habits, they need have no fear for the future, for their services will always be in demand."

Such trades were a middle ground between skilled workers

employed by large enterprises and the future of the Carnegies and Fricks taking away these skills and reducing workers to common labor inside and outside the mills. The *Sewing Machine News* argued that expanding skilled trade schools would also be a way to get around piece rate wages for the unskilled that were being imposed in the shops, insisting that this was caused by "the trade unions by too many and too heavy restrictions regarding the number of apprentices which [sic] will be employed or allowed."

This paper surmised that young workers preferred to stand behind a store counter, or canvass for some agency, or work at factory piece work to "earn money, at first, per week, then if put to learning a trade." But it concluded this was a very shortsighted preference: "This would all be very well, provided he could be sure that such employment would last as long as he needed to work for a living."[242]

Since everyone who worked for wages in Homestead knew there was no such thing as a lifetime guarantee of employment, even as a member of the Amalgamated Association of Iron and Steel Workers, the argument to take up a craft made sense. Besides, Homestead had developed its own crop of builders, bricklayers, plumbers, tinsmiths and other trades, so the chances for apprenticeships were many.

Most young men in Homestead did not enter these crafts, however. The boys were usually needed to help support their families, and the jobs they sought were mainly found in the Valley's mines, coking plants and mills. As the mills were the principal employers in the area, a brief technical and mechanical education was to remain the only prerequisite for industry in Homestead, and that is what the young men received. Carnegie and the other steel employers meanwhile offered them hope of security and high initial wages on the piece-rate tonnage system. And with the near impossibility that large numbers could be upwardly mobile, most roads led Homestead's young men into mill work, setting the daily rhythms of the town.

* * *

Even the sports they played became vehicles for the display of the generosity and message of the mill owners: "It is announced that when the Homestead and Braddock baseball teams meet at Braddock this afternoon, to play a match game," the *Local News* noted somewhat sarcastically, "that Superintendent Schwab of the Edgar Thomson Steel Works, will make the address to them. Since that gentleman left Homestead, he seems to be taking quite an interest in the matters pertaining to the town. We believe in this, as a man of his position can exert great influence for good, if disposed to do so, but just whether or not baseball is the best field in which to display his efforts, is a disputable point. However no harm can be done to baseball."[243]

The double *entendre* of the last phrase was a warning that extended beyond the realm of sports. Schwab had already brought his message to the Braddock ball club and beyond. The "Young Americans," as the Braddock club was called, had had their new uniforms bought by the mighty Schwab[244] – whatever else being purchased lying within the inner reaches of the players' psyche. But after the *Local News* report, Schwab failed to appear with his message for the ball players. The *News* offered ironic condolences: "It is a matter of regret that he failed to deliver the address referred to for some reason or other. It would certainly have been a novelty upon a baseball field."[245] Immediately after their regrets, the editors of the *News* ran an important editorial that may well explain the reason Schwab had lost his voice on this occasion:

> The changes proposed in the management of the Carnegie interests in Homestead and elsewhere is looked upon as foreshadowing a labor conflict. There are some who surmise that an attempt will be made to bring the various plants in the different localities under one general system of employment. So far the new combination has in nowise intimated what will be their policy in this regard. It is reasonable to suppose, however, that things will continue to move along smoothly so far as Homestead is concerned, since the firm made overtures early in the spring which indicated a speedy and satisfactory adjustment of the scale. The Amalgamated Association still have the firm's assurances before them. It is altogether probable that before the new combination to be known as the Carnegie Steel Company, Limited, springs into actual existence on July first all labor questions will be satisfactorily settled.
>
> Moreover, in view of the fact that a presidential campaign is approaching, it is exceedingly improbable that any step would be taken which might involve a disturbance in labor circles.[246]

Clearly, some harm could be done if Schwab, Frick and other Carnegie officials moved non-union labor from Braddock or elsewhere to the organized Homestead Works. With the planned July 1st consolidation of the Carnegie's combined interests capitalized at least $50 million, and Frick's new position as central manager over this entire empire,[247] we will see that in the ninth inning of labor conflict Schwab would reach into the bullpen of scab workers to try to retire the Amalgamated Association of Homestead.

## Notes

### Part I

[1] Alice Alfonsi, " The Blue Collar World," *Pittsburgh Post-Gazette*, February 3, 1983.

### Chapter I

[2] Emma Goldman, *Living My life* (Salt Lake City, Utah: Peregrine Smith, 1982), 122-123.
[3] Harold C. Livesay, *Andrew Carrnegie and the Rise of Big Business*, Boston: Little, Brown & Co., 1975), 133-134.
[4] Leon Wolff, *Lockout: The Story of the Homestead Strike of 1892: A Study of Violence, Unionism, and the Carnegie Steel Empire* (New York: Harper & Row, 1965), 28-29.
[5] A. G. Burgoyne, A Complete History of the Stuggle of July, 1892, Between the Carnegie Steel Company, Limited, and the Amalgamated Association of Iron and Steel Workers (Pittburgh: Self-published Edition, 1893), 3.
[6] *The Local News* (Homestead, Pennsylvania), May 21, 1892, 3.
[7] Burgoyne, 6-11.
[8] *The Local News*, May 21, 1892, 3; Burgoyne, 4; M.R. Stowell, *Fort Frick* (Pittsburgh: Self-published Edition, 1893), 20-21; *The Cleveland Leader* (Cleveland, Pennsylvania), July 11, 1892, 3; *The Pittsburgh Times* (Pittsburgh,Pennsylvania), June 7, 189. Special Edition.
[9] Stowell, 20-21.
[10] Burgoyne, 4.
[11] Stowell, 24; Wolff, 44
[12] Stowell, 24.
[13] Burgoyne, 6-11.
[14] Stowell, 21.
[15] Livesay, 20.
[16] *Ibid.*, 126.
[17] Burgoyne, 5.
[18] *The Local News*, June 18, 1892, 3; Burgoyne, 21-22.
[19] *The Local News*, May 7, 1892, 3.
[20] Burgoyne, Chapters II and III, *passim.*
[21] *Ibid.*, 4-5.
[22] *The Local News*, May 21, 1892, 3.
[23] *Ibid.*, June 11, 1892, 4.

[24] Margarette F. Byington, *Homestead: The Households of a Mill Town* (New York: Charities Publication Committee, The Pittsburgh Survey, Russell Sage Foundation, 1910), 23-24 ff.
[25] *The Local News*, August 13, 1892, 3.
[26] *Ibid.*, June 11. 1892, 2.
[27] *Ibid.*, 4.
[28] *Ibid.*, June 18, 1892, 3.
[29] *Ibid.*, July 23, 1892, 2.
[30] *Ibid.*, May 14, 1892, 2.
[31] *Ibid.*, 3, and June 11, 1892, 3; see Byington , *passim*; see also Howard Waitzkin, "A Marxist View of Medical Care," *Annals of Internal Medicine*, Vol. 89, No. 2 (August, 1978).
[32] *The Local News*, August 13, 1892, 2.
[33] *Ibid.*, May 14, 1892, 3.
[34] *Ibid.*, May 7, 1892, 3.
[35] *Ibid.*, May 21, 1892, 4.
[36] *Ibid.*,, June 4, 1892, 3.
[37] *Ibid.*, May 14, 1892, 3, and May 21, 1892, 4.
[38] *Ibid.*, June 11, 1892, 2.
[39] *Ibid.*, June 4, 1892, 3.
[40] James H. Bridge, *The Inside Story of the Carnegie Steel Company* (New York: Aldine Press, 1903), 124 ff.; *The Pittsburgh Times*, June 7, 1892, Special Edition.
[41] *The Local News*, May 21, 1892, 1.

[42] Company towns in Pennsylvania's iron and steel industry were designed by the employers along the same lines as those on the frontier: The cost of the the maintenance of the workers and their families was controlled by the employer through company-run or company-influenced stores, banks and housing facilities. Thereby, wages could be set to cover these costs, and these wages as a cost of production could be controlled, James D. Norris, *Frontier Iron: The Maramec Iron Works, 1826-1876* (Madison, Wisconsin: The State Historical Society of Wisconsin, 1964), 25, 36-69. See also Edward Bemis, "The Homestead Strike," *Journal of Political Economy*, 1893-1894.

[43] *The Local News*, May 7, 1892, 1-2.

[44] *Ibid.*, 4.

[45] *Ibid.*, July 30, 1892, 1, 2, 3.
[46] *Ibid.*, May 7, 1892, 1-2.
[47] Bridge, 194; Byington, *passim*.
[48] John A. Fitch, *The Steel Workers* (New York: Charities Publication Committee, The Pittsburgh Survey, Russell Sage Foundation, 1911), 25, 226-228.
[49] Bridge, 194.
[50] *The Local News*, May 7, 1892, 1.
[51] *Ibid.*, May 7, 1892, 1; May 28, 1892, 1.

[52] *Ibid.*, May 7, 1892, 2; May 28, 1892, 1.
[53] *Ibid.*, May 7, 1892, 4.
[54] *Ibid.*, May 14, 1892,1.
[55] *Ibid.*
[56] *Ibid.*, May 21, 1892, 2; and May 28, 1892, 1.
[57] *Ibid.*, May 21, 1892, 1.
[58] *Ibid.*, May 28, 1892, 2.
[59] *Ibid.*
[60] *Ibid.*, June 11, 1892, 2.
[61] *Ibid.*, June 18, 1892, 1.
[62] *Ibid.*
[63] *Ibid.*, 1-2.
[64] *Ibid.*, 2.
[65] *Ibid.*, June 4, 1892, 2.
[66] *Ibid.*, May 28, 1892, 2.
[67] *Ibid.*
[68] *Ibid.*, May 14, 1892, 2.
[69] *Ibid.*, 3.
[70] *Ibid.*, May 7, 1892, 2.
[71] *Ibid.*, 1.
[72] *Ibid.*, 2.
[73] Fitch, 129.
[74] *Ibid.*, 193-194.
[75] *Ibid.*
[76] *The Local News*, May 14, 1892, 2.
[77] *Ibid.*, 1.
[78] *Ibid.*, 3.
[79] *Ibid.*, January-June, 1892, *passim.*
[80] Fitch , 192-193.
[81] *Ibid.*, 193.
[82] Robb v. Carnegie Bros. & Co., Limited, *Pittsburgh Legal Journal*, Number 13, October 28, 1891, 108-109; Lenz v. Carnegie Bros. & Co., Limited, *Pittsburgh Legal Journal*, Number 42, May 18, 1892, 383-385.
[83] *The Local News*, June 11, 1892, 3.
[84] *Ibid.*, September 17, 1892, 1.
[85] *Ibid.*, October 15, 1892, 1, 2.
[86] *Ibid.*, June 11, 1892, 3.
[87] *Ibid.*, October 1, 1892, 3.
[88] *Ibid.*, May 28, 1892, 2.
[89] *Ibid.*
[90] Wolff, 46.
[91] Burgoyne, 21-22; "True to the Few," *The San Francisco Chronicle*, July 2, 1892, 1; *The Local News*, June 18, 1892, 3, quoting the_*Pittsburgh Post.*
[92] See Samuel P. Hays, *The Response to Industrialism: 1885-1914* (Chicago: University of Chicago Press, 1957);. see also Amelia E. Barr, "The Servant

Girl's Point of View," *The North American Review*, June, 1892.
[93] *The Local News*, May 7, 1892, 1-3.
[94] *Ibid.*, May 28, 1892, 3.
[95] "A Good Workman is Known by His Tools," *Popular Educator*, Vol. VIII, No. 10 (June, 1891), 321.
[96] *The Local News*, May 21, 1892, 3.
[97] *Ibid.*, May 28, 1892, 3, original emphasis.
[98] *Ibid.*, May 7, 1892, 4.
[99] *Ibid.*, May 28, 1892, 3.
[100] *Ibid.*, June 11, 1892, 1.
[101] *Ibid.*
[102] Ibid.
[103] *Ibid.*, June 4, 1892, 1.
[104] *Ibid.*, May 28, 1892, 3.
[105] *Ibid.*, June 11, 1892, 3.
[106] *Ibid.*
[107] *Ibid.*, May 7, 1892, 4; December 22, 1892, 4; January 11, 1893, 4.
[108] *Ibid.*; Hays, 9.
[109] *Ibid.*, May 14, 1892, 3.
[110] *Ibid.*, May 7, 1892, 4.
[111] *Ibid.*, May 14, 1892, 3; May 28, 1892, 3; June 4, 1892, 2; June 11, 1893, 3; L. Mario, "The Winner in the End," *C. S. Greer's Monthly Musical Journal*, (Fall River, Mass.) January, 1895, 3-5.
[112] *The Local News*, May 28, 1892, 3.
[113] *Ibid.*, June 4, 1892, 2.
[114] *Ibid.*
[115] *Ibid.*, June 11, 1892, 3.
[116] *Ibid.*, May 14, 1892, 2.
[117] *Ibid.*, June 4, 1892, 2-3; Goldman, 122-123; Mario, 3-5.
[118] See *passim* at Bridge, Byington, Fitch, Burgoyne, and Stowell.
[119] *The Local News*, May 7, 1892, 4.
[120] *Ibid.*, 3.
[121] *Ibid.*, May 21, 1892, 1.
[122] *Ibid.*, June 4, 1892, 2.
[123] *Ibid.*
[124] *Ibid.*, June 18, 1892, 3, quoting the *Pittsburgh Dispatch*.
[125] Fitch, 235-236.
[126] *Ibid.*, 19.
[127] *The Milwaukee Advance* (Milwaukee, Wisconsin), June 25, 1892, 1.
[128] *Ibid.*
[129] *Ibid.*, July 30, 1892, 1.
[130] Livesay, 134.
[131] Fitch, 18, 216. In 1908, one steel worker insisted that it was no use for the unions "to try to regulate wages, anyhow, for labor is a commodity and its price is regulated by supply and demand," adding: "The only way out for the

laboring men is to get together in a labor party." which, in his view was the Socialist Party. But this was not the predominant view in Homestead in 1892.
[132] *The Local News*, May 7, 1892, 3.
[133] *Ibid.*, 4.
[134] *Ibid.*, June 11, 1892, 2.
[135] *Ibid.*, June 4, 1892, 3.
[136] *Ibid.*
[137] *Ibid.*, June 18, 1892, 2.
[138] *Ibid.*, June 18, 1892, 1.
[139] *Ibid.*
[140] Andrew Carnegie, "An Employer's View of the Labor Question," *Forum*, Vol. I, 114-125.
[141] Andrew Carnegie, "Results of the Labor Struggle," *Forum*, Vol.I, 538-551.
[142] *The Local News*, June 18, 1892, 2.
[143] *Ibid.*, June 11, 1892, 2.
[144] Fitch, 226-227.
[145] *Ibid.*, 227.
[146] *Ibid.*
[147] *Ibid.*, 228.
[148] *Ibid.*, 15.
[149] *The Local News*, May 7, 1892, 1.
[150] *Ibid.*
[151] Fitch, 203.
[152] *The Local News*, May 28, 1892, 4.
[153] *Ibid.*, October 29, 1892, 2, citing the *American Farmer*.
[154] *Ibid.*, August 20, 1892, 2. See also *Ibid.*, October 29, 1892, 2;.
[155] *Ibid.*, November 12, 1892, 2.
[156] Fitch, 203.
[157] *The Local News*, June 11, 1892, 4.
[158] *Ibid.*
[159] *Ibid.*, May 7, 1892, 3.
[160] *Ibid.*, May 14, 1892, 2.
[161] *Ibid.*, June 18, 1892, 3.
[162] *Ibid.*, June 4, 1892, 4.
[163] See Dan Barry, "Preserving a Buried Legacy," *The New York Times*, March 6, 1999, A10. One of many inscriptions on temperance movement coins struck in the 19th century.
[164] Fitch, 12.
[165] *Ibid.*, 203-204.
[166] *Ibid.*, 204-205.
[167] *The Local News*, May 7, 1892, 3.
[168] *Ibid.*
[169] *Ibid.*, May 21, 1892, 1.
[170] *Ibid.*, May 7, 1892, 3.

[171] Lucinda Franks, "Where Carnegies Rusticated," *The New York Times*, May 1, 1983, Sec. XX, 27.
[172] *The Local News*, May 7, 1892, 2.
[173] *Ibid.*, June 4, 1892, 1.
[174] *Ibid.*, December 13, 1892, 4.
[175] *Ibid.*, December 14, 1892, 1.
[176] *Ibid.*, June 18, 1892, 2.
[177] *Ibid.*, May 21, 1892, 3.
[178] *Ibid.*, 4.
[179] *Ibid.*, May 28, 1892, 4.
[180] *Ibid.*, May 28, 1892, 4; May 21, 1892, 4.
[181] *Ibid.*, May 14, 1892, 2.
[182] *Ibid.*, May 7, 1892, 4.
[183] *Ibid.*, 2.
[184] *Ibid.*, May 14, 1892, 2.
[185] *Ibid.*, May 7, 1892, 4.
[186] *Ibid.*, 3.
[187] *Ibid.*, May 28, 1892, 3.
[188] *Ibid.*, June 4, 1892, 3.
[189] *Ibid.*, June 11, 1892, 2.
[190] *Ibid.*, June 4, 1892, 2.
[191] Fitch, 223.
[192] Ibid.
[193] *Ibid.*, 223-224.
[194] *Ibid.*, 225.
[195] *The Local News*, July 9, 1892, 1-3.
[196] Fitch. 70-71.
[197] *Ibid.*, 68-69.
[198] *Ibid.*, 67.
[199] *The Local News*, May 14, 1892, 3.
[200] *Ibid.*, May 7, 1892, 3.
[201] *Ibid.*, June 4, 1892, 3.
[202] *Ibid.*, June 4, 1892, 3.
[203] See Childs v. Pennsylvania R. R. Co., *Pittsburgh Legal Journal*, No. 50, July 13, 1892, 466-468; Tunney v. Carnegie Bros. & Co., Limited., *Pittsburgh Legal Journal*, No. 24, January 13, 1892, 219-221.
[204] *Ibid.*, January 11, 1893, 1.
[205] *Ibid.*, January 11, 1893, 1.
[206] *Ibid.*, May 21, 1892, 4.
[207] *Ibid.*, 3.
[208] Tunney v. Carnegie Bros. & Co., Limited., *Pittsburgh Legal Journal*, No. 24, January 13, 1892, 219-221.
[209] R. C. Metcalf, "A Few Problems in Education," *Popular Educator*, Vol. VIII, No. 1 (June, 1891), 309.
"Primary Manual Work," *Popular Educator*, Vol. VIII, No. 1 (June, 1891), 297.

[211] *The Local News*, May 7, 1892, 4.
[212] *Ibid.*, 3.
[213] Charles P. Steinmetz, *America and the New Epoch* (New York and London: Harper & Brothers, 1916), 45-46.
[214] *Ibid.*, 46.
[215] Byington, 112.
[216] Ray Stannard Baker, *The New Industrial Unrest: Reasons and Remedies* (Garden City, New York: Doubleday, Page and Company, 1920), 106-107.
[217] *Ibid.,* 106.
[218] Fitch, 10.
[219] See A. Austin, *The Labor Story* (New York: Coward-McCann, 1949); Anthony Bimba, *History of the American Working Class* (New York: International Publishers, 1929); Richard O. Boyer and Herbert M. Morais, *Labor's Untold Story* (New York: Cameron & Kahn, 1955); Richard Krooth, ed., *The Great Social Struggle*, Vol. II (Santa Barbara: Harvest Publishers, 1979).
[220] *The Local News*, May 28, 1892, 3.
[221] *Ibid.*, 1, 3.
[222] *Ibid.*, 3.
[223] *Ibid.*,
[224] *Ibid.*, 4.
[225] *Ibid.*
[226] *Ibid.*
[227] Ellen LeGarde, *Popular Educator*, Vol. VIII, No. 10 (June, 1891), 324.
[228] *The Local News*, May 28, 1892, 4.
[229] *Ibid.*, 1.
[230] *Ibid.,* June 4, 1892, 4, citing the *Elizabeth Herald.*
[231] *Ibid.*, May 28, 1892, 3.
[232] Le Garde, *Popular Educator*, Vol. VIII, No. 10 (June, 1891), 326.
[233] Upton Sinclair, *The Goose Step* (Pasedena, Cailifornia: Self-published .Edition, 1923).
[234] *The Local News*, June 11, 1892, 3.
[235] E. C. Allen, "Advertisement," *Popular Educator*, Vol. VIII, No. 10 (June, 1891), 326.
[236] *The Local News*, June 4, 1892, 2.
[237] *Ibid.,* June 11, 1892, 3.
[238] *Ibid.*, 2.
[239] *Ibid.*, June 18, 1892, 2.
[240] Ibid., May 14, 1892, 1; May 21, 1892, 1; June 18, 1892, 3.
[241] *Ibid.*, May 14, 1892, 2.
[242] *Ibid.*, 3.
[243] *Ibid.*, May 21, 1892, 2.
[244] *Ibid.*; see Fitch, Chapters I and II, *passim.*
[245] *The Local News*, May 21, 1892, 3.
[246] *Ibid.*

[247] Stowell, 21, 24; Burgoyne, 6-11.

**A Breathing Spell**
**(Drawn by Joeseph Stella, circa 1892)**

# McKINLEYS FRAUD ON LABOR!

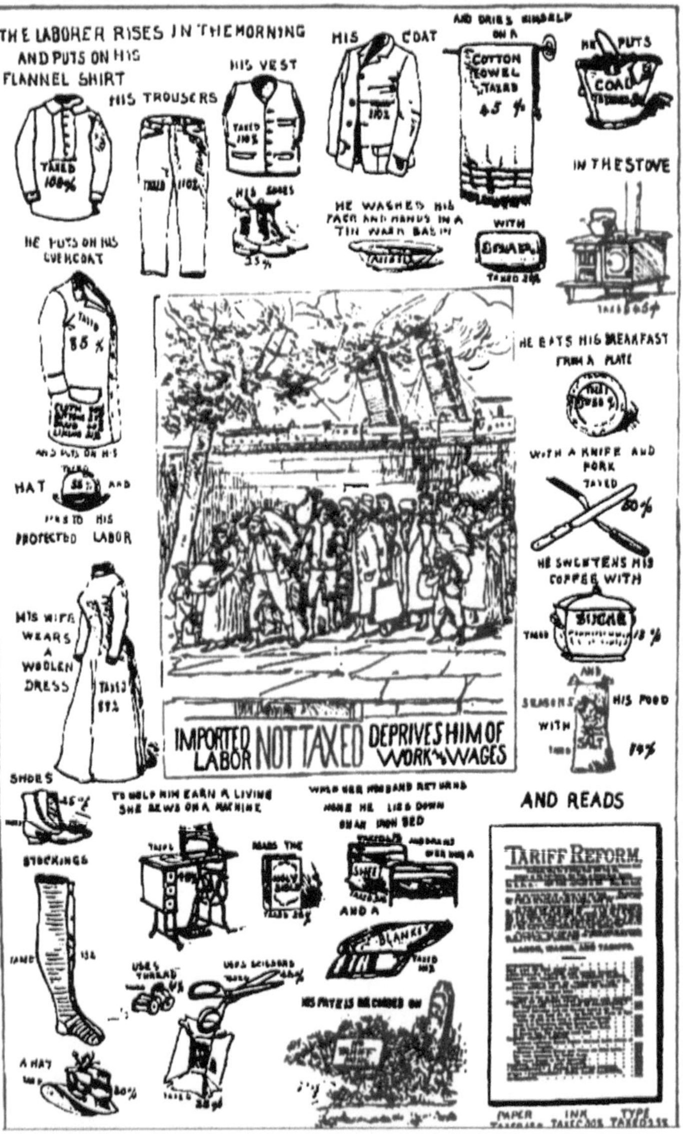

# Part II

# The Hidden Dynamics of Homestead

The demand will just fall short of the capacity to produce, therefore a struggle must ensue among producers for orders. . . . The sooner you scoop the market the better.

It has never failed that [in times of recession] the lowest price given has proved to be the highest price at time of delivery on a falling market. When you want to capture a falling stone, it won't do to follow it. You must cut under it, and so it is with a falling market.

– Andrew Carnegie's advice to his partners
during the 1893 depression

Of course you will win, and win easier than you suppose, owing to the present condition of the market.

– Carnegie letter to Frick during the
Homestead lockout, 1892

Hardly an industry in the United States could ignore competition in the last two decades of the nineteenth century. The old Napoleons of Europe had been displaced in part by the industrialists. They had the financial resources and factory facilities to master production. They had the worldwide connections to grab markets for themselves and exclude others. They also had a network for collecting information about potential markets and potential competitors. And they viewed the upstarts in America as dangerously likely to impinge on their sales areas, seeing the growth of monopolies and trusts in the United States as particularly aggressive in introducing new technology, lowering costs, and advancing on new markets. More than this, they witnessed with trepidation the attempt of the iron and steel industry in America to erect tariff fences against European metal and, at the same time, to introduce those technological advances that would replace workers with machines and lower the labor cost of production.

The U. S. iron and steel mill owners reasoned that there were two ways to hold off European competitors. Tariffs could be placed on foreign steel and its price thereby raised, when that steel was able to undersell U. S. steel. Or such steel could be undersold by simultaneously allowing the tariff to be reduced and scaling down wages as the market price of steel fell in response to cheaper imports. If foreign steel was more expensive than U. S. steel, then it mattered little if the tariff was high or low because foreign steel could not undersell the American product. But this situation could be and was taken advantage of, and in 1890 Andrew Carnegie applied pressure on Congress to have the tariff on steel billets cut when competition did not threaten.

At Carnegie's flagship plant at Homestead, wage agreements had historically tied the workers' pay scales to the market price of billets, the starting form from which beams, rails, plate, etc., were forged, although nearly all Homestead steel was sold in some structural form. By lowering tariffs on billets (but not on other forms of steel) Carnegie justified Homestead wage-cuts on the grounds that he had to undersell foreign competition. Of course, his main production was not billets, but beams, plate, rails and other structural steel. This slick scheme to cut wages was the root of the conflict at Homestead.

The hidden dynamics of the struggle included other interrelated factors as well. Empire builders and arms contractors had pressured the government to grant military contracts for the major steel works to turn out vessel fittings, with the purpose of building a stronger Navy to control the Pacific Ocean. These contracts permitted Carnegie's Homestead Works and the Bethlehem Steel Company build up an inventory of steel when their workers' wages were low and later to use that inventory in attempting to meet the government demand when the workers went out on strike. Guaranteed government orders also allowed the Homestead Works to continue production under these military contracts with non-union, "scab" laborers. And since arms production required highly sophisticated equipment, the technical advances introduced at Homestead were designed to lower production costs and undercut many skilled tasks the workers performed in the past.

Together, foreign competition and arms contracts fostered a new balance of skilled labor and technical machines; and Carnegie planned to introduce still more technology that would displace certain skilled workers, take away their production-line controls, lower the wage scale and reduce the cost of production. To do this, Carnegie felt he had to break the Amalgamated in order to replace part of his expensive work force with the latest machinery, pressure the remaining skilled workers to fully use these machines, and thereby speed up production. Thus the battle against the Amalgamated at Homestead was a conscious design to eliminate the old technology and skilled tasks in steel-making and to replace them by new ones, as well as to replace expensive skilled workers with inexpensive unskilled laborers.

But this grand scheme did not proceed as the Carnegies and the other industrialists had hoped. When labor went out for better wages and job security, protection of the Amalgamated and the union's right to bargain, a full-scale war began. Carnegie fell short in meeting both his obligations to the government and his competitors in the market. This shortfall added pressure on Carnegie to rapidly upgrade his technology, to replace skilled laborers with new machinery, and to stop at nothing in defeating his workers at Homestead. These became the underlying pressures and hidden dynamics of Homestead that shaped the legal steps taken, the ensuing military standoff, the readjustment of the relations between Carnegie and his work force, and the balancing of labor and machines deployed in production.

**Gun Forging**
**(Credit: *Harvest Quarterly*)**

# Chapter 2

# Empire Builder & Arms Contractors

> The Report found that the Navy, in its haste to meet a deadline, improperly restricted competition for the $30 million contract. . . . The G.A.0. [General Accounting Office] has been investigating accusations made . . . to . . . the Secretary of Defense. The companies accuse the Navy of using such biased procurement practices as writing unfair contract specification and manipulating the bidding process to favor [the largest producer] on contracts worth billions of dollars.
>
> - Summary of a General Accounting Office Report, 1989[1]

With market size, a primary concern to owners of steel mills in America, Andrew Carnegie and others had sought to ensure a steady demand in hopes of increasing production and thus profits. Toward this end, Carnegie had sought the best of all possible markets:

Government military contracts. He saw them as a way to ensure sales, upgrade his technology, and guarantee his future markets.

He was also aware of the special position of such guaranteed sales. Government contracts provided him the mobility to speed up production while the workers were compliant, but to lock out the workers when they demanded wage security by using (or trying to use) his accumulated inventories to meet government obligations. In the case of his lockout at the Homestead Works, Carnegie also switched output to his Duquesne Works, allowing a continued accumulation of profits even during the all-out battle between the workers and the company.

Carnegie, moreover, had expanded his business empire at precisely the time the United States sought a military empire in the Pacific. The Navy's need for steel vessels made it look to Carnegie's technically advanced Homestead Works. So there was a direct relationship between U. S. foreign policy, a strong Navy, military contracts awarded Carnegie's mills, his upgrading of technology, and the great strike at Homestead.

**American and Pacific Navies**

In the 1880s, the weakening British Empire in the Americas was being replaced by a U. S. sphere of influence. England, determined to drive out the last traces of Spanish domination in the Americas, was itself too weak to enforce its military control. As the United States prepared to step into any area lost by either Spain or Britain, U.S. leaders reasoned a powerful Navy was required.[2]

The fighting force of any Navy depended, in turn, on its fire-power to sink enemy ships. So the objective of naval tactics was to mobilize more "capital ships" than the enemy, that is ships able to engage in battle with enemy fleets by virtue of sufficient armor and superior gun-power, commonly termed "weight of metal."[3]

When the work of building up the new Navy began in the '80s there was not a properly equipped shipyard in the nation where the requisite materials were available or the assembly could be done. There was no armor plate made in the whole country, and no plant existed capable of handling such heavy work. No one was prepared to forge the shafting for ship propellers or to make the great forgings required to build a modern high-powered naval ship. In fact, until Secretary of the Navy Whitney came into office in 1888, gun forgings and shafts were bought in England from Whitworth, armor plates from Cornnell and other materials elsewhere abroad.[4]

Under President Harrison's policy of supporting the Navy,[5] Whitney stopped all these foreign purchases. He adopted a plan for buying everything from American manufacturers, even if there was an initial delay. He purposely awarded large contracts to justify increases in the size of privately-owned U. S. mills. According to one report, he provided the original stimulation for the growth of Bethlehem's Pennsylvania plant into "the greatest steel fabricating plant in the world in some respect, and one that, when it had reached full development, will be fully as extensive as the famous Krupp establishment at Essen, Germany, and of the Schneider establishment at Creusot, France."[6] At Bethlehem, wrote the Baltimore *Sun*, the heaviest gun forgings, propeller shafts and all other heavy forgings are turned out, and there the best armor plate in the world has been produced. A complete machining plant is being put in, and soon this establishment will be at work on their order for a hundred guns, forged and assembled and complete in every part, for the army for coast defenses.[7]

Clearly, then, government contracts and subsidies to Bethlehem's steel mills had helped build its competitive stature with giant foreign mills specializing in armaments. Near San Francisco, meanwhile, brothers Irving Murray and Henry Scott held an inside political track with Secretary of State James G. Blaine, their Union Iron Works landing the $3 million contract for the battleship *Oregon*, though its bid was not the lowest one received. By 1891 Irving Scott boated that his plant ranked among the most complete in the world – covering twenty-five acres, employing thirteen hundred men – predicting that an extra $5 million in capital would make it equal to Krupp, Cresot, Bethlehem, or any other arms-making behemoth of Europe or America.[8]

"Another great forging establishment has been built at the Midvale Steel Works," added a reporter, "and under the same policy carried out by Secretary [of the Navy] Tracy the Carnegie establishment has taken up the manufacture of armor plates."[9]

## Political Contributions and Government Contracts

Carnegie's contracts were, in fact, return favors based on his organization's contributions to the Republican Party. *Sub rosa*, Andrew Carnegie himself had contributed $100,000, H. C. Frick $250,000, and the Carnegie Steel Company of Pittsburgh $10,000. *The New York World* denounced this funding of pliable politicians, headlining its message: "Frick found out!" Frick had plunked down $250,000, rounding out $1 million in contributions from Pennsylvania business interests, put at the disposal of Republican Party leader Matt Quey and

Party chairman Tom Carter.[10]

Tariff-seeking Republicans backed by the manufacturers had then acted as a counterweight to Democratic President Cleveland's firm attitude against excessive tariff protection. The tariff issue was also used by the Republican organization to "fry all the fat out of the manufacturers," as its confidential circular declared: "We want money, and we want it all at once."

As a consequence of such pressure, manufacturers everywhere, and especially those of Pennsylvania, got together very substantial sums for the presidential campaign of both 1888 and 1892; and a very large portion was then used to buy votes. By "dint of corruption," Ostrogorsky wrote in his classic study on American political parties, "they brought the Republicans back into power. The latter hastened to repay the advances with some coin of their own making: Mr. Wanamaker [the political fund raiser] received a post in the Cabinet for his brokerage, and the manufacturers were presented with a new tariff still more prohibitive than the preceding ones, [one], which became famous under the name of the 'McKinley tariff.' "[11]

The tariff represented government interference and protection for the American trusts from foreign competition. It applied to most manufactures, though it was lowered on steel billets through Carnegie's personal Congressional lobbying. It functioned as a "constructive supervisory power," like that of the central government in Germany. Though in the U. S., governments from the federal to the municipal levels were not officially organized for such activity and, where they intervened were "largely inhibitory, liable to disorganize by interference," still: "The tariff wall excluded the check afforded by competition with other nations" - so the price of American steel could be raised. Tariffs also stimulated absorption of smaller concerns at exorbitant prices above the value of their assets to prevent costly, unrestrained competition at home[12] – again keeping up steel prices.

At the time, the Frick-negotiated political contribution fit within the frame of these pressures and expected returns, facilitating Carnegie's take-over of smaller concerns and the award of government contracts. The Republican Party "Secretary Elkine is credited with having put the screws on Mr. Frick," the *World* reported. "It is known that the Carnegie people contributed $50,000 to the corruption fund and in return for that contribution and to assume like liberal contributions in this campaign the firm of Carnegie, Phipps & Co. received from the Navy Department the largest contract for armor plate that has ever been given out, and at a price much higher than many other steel manufacturers would have been glad to accept."[13]

Charles Harrah, a manufacturer who contributed $10,000 to the $60,000 political fund raised by the Manufacturers' Club of Philadelphia in the national election in 1888, had made this incontrovertible public statement, and the first of many contracts subsequently given to Carnegie and his partner H. C. Frick was for 7,000 tons of armor plate at as previously unheard of price of $540 a ton. The contract had been awarded without any competition or any information to other manufacturers to allow them to bid for the contract.

Later, as the 1892 struggle at Homestead reached a head, Frick was roundly criticized by the Republican Party because, after having received this lucrative military sinecure, he precipitated the reduction of workers' wages just before the election. The resulting tumultuous labor strikes, linked by happenstance to the growing Farmers' Alliance which joined issues with the newly-formed People's (populist) Party with General James B. Weaver of Colorado as their presidential nominee, drew off many votes from General Harrison, the Republican nominee – ensuring the election of Cleveland by the largest plurality to that time received by any presidential candidate.[14]

Before then, however, Frick was unable to meet his government steel delivery obligations. Using scab workers, the firm produced an inferior quality of armor plate. And even before such failures, to keep his government contracts under Republican patronage, Frick had had to contribute to the party. "When H. C. Frick was summoned to New York with Secretary Elkine to meet him and explain the necessities of the party and with the power of revoking the most profitable contract ever given out under the Harrison Administration," wrote the *New York World*, "it is not difficult to understand how H. C. Frick was compelled to stand and deliver $250,000."[15] This sum comprised about 4 percent of the entire cost of $6,000,000 spent by both parties in the previous 1888 election, with the manufacturers protecting their interests and honor by generous contributions as "often as the Republican hat went round."[16] The *World* reasoned, then, that Carnegie's $250,000 had effectively "paid for a monopoly" – an armor plate contract – without competitive bidding.[17]

**Patriotism, National Defense & Spheres of Influence**

Later, in his autobiography, Andrew Carnegie attempted to make it appear that Homestead had accepted the role of arms maker out of patriotic zeal and that the contract was inherently unprofitable. He claimed his firm had twice declined to manufacture armor for the

U.S. Government, but had accepted the contract because it was "urgently needed."[18]

The truth in Carnegie's claim was that the need for national defense was indeed urgent and war contracts were lavished on his firm. Previously, the Homestead Works under Carnegie, Phipps & Co. had won contracts for two warships, the *Maine* and the *Monterrey*. And by 1892, the Homestead Works had become the nation's *top manufacturer of armor plate* and had exceeded Bethlehem's rival plate mill[19] Carnegie's autobiography coyly neglected to recall Navy contracts automatically secured his profits, as did those awarded at Homestead when war with Chile loomed in 1891.

Naval Secretary Tracy, acting under Congressional and Executive orders, had become America's foremost hawk in 1891. Tracy and the Administration saw clearly that Britain was attempting to use the Chilean Navy as its foil for controlling the Pacific Ocean. They believed the British were scheming to force Spain completely out of Latin America and to absorb her territory, while excluding the United States from that region's trade, resources and investments.[20]

Both Britain and the United States had initially favored the independence of all of Latin America from Spain to free up raw materials and markets for themselves. In the post-Napoleonic period after 1815, both had actively prevented other European nations from aiding Spain militarily so as to ensure victory of many Latin American rebellions. Britain's Lord Cochrine had led the first Chilean Naval Squadron against the Spanish in Peru, for example, and British interests in Chile had made that nation an economic subject, so much so that by 1875, 58 percent of Chile's exports were going to Britain while 41 percent of her imports came from there.

Besides capturing Chile's copper resources and indebting the nation to English bankers, Britain forced Chile to adopt free trade, so that England could undersell all competitors. British speculators also gained the advantage of Chile's nitrate wars with Peru and Bolivia from 1879-83.

When Chile's own industrial bourgeoisie developed and demanded protection for national industries, however, Britain took alarm. Chilean President Balcemeda demanded that the British nitrate monopoly be broken and the industry become nationally owned. He wanted a Chilean tariff to protect home industries, so that the mines of Chile could supply them.

He was strongly opposed by the aristocratic classes in Chile, who had done well under British neocolonialism. With their backing, insurrection against Balcemeda began in 1891, led by a Navy organized along the lines used by the English Royal Navy. British firms also contributed liberally to the "revolutionary" fund mobilized to checkmate Balcemeda's political strength. English vessels transported opposition leaders, to say nothing of coal and food to the conservative "insurrectionists" headquartered in the Iquique nitrate country. Victory for the insurrectionaries in August 1891 was thus also a British success.[21]

The key to the struggle had been the British-backed Chilean Navy. On January 5, 1891, the Navy had declared itself in favor of the Legislature controlled by the upper classes. When the Navy seized every steamer carrying the Chilean flag, President Balcemeda was left without a seagoing warship on the coast. This enabled the insurrectionaries to capture the various nitrate ports; thereafter, Rear-Admiral Hotham of the British Pacific Squadron invited rival commanders to confer aboard his flagship in preparation for an armistice that gave the insurrectionaries crucial control of Iquique.[22]

At this point the United States worried that Britain would be the real victor of Chile's Civil War. Balcemeda still had a following, and as late as April 23rd of 1891, torpedo gunboats entered Caldera harbor to sink the "rebel" ironclads, for the first time in world history successfully using a Whitehead torpedo against such a vessel.[23] In the meantime, the U.S. sided with Balcemeda, and U.S. Minister Egan located at the Chilean capital requested that the Itata, the insurrectionaries' cruiser, be seized on the ground that she was about to carry a cargo of arms from the U. S. to the revolutionists. Evading the U. S. federal authorities at San Diego, California, on May 6, 1891, she nonetheless surrendered to the U.S. Squadron at Iquique. This earned the U.S. the wrath of the Chilean aristocrats, as well as the anger of the Balcemeda party which insisted the U.S. actions were insufficient to protect them. Only after the end of August, when Balcemeda's forces were completely routed at Placilla near Santiago, did the U. S. stiffen in determination to protect its rights.

When United States sailors fought with a Chilean mob in Valparaiso, leaving two U. S. sailors dead and eighteen injured,[24] U.S. Secretary Blaine demanded indemnity and apology, the Chilean Government insisted the matter was in Chile's sole jurisdiction and authority[25] – and the U. S. Navy was readied for action, Secretary of the Navy Tracy immediately awarding contracts to the nation's leading steel mills. Despite Chile's subsequent apology and its U.S. acceptance in January 1892,[26] the U. S. viewed the whole affair and its military build-up as a way to keep the Chilean Navy from expanding in the Pacific and acting as a wedge for British imperial domination.

At first the nation's leading mill owned by Bethlehem gained the greatest benefit. "We also have a contract with Carnegie and Co. to deliver 500 tons [of steel] a month," Secretary Tracy told reporter Julian Hawthorne in early 1892, adding: "As it is, our armored cruisers and line of battleships will be the best of the kind in the world."

"But until they are completed – " Hawthorne started and Tracy finished: "until they are complete, we are practically defenseless against any nation that attacks us."

"Then if Chile - " Hawthorne began again, the Secretary cutting him off, weighing his words: "You may say, too, . . . . : Four months from the time those sailors were killed at Valparaiso, Chile, if we had declared war, we would have had our Pacific coast entirely at her mercy. With the ships she has, together with those that will be ready shortly, she could have done anything she pleased with us. It has been her ambition to be mistress of the Pacific."[27]

**Government War Contracts, Speedups and Temporary Delays**

As government contracts were sped up, there was unusual production activity in Carnegie's Homestead mills. "Work for the United States Government is being pushed at the Homestead mills of Carnegie, Phipps & Co.," noted the *New York World*: "Not an hour is being lost in the armor and plate mills, and not another man could be given work. A train load of plates and structural iron is being shipped daily from the Homestead plant to the Atlantic coast."

A tremendous speedup had taken place during the holiday season. "For years," the *World* recalled, "it has been the custom at the steel works to close down entirely for twenty-four hours and then operations were resumed in the armor plate mill and other departments where Government work is being done. About five hundred men are working at Homestead day and night on the contract from Washington. Piles of steel ingots high as houses have been made ready in the other departments, so that in an emergency there will be materials on hand to put five additional mills on Government work. There are five mills at Homestead that can be put on plates for warships, and [in which] a special device for speedy work has been introduced. This is the new beam mill."[28]

The work force at Homestead had demanded control over the speed of production, especially when government contracts stretched out their efforts in longer hours and made the work more intense. Workers nationwide had been under comparable pressures in 1892, and the People's Party had tried to raise the issue at its convention in Omaha, resolving: "That we cordially sympathize with the efforts of organized workingmen to shorten the hours of labor, and demand a rigid enforcement of the existing eight-hour law on government work and ask that a penalty clause be added to the said law."[29]

But Frick had not been amenable. During the countdown at Homestead he proceeded on the belief that labor troubles would not affect military contracts, telling the press on July 4th that: "Our contracts to furnish the government with armor plate contain clauses governing strikes and other labor disturbances. Should the Homestead mills be closed a year it would not hurt us, as far as those contracts are concerned. Armor plates must be made to fit as exactly as the parts of a sewing machine or a typewriter. They are subjected to the severest tests by the government. The finishing machinery is operated by specialists. The mechanics are high-priced men. We take the risk of having our work rejected. Under these conditions the manufacture of our armor plate does not seem so lucrative as has been claimed."[30]

Frick was making this argument just then because Carnegie was also in a hot squabble with President Harrison, Carnegie understanding the crucial place – and power – he held in the nation's defense system. "When he was in Washington last winter," the *Milwaukee Daily News* editorialized in July, 1892, of Carnegie's resolve, "Mr. Carnegie was greatly exercised lest some blood be shed by the United States upholding its dignity in the controversy with Chile. He undertook to tell the President that the country wouldn't stand it and was given a sharp lecture by Gen. Harrison. He has not been to Washington since that time."[31]

A vain man and allegedly a pacifist, Carnegie repaid President Harrison and the Republicans by locking out his workers, cutting steel output during the government's naval buildup. The President and his administration then planned to retort in kind: "There is not the slightest possibility of the War Department having anything to do with the struggle between Carnegie and his men," the pro-labor *News* reported from Washington. "It would only act when the Governor of Pennsylvania applied directly for federal help in quelling the disturbance and stated specifically that the state authorities were unable to preserve order. No such appeal from Governor Pattison is looked for here."[32] Pattison, a Democrat aspiring to be his Party's presidential nominee, certainly had no intention of asking a Republican President for military assistance.

Frick had already sought a way out of the obligation to supply specialized parts under the Government contracts; by July 8th the lockout and strike were in full swing and the Homestead Works could never fill the contracts an time. In Philadelphia at Cramp's shipyards, construction of three vessels had to be halted because of the lack of armor plate coming from the Homestead mills. Of the five warships underway at that yards, armor plate contracts held by Carnegie covered *Battleship No. 2*, *Cruiser No. 12*, and armored *Cruiser No. 2* (better known as the New York), launched in the winter of 1891, but now ready for her armor finish. Due to the suspension of work at Homestead her completion was delayed. Camps notified the Secretary of the Navy that Homestead could not fill its contracts due to the labor dispute, and a member of the Camps firm anticipated more serious consequences if the strike lasted much longer.[33]

"In order to protect ourselves from being called upon to pay possible demurrage through inability to complete the ships for which the Carnegie Works have contracts to make armor, within the time required by our contracts," said Henry Cramp, "we wrote to the Secretary of the Navy notifying him of the strike and stating that it might cause us considerable delay unless it was speedily settled. . . . None of the 14-inch and 17-inch armor for *Battleship No. 2* has been received, nor has any of the conning tower armor for *Cruiser No. 2*, the *Pirate*, been sent forward."

As for the lockout, Camps said he did "not believe this trouble can last much longer. The firm will in my opinion man its works with other men and start up in the not distant future. Mr. Frick is a very determined and strong-minded man, and I feel certain he will never give in. He would never have taken the stand he has if he intended to recede from it."[34]

## The Military Madness

Frick's determination was in part based on the powerful connection he had with the government, whatever the momentary disagreement, and its need for timely performance of the war contracts. Two days after the battle of Homestead, a rumor floated that the Carnegies would request the National Government to station United States regular soldiers at the Homestead plant, so that the firm would not be hindered in filling their contract for steel for making the armed cruisers. Though it was only one of the stories of the street, and no foundation for the rumor could be discovered,[35] the Government indeed had a key interest in Homestead production.

Military officers suggested that once the Homestead troubles began the National Government might interfere to suppress the riot on the ground that the completed armor plate in Carnegie's shops was government property.[36] They also argued that the rioters were interfering with government work upon which the company was engaged.[37]

Undoubtedly, Carnegie was falling short on his contracts, but there was nothing the National Guard could do *directly* to prevent it. The railroad strikes of 1877 had effectively put the National Guard out of the business of acting as the first line of repression, save in a national emergency, which this was not.

Besides, there were other ways of dealing with the lockout at Homestead. Governor Pattison was in close touch with the federal authorities and their advice was to allow the local authorities to expend all means at their disposal to suppress the "rioters." As the Pennsylvania Guard was effectively part of the National Militia, 8,000 strong, they could easily put Homestead under martial law. Thus, once the legal scenarios were played out, the State Militia could take charge and ensure non-unionists would be brought into the mill and put to work on government contracts.

Aside from the legalities designating which authorities would first go into action, so far as the Federal Government was concerned, these contracts had to be filled in a timely manner as a necessary part of the nation's military preparedness on the high seas. So even before the Militia occupied Homestead, the Federal Government had assessed the mill's ability to deliver on its contracts using both its old and new technology. "There is the old beam mill," Secretary of Navy Tracy inventoried the plant. "In addition there is a universal mill, a blooming mill and 119-inch plate mill. By changes in these mills that can be effected in a few hours they can be made ready for the milling of Government plates."[38]

Nor was military production limited to supplying ships for the incidental Valparaiso affair, as the United States was now interested in building its Pacific Empire with a full line of war vessels. "There is the highest authority for saying that the activity in naval circles, which has been interpreted in sensational quarters to mean certain war with Chile, will not cease even if within the next few weeks Chile should come forward with the amplest reparation for the assault upon the Baltimore's men at Valparaiso," wrote the *New York World*. No war today clearly did not mean peace tomorrow, this paper added: "The Chilean *imbroglio* has served to emphasize the importance of the steps taken months ago to provide adequate naval equipment. There will be no cessation of activity, therefore, now that the original program is on foot, but regardless of all abstract issues, the work cut out in the interest of the general need of the Government under the new policy upon which it is entering will be rapidly and generously pushed to completion."[39]

Generous were the contracts pushed to completion at Homestead. "In a calculation made today by one familiar with the product of the steel mills," the *World* reported on January 4, 1892, "the conclusion was reached that the Homestead mills are in a position to make all the iron and steel necessary in the construction of the average size gunboat in a week at the very longest. The firm's plate mills at the Upper and Lower Union Mills can also be changed to roll ribs for a man-of-war. The Duquesne plant of Carnegie Bros. and Co., as well as the Edgar Thomson plant of the same firm at Braddock have machinery for the making of the lighter sorts of work. This would give Carnegie a chance to keep the Homestead Works an the heaviest work."[40]

There was no end to the potential contracts Carnegie's Works might receive, for the military madness had seized the imagination of the masters of war.[41] "The success attained with high explosives in powder guns and with the submarine gun have made it possible to devise a vessel embracing so many Naval features as to make an almost entirely new type, and the Bureau has prepared and submitted to the Secretary [of the Navy] . . . plans for such a vessel," the Baltimore *Sun* reported:

> The vessel is a modification of the ram. Her deck is nickel-steel. . . . Armored structures rise fore and aft, each surrounded by barbettes containing bored rifle mortars, firing projectiles of drawn or rolled nickel-steel containing bursting charges of 200 pounds of high explosives. . . . In this new vessel it is proposed to carry the heavy side armor clear down to the bottom of the ship as a protection against torpedoes. . . . The heaviest armor, as it is now applied to vessels, would be of no more avail against this ship, as the heavy mortars could drop charges of 200 pound of high explosives on the deck of an enemy or send a couple of shells containing 500 pounds each of Emnesite into her vitals below the armor plate on her sides. If for any reason these projectiles failed to dispose of the enemy, this novel craft would keep right on and ram her enemy.[42]

Such an advanced armada envisioned military budgets on the rise, profitable contracts being promised by the government to the Carnegie mills. The pattern was clear enough: U. S. Government outlays on arms, already 33.5 percent of the budget in 1875, were to rise to 56.9 percent by 1907-08. In the interim, for the fiscal year June 30, 1894, the Navy had provided $2 million for the armaments of authorized vessels. Carnegie's armor plate, nickel-steel, was the wave of the future for his Homestead mill.[43]

## Competitive Arms Sales and Government/Industry Alliances

A world traveler, Andrew Carnegie could not miss the great advance Germany's Krupp had initiated in armaments made of nickel-steel.[44] Alfred Nobel's new gunpowder, with its picric acid base, had proved so powerful in shattering both bronze and cast steel that Krupp had turned to nickel-steel cannon by the early '90s, proving to the German crown that unalloyed steel was as obsolete as bronze, that nickel-steel had to become the basis of the nation's armor. Taking his advice, the Kaiser appointed Fritz Krupp to the post of Privy Councilor (*Geheime Kommerzienrat*), conferred on him the title *Exzellenz*, and, thanks to the personal patronage of *Seine Majestat*, *Exzellenz Krupp*_acquired the absolute Crown monopoly to provide guns to the government. Fritz's annual income jumped from 7 million marks to 21 million mark as the firm's business with Berlin rose from 33 percent of its gross product to 67 percent.[45] The significance of Krupp's technology and his tie with the government was not lost to Andrew Carnegie and other military contractors in Europe.

In fact, the arms merchants were at loggerheads. Each new invention had increased the competitive struggle to gain government contracts, and five great arms merchants battled for their share of the world's armament treasuries. Krupp in Germany, Schneider in France, Armstrong and Vickers in England and Mitsui in Japan were soon joined by Emil von Skoda in Czechoslovakia. Krupp looked to the American market as well, expending a small fortune at the Chicago Fair of 1893, where he lavished some $1,500,000 on an exhibit with drawings illustrating a cast steel bow frame for a new German ironclad, one of Krupp's traveling cranes used for slinging and moving the great Krupp guns, three 16.24-inch cannon mounted on hydraulic carriages, and a gun capable of firing thirteen miles.[46]

But the time had passed when U. S. buyers had to turn to Germany for Krupp's steel. Bethlehem and Homestead were operating full steam, and continued to do so, equaling Vickers weaponry sold by Sir Basil Zaharoff to Spain and Krupp armaments contributed to the insurgents in Cuba. "America was satisfied with its hardware, and rightly so," says historian William Manchester; "five years later United States guns demolished Zaharoff's merchandise at Manila Bay, El Caney, San Juan Hill, and Santiago."[47]

The foundation of these victories for the American Empire over those of Spain and Britain was the Carnegie Works. The U. S. Naval Squadron in the Pacific had become dependent on the quality of structural steel made at Homestead and Bethlehem.

There were other mills also building for the Squadron under government orders, so by the close of 1892, the *Monterey*, originally contracted with the strife-ridden Homestead mills, would be completed elsewhere.[48]

Intermeshing of government contracts and armories in America repeated the Krupp/*Seine Majestat* alliance in Germany. The *Sun* reported on November 17th that:

> One of the objects of the visit of Mr. Frick of the Carnegie Steel Company to the Navy Department yesterday, was to discuss the matter of future armor for the new ships. It is understood the Department will soon be ready to let contracts for all the armor needed for all the ships now authorized. Under the present law it will not be possible for a contract to be awarded, as the present Carnegie contract was without competition, but bids must be invited from all firms willing and able to undertake the work. The *only establishments in the country now equipped for armor work are the Carnegie Company and the Bethlehem Iron Works*, though it is possible some other firms might enter into competition and show they were prepared to erect the necessary plant at once and begin work in time to deliver the armor when needed. Owing to the time required to build the machinery needed, however, it is not likely that any new firms will compete, and it is almost certain that the only bidders will be the two firms mentioned. It is probable that the specifications for the new armor will require most of it, or probably all, to be treated by the Harvey process of superficial hardening. The Bethlehem establishment is at present the only place equipped for this work on a large scale, and if the contract is awarded to the Carnegie Company they will have to put in a Harveyizing plant in addition to the new machine tools required.[49]

To secure the new government orders, Henry Frick and Andrew Carnegie used their political influence and promised to install the Harvey machines at Homestead in time. Despite the lockout at Homestead, despite the fact that Carnegie's steel fell short of government standards on inspection, despite a delay in securing the new machines promised, and, finally, despite the employment of unskilled strike-breakers operating old technology, Carnegie received sufficient government contracts to enable him to reap substantial profits in 1892, the year of the great Homestead battle.

True, Carnegie had locked out the workers while his government contracts remained unfilled, but both early in 1892 and after the battle of July, the government used its multiple leverages to protect the mills so essential for national defense against the strikers.

Even before Homestead, precedents had been laid for speedups and lockouts. In the case of the Midvale Steel Works, for example, the Department of the Navy had contracted that the work be finished on time and that laborers work night and day. In January 1892, when the potential war with Chile pressed and the Navy was anxious to secure timely production to offset the Chilean Armada, the U. S. Bureau of Ordnance of the Army and Navy telegramed Midvale to keep the workers operating night and day in order to complete gun forgings in the shortest possible time, as well as to finish steel castings for the Ammen harbor defense at Bath, Maine.[50] Throughout the nation, preparations for war demanded emergency production of munitions. In the army gun factories, orders were given for workers to operate at utmost capacity, 300 machinists and workers at the Waterulies Plant starting early and finishing late.[51]

The requirement of speed had meanwhile led the government to favor the largest enterprises that produced steel. In Congress, Representative Hale insisted that the Republican Party and its protectionist policy had been responsible for the building up of the works at Homestead and Bethlehem, in either of which could be constructed the finest ordnance material and the best armor plate in the world. If in power, he continued, the Republican Party would duplicate Bethlehem and triplicate Homestead, instead of allowing competitive establishments to grow up at Newcastle in England, Essen in Germany, and elsewhere in Europe.[52]

Following this policy, despite the lockout by Carnegie at Homestead, Government contracts were guaranteed the firm: "The strike at the Carnegie Steel works will not interfere with any contracts which the firm at present have [sic] with the Navy Department for furnishing armor," the *Sun* stated. "They are now under contract to furnish 6,000 tons of armor, aggregating nearly $4,000,000. The prices are the same as paid the Bethlehem Iron Works under their contract made two years previously. A clause in this contract for armor provides that the Carnegies shall not be held responsible for delays in the delivery of material when such delays are caused by 'fire or water, or by any strike or stand-out of workmen employed in or about the work, or by other circumstances beyond their control.'"[53]

This was already "old hat" in the war contracting business. When workers had struck a few years earlier at the Union Iron Works in San Francisco, the government allowed the company to break its contract, teaching the workers the bitter lesson that they could not hope their strike would force the employer to give-in in order to meet the government contract deadline.[54]

**Military Production During the Homestead Lockout**

But how explain the lockout at Homestead, when the government had big contracts that could be fulfilled nowhere else and that required substantial investments in machinery, undercutting old skilled crafts?

Of Homestead, wrote the *Sun*:

> It is here that armor plate for most of the vessels of our new Navy has been made. The defensive covering of the steamship Monterey have only been just completed. Huge ingots measuring 55 inches square by 14 inches wide and weighing 160,000 pounds are lifted from the pits, charged into the heating furnaces, and rolled into plates by machinery that is unequaled by any other in the world. Appliances imported from Scotland, plane and mold and fashion the plates to the exact shapes required by the contour of the ship's sides, so that when the material is placed in the cars it is ready for position on the vessel. In perfecting the armor-making branch of the business the firm has invested over $1,000,000. The other departments have manufactured massive beams up to forty inches high, plates of every kind, and structural steel of all descriptions."[55]

In turn, this required the introduction of still more technology: "Everywhere throughout the Works labor-saving appliances have been introduced at great cost, and within the last three years the necessity of human labor has been greatly reduced," the *Sun* elaborated. "The latter is one of the chief reasons why the firm insists on cutting the rates for work."[56] Indeed, Carnegie planned to lock-out the workers while retaining his government contracts; he then sought to do away with skilled workers by replacing them with low-price, unskilled workers to operate the newly installed technology.

By the beginning of August, the 2,000 striking skilled and semiskilled specialists Frick needed to produce steel could not be replaced, and the unskilled scabs still could not make quality steel. But by the end of the month, the last slab of armor plate for the cruiser *Monterey* was somehow finished; and it was clear that government orders would ultimately be filled, since the government contracts allowed an extension of deadlines due to labor troubles.[57]

Speculation that the slabs had been worked on by the Homestead regulars before the strike, or that skilled workers from the Duquesne or other mills had trained the unskilled to produce them, either did not square with the confidence of the striking Homestead workers that Carnegie could not meet the government deadlines, or seemed incongruous with the fact that the Duquesne and other Carnegie employees were in sympathy with the Homestead workers. What is known, however, is that the steel produced by the scab workers was of low quality and that the only way the government might have felt satisfied was to wink at its previous standards and to allow Carnegie to shift production elsewhere.

In fact, during the lockout both Homestead and Bethlehem failed to meet the strict government contract standards of quality. Nonetheless, they were allowed to continue as the nation's two leading armories, for indeed there were no others who could supply vessel fittings, as they, hand-in-glove with the government, had long since planned. The government had initially established stringent tests for weaponry, one report noting the process involved:

> A remarkably successful government armor-plate test took place at Redington proving ground of the Bethlehem Iron Works today. it was a ballistic test, and by its results 480 tons of armor were accepted by the Navy Department. it tested twenty-three barbette plates of the Maine, in a building at the Brooklyn Navy yard. They are worth $350,000.
>
> The test was made in the presence of Commodore Folger, chief of the ordnance bureau. The plate is 10 inches thick and 12 feet high. . . . Three shots were fired from an eight-inch gun. The projectiles weighed 250 pounds and the velocity was 1,400 feet a second, with a charge of 60 1/2 pounds of Dupont brown hexagonal powder. The penetration was 10 inches, 11 inches, and 10 inches for each shot respectively.
>
> Commodore Folger was delighted, and announced that the test would pass the group of twenty-three plates for which the test plates stood as representative.[58]

Applying such tests as these, the Carnegie firm fell short on several contracts. The defectiveness of Carnegie steel was in part due to the output made with non-union workers after the lockout began. But by September, Secretary of the Navy Tracy was bending over backwards to deny charges that contract specifications had been changed to benefit Carnegie & Phipps. Rather, Tracy declared, the contractors had changed the specifications to increase the fighting efficiency of the ships and to expedite their completion.[59]

Regardless of the real reason, by the next month, the Navy Department had become dissatisfied with the inadequate deliveries of structural, armor and gun steel by Homestead and Bethlehem for new vessels. Sufficient domestic steel could not be procured to prosecute the work quickly, the slow deliveries being due to the Homestead strike on one hand and Bethlehem's inadequate machinery on the other. "The department has now determined to cure this state of affairs and bring the contractors to terms," wrote the *Sun,* "and notice has been sent in at least one case that the promises made when the contract was signed that a proper plant would be established must be fulfilled, and that the deliveries of steel must be brought up to the contract requirements or the contractors will be dealt with strictly according to the terms of the document."[60]

The new Secretary of the Navy, H. A. Herbert, had come in with the unseating of President Harrison by Cleveland in the 1892 elections, and the Democratic administration was of a different frame of mind than the Republicans backed by the substantial industrialists and military contractors.[61]

Secretary Herbert thus flatly reported that the Carnegie Company was remiss in that it "had failed to temper armor evenly . . . had plugged and concealed blow-holes, which probably would have caused a rejection . . . had re-treated, without knowledge of the inspectors, plates which had been selected for ballistic tests. . . ." Once President Cleveland reviewed and upheld these charges, the Carnegie firm was fined $140,000.[62] But even then, every effort was made by the Navy Department to let its principal suppliers off the hook. As one reporter noted:

> The department has therefore waived some of its rights in the interest of expedition. The contracts expressly forbid any transfer of the work, yet one company was allowed to exchange with the other a large part of its obligations upon reports made to the department that the work could be better and more quickly done. This was the case where one company [Bethlehem] was better fitted to make thin boiler and structural steel and the other [Homestead] heavy armor and gun steel. The first was to be rolled and the latter to be forged. But these concessions by the department have not been effectual. The new ships are being delayed, and hereafter the contract stipulations will be enforced unless an improvement is noted.[63]

## Outline of the Future

No end to the state-fostered, military-industrial complex was in sight. The military contract business was strengthened, even after the potential battle with the Chilean Navy faded. U. S. steel mills continued to be called upon to provide ship sides and weaponry, for now the United States was committed to competition with England and her pawns in the Pacific, be they Chile or other nations.

The outline of a new struggle for empire was just emerging. When on January 25th, 1892, Chile had purchased still another cruiser from Britain's Armstrong Shipyards at Newcastle, the Secretary of the Navy was adamant that the U.S. had to prepare to take on one of the most efficient men-of-war afloat. He cited the *25th de Mayo*, which was the fastest vessel sent down the ways in 1891, as well as the best armed for a vessel of the 3,000 ton class.[64] Nor did the situation change when the war threat subsided on January 28th.[65] The Navy Department still did not cut back its immediate orders, so that of some $2 million in preparatory expense for war, much was assigned to coal provisions and the balance was to speed up work on vessels under contract and to award more work to yards and shops.[66] The future in part was being planned by the American Navy.

By these means, the United States Government elevated the arms race for control of the Pacific Ocean, planning and contracting a Pacific fleet to replace Spain's far-flung sphere of influence. Between 1891 and 1898 there was an enormous U.S. naval buildup, culminating in 1898 as the U. S. deployed its power in the Pacific and the Caribbean to push Spain out of Cuba, Hawaii, and the Philippines – planting its flag, building new deep-water ports and dry docks, shipyards and armories, coaling stations, colliers and powerful fortification – to make these areas safe for trade and investments.[67] In the process of preparing American Naval forces the arms contractors, led by Carnegie and Bethlehem, received lush rewards and the power to subjugate their labor force with aid from the State Militia. It was not only Carnegie's wealth that defeated the workers at Homestead, then, but also the military requirements of a government bent on establishing a Pacific empire.

---

## Notes

[1] Calvin Sims, "G.A.0. Says Navy Contract Unfairly Favored I.B.M.," *The New York Times*, June 23, 1989.

[2] See Daniel del Rio, *Simon Bolivar* (Clinton, Massachusetts: The Colonial Press, Inc., 1965); Scott Nearing and Joseph Freeman, *DollarDiplomacy* (New York: Monthly Review Press, 1966); William Appleman Williams, *The Tragedy of American Diplomacy* (New York The World Publishing Company, 1959).

[3] *War Supplement to Compton's Pictured Encyclopedia* (Chicago: F. E. Compton & Co., 1939), 59.
[4] *The Sun* (Baltimore, Maryland), November 23, 1892.
[5] E. Benjamine Andrews, *History of the United States* (New York: Charles Scribner's Sons, 1927), Vol. V,, 49
[6] *The Sun* (Baltimore, Maryland), November 23, 1892.
[7] *Ibid.*
[8] Gray Brechin, *Imperial San Francisco: Urban Power, Earthly Ruin* (Berkeley: University of Califoirnia Press, 1999),128.
[9] *The Sun* (Baltimore, Maryland), November 23, 1892.
[10] *The World* (New York, New York), November 2, 1892.
[11] Ostrogorsky, M. *Politics in Nineteenth Century America.* Vol. II (Santa Barbara: University of California, Santa Barbara: Special Collections, bound, undated manuscript), 100-101.
[12] Charles P. Steinmetz, *America and the New Epoch* (New York: Harper & Brothers Publishers, 1916), 126.
[13] *The World* (New York, New York), November 2, 1892.
[14] Andrews, Vol. V,, 45; Edwin Emerson, Jr., *A History of the Nineteenth Century* (New York: P. F. Collier & Son, 1900), Vol. III, 1746; see also *The Labor World* (Pittsburgh, Pennsylvania), 1892, passim.
[15] *The World* (New York, New York), November 2, 1892
[16] *Ibid.*
[17] *Ibid.*
[18] Andrew Carnegie, *Autobiography* (New York: Houghton Mifflon Co., 1920), 220 ff.
[19] Wolff, 44.
[20] The Social Economic Union of Cuba, *Commercial Relations Between Cuba and the United States of America* (Havana, 1936), 23; see also Henry Steele Commanger, *Documents of American History*, 7th ed. (New York: Meredeth Publishing Company, 1963), 234-237.
[21] Edward Boorstein. *Allende's Chile: An Inside View* (New York: International Publishers, 1997), 18-22.
[22] Emerson, Jr., Vol. III, 1720-1721.
[23] *Ibid.,* 1721.
[24] Andrews, Vol. V,, 64.
[25] Emerson Jr., , Vol. III, 1722.
[26] *Ibid.*, Vol. III, 1738.
[27] *The World*, January 3, 1892.
[28] *Ibid.*, January 4, 1892.
[29] *The Milwaukee Daily News* (Milwaukee, Wisconsin), July 5, 1892. The National Labor Union and other trade unions had originally pressured Congress to pass this legislation in 1868, defining eight hours as "a day's .work for all laborers, workingmen, and mechanics now employed, or who may hereafter be employed by, or on behalf of the government of the United States." Though Secretary of the Navy, Adolph E. Borie, proceeded to cut the wages of

government workers one-fifth to correspond with the reduction of the 10-hour day, four years later, on May 18, 1872, Congress again enacted the eight-hour day and demanded that the old rate for ten hours be paid. See also *The Cleveland Leader* (Cleveland, Ohio), July 5, 1892.

[30] T. V. Powderly, *Thirty Years of Labor, 1859 to 1889* (Columbus, Ohio: Excelsior Publishing House, 1889), 473-476.

[31] *The Milwaukee Daily News*, July 7, 1892.

[32] *Ibid.*

[33] *The Cleveland Leader*, July 10, 1892.

[34] *The Philadelphia Press* (Philadelphia, Pennsylvania), July 10, 1892.

[35] *Ibid.*, July 9, 1892.

[36] *The Cleveland Leader*, July 8, 1892.

[37] *The World*, January 4, 1892.

[38] *Ibid.*

[39] *Ibid.*

[40] *Ibid.*

[41] *The Sun* (Baltimore, Maryland), November 23, 1892.

[42] *Ibid.*

[43] *Ibid.*, October 29, 1892.

[44] William Manchester, *The Arms of Krupp, 1587-1968* (New York: Bantam Book, 1968), 242-243.

[45] *Ibid.*, 242-243, 247-248.

[46] *Ibid.*, 247- 248.

[47] *Ibid.*, 248.

[48] *The Sun* (Baltimore, Maryland), November 17, 1892.

[49] *Ibid.* Emphasis added .

[50] *The Philadelphia Press,* January 25, 1892.

[51] *Ibid.*

[52] *Ibid.*, January 26, 1892.

[53] *The Sun* (Baltimore, Maryland), July 8, 1892.

[54] San Francisco Maritime Museum, Permanent Exhibit, Second Floor. (San Francisco, California); see Brechin,, *passim.*

[55] *The Sun* (Baltimore, Maryland), July 7, 1892.

[56] *Ibid.*

[57] Burgoyne, 191; Wolff, 191.

[58] *The Sun* (Baltimore, Maryland), December 1, 1892.

[59] *Ibid.*, September 14, 1892.

[60] *Ibid.*, October 13, 1892.

[61] Ostrogorsky, Vol. II, 101; John and Alice Durant, *Pictoral History of American Presidents* (New York: A. S. Barnes and Company, 1955), 183-187, 195; Emerson, Jr., Vol. III, 1746.

[62] J. K.Winkler, *Incredible Carnegie* (New York: Vanguard Press, 1931), 227.

[63] *The Sun* (Baltimore, Maryland), October 13, 1892.

[63] *The Sun* (Baltimore, Maryland), October 13, 1892.
[64] *The Philadelphia Press*, January 25, 1892.
[65] *Ibid.*, January 28, 1892.
[66] *The Sun* (Baltimore, Maryland), October 13, 1892.
[66] *The Philadelphia Press*, January 25, 1892.
[66] *Ibid.*, January 28, 1892.
[66] *Ibid.*, January 29, 1892.
[67] See Nearing and Freeman, Chapters I, IV, V, VII; Richard Krooth, *Arms & Empire, passim*; Brechin, 140, 166-167, 338 n35..

# Chapter 3

# Tariffs and Employers; Machines and Workers

The American People know a good thing when they see it. Heartiest congratulations. You deserve this triumph.

– Andrew Carnegie's cablegram to President Benjamin Harrison on lowering the tariff on steel billets, Sunningdale Scotland, June 12, 1892.

At Homestead wages were tied to the price of steel billets; wages rose if prices did. The reverse applied, however, only until billets dropped to $25 per ton. At that point wages froze, no matter how low the price went. The company wanted to reduce the minimum to $22.

– Prof. Harold C. Livesay[1]

Homestead's military contracts were only one of the forces shaping the development of Carnegie's mills and his relationship with his workers. More important still were his influences over tariffs and technology, with H. C. Frick, Carnegie's implacable right-hand man at

Homestead, acting as his principal tariff negotiator with the Republican administration.

Frick, in the vernacular of the age, controlled the "boodle" – the slush fund – for the protection of manufacturers in the iron and steel industry. In 1888 Carnegie, Frick and their fellow manufacturers had already helped to buy the Presidency and its tariff policy. Charles Harrah of the Manufacturers' Club of Pittsburgh boldly bragged after the election of President Harrison: "This tariff belongs to us; we bought it; we paid for it; and it is ours; we did not put up our money to raise the price of labor, to raise wages . . . we put up the money to buy the legislation we wanted and we got it. It is ours, we bought it and paid for it and that is the whole story."[2]

**The Politics of Tariffs**

At least it was part of the story behind the McKinley Tariff. Led by Frick, the plutocracy of manufacturers able to dispense money to the Republican Party acquired the power to shape tariffs imposed on foreign steel.

"He showed how the protective tariff in McKinley's nimble fingers was juggled so that State elections could be bought every four years – or every two years or every year – it doesn't matter to the tinkers," a critic said of Frick. "So long as some millionaire is willing to pay for more bounties, so long will the Republican protectionists give him the legislation he wants, and the plain people are 'whipsawed' every time, as the professional gamblers say. They pay higher taxes on all they buy to enrich an individual. He [Frick] pays a moiety of his increased bounty to debauch the ballot box and deface those unfortunate voters who have a price even on their manhood and their right of suffrage."[3]

Polemics aside, Frick was able to lobby for the tariff he and the other steel manufacturers wanted. As Republicans determined to reduce wages, they were able to influence the policy of their party, first to aggrandize wealth and power by raising tariffs to keep out foreign competitors, then by lowering the tariff on steel billets momentarily, double-crossing the steel workers who voted for that party in expectation that it would keep the tariff on billets high to protect their wages!

Not only were wages cut at Homestead and the steel workers locked out, but the introduction of H. C. Frick into the 1892 presidential campaign "as a dispenser of Republican boodle . . . stirred up the indignation of honest citizens all over the country," said the *New York World*. "It is no longer a secret that the Republican manufacturers base their claims of success on the amount of money they have to dispense."[4] The rich manufacturers had not only put Harrison in the executive mansion by contributing the largest campaign fund in history up to that time, but they had also received the tariff and other benefits they expected from him.[5]

What objective each manufacturer had was unique in a sense. Yet there was the common purpose in keeping out foreign goods, and the *New York World* compiled a list of the 100 greatest trust combinations which were made possible under the protective shield of the McKinley tariff. "Of course, such omnibus collection of names have their uses in showing to the eye at a mere glance in gross the gigantic grip which the trusts have upon the pockets of the people," the *Milwaukee Daily News* editorialized. "A clear inside history of one individual trust is worth more, however, as a means of illustration than any list a mile long."[6]

Republican money had, over time, positioned the major steel companies to use the tariff flexibly. When the steel manufacturers needed the tariff, they were able to propagandize among the smaller mill owners and steel workers that the tariff was essential for both high profits and secure wages. The steel workers had come to believe this line. But the employers knew better: It was not a principle, but their own policy in using the tariff for certain objectives. In fact, before 1890, the tariff had played an important purpose in strengthening U. S. steel manufacturers to take on foreign competitors. It had acted as the first line of defense by increasing the price of foreign steel sold on the U. S. market. Behind its protective barrier, Carnegie and others were able to expand their steel mills, upgrade their technology, outmode old labor skills and replace them with new ones.

The tariff could be used by the U. S. manufacturers as a weapon of offense or defense. If, on one hand, foreign competitors showed signs of capturing a substantial share of the U. S. market, even higher tariffs would keep foreign steel out and domestic prices up, thereby protecting profit margins for Carnegie and other steel masters. If, on the other hand, foreign nations defended by raising their own tariffs on American steel, the U. S. tariff could be negotiated with them to reciprocally cut tariffs, encourage foreign imports into the U.S. market, and – to protect profit margins – lower both steel prices and the wages of American steel workers.

But what would happen if tariffs were lowered and foreign steel still could not underprice U. S. steel on the American market? This question was answered by Andrew Carnegie in 1892.

In that year, markets were slack, foreign competition did not present a major obstacle to U. S. mills, and Andrew Carnegie planned to reduce wages at Homestead.

## Tariffs, Steel Mills and Skilled Workers

Before 1890, the iron and steel manufacturers had lobbied for a high tariff to keep out foreign steel. The tariff protected their profit margins between costs and selling prices. High wages for American steel workers were also based on the high tariff that permitted Carnegie and other iron and steel mill owners to monopolize the home market, keep prices up, and thereby maintain substantial returns on invested capital. If tariffs declined and wages could, not be reduced, however, that return would have declined.

This was the backdrop of the national elections of 1888, resulting in the Republican Party victory and economic policy that fostered the 1890 McKinley tariff, which increased the duty on many commodities from 38 to 49.5 percent. During the debate on the bill, it was argued that American labor would be "pauperized" if it were not protected on the job and in its wage scale by a high tariff keeping out goods made by cheaper, foreign workers. Accordingly, the bill imposed higher protective duties on wool, the finer grades of woolen and cotton goods, cutlery and tin-plate, and also extended duties to cover a number of agricultural commodities.[7] But on iron and steel the tariff was cut on many products, and, as we shall show, the owners of ironworks used the reduced tariff as an excuse to reduce wages, break the workers' Amalgamated Association, and deunionize the entire industry in preparation for a new market struggle with European competitors – Britain, France and Germany being especially large producers and exporters.[8]

The secretary of the American Iron and Steel Association praised the newly passed McKinley tariff signed by the President on October 1, 1890:

> The metal schedule as it was finally adopted is, in my judgment, the most harmonious and most completely protective of all metal schedules that have formed a part of our tariff legislation, notwithstanding the fact that it embodies many reductions on duties. But these reductions which would have been disastrous ten or twenty years ago, our iron and steel industries can now afford to accept.[9]

The tariff could not at the same time be the most protection ever offered steel mills and also reduce duties that would let in foreign steel to undercut the sale of U. S. steel. If the cuts in duties on steel rails (from $17 to $13.40 a ton) and on iron and steel structural shapes (from $28 to $20.16 per net ton) really diminished the industry's return on capital, then wages would have to be cut to maintain the old return.[10] But if these reductions in tariffs did not in fact bring forward competitors from abroad, then wages would not have to be reduced to maintain profit margins, and any cut in wages that was effected would automatically raise profits. Since it had been the policy of lobbyists in Congress to lower tariffs on items for which no foreign competition threatened, they promoted the cutting of tariffs in these areas and used the cuts as an excuse to lower the wages of the work force.[11]

This is likely what happened in the iron and steel industry when the American Iron and Steel Association pressured Congress in the late 1880s and early 1890.[12]

It had been freely charged that, when the McKinley tariff bill was being prepared, Mr. Andrew Carnegie himself had waited on the conference committee which put the finishing touches to the measure. In return for his generous contribution to the Republican campaign funds, he secured a reduction in the duty on steel billets, the price of this product being the single standard for calculating wage payments at his Homestead Works. Since the Carnegie firm also controlled the nation's billet market, he could lower prices as a seeming consequence of a lower tariff. This was to serve as a cover-up and justification for the new lower wage scale to be imposed at the Carnegie Works by ultimatum.[13]

Towards this end, the Republicans carried on a massive public relations campaign, largely designed to appeal to the iron and steel workers who voted for their party.[14] Hearst's *San Francisco Chronicle* insisted that it was the Democrats' free trade policy that posed the real danger to labor:

> Whatever may be the outcome of the contest between the competitor iron industry in this country, it certainly is absurd to assume that free trade would help the situation of either employer or employed. A [low] tariff for revenue [as proposed by the Democrats] would only open wide the door to the tremendous surplus stocks of England, and close down all the iron works in the United States or compel them to adjust their scale of wages to those in force in Great Britain at present, and that would mean an average reduction of at least fifty-five percent in every branch of the industry.
>
> The favorite argument of the [Democrat] free trader, that we could find a market for our surplus iron if we were on a revenue-tariff basis, will not be urged with much persistence at a time like the present, when all of the English iron manufacturers are suffering from depression

> owing to their failure to find customers for their products. If free-trade England is unable to find markets, or make them by exploring and opening new countries, the United States can hardly do so . . . . Until the process of [market] contraction is reversed enterprise is sure to remain paralyzed and employers will have to put up with less profits and the wage earners less wages.

This Republican soliloquy also insisted that the Democrats had seized upon the trouble between the employers and the workers in the iron district of Pennsylvania and misrepresented the operation of the McKinley bill, which reduced the tariff on steel rails and many of the primary forms of iron and steel. The real fault, it continued, was not the tariff, but the small domestic market for steel and competition from America's southern manufacturers who were "actively threatening the supremacy of Pennsylvania in that industry."[15]

Hearst's *San Francisco Chronicle* assessed the skilled workers' need for some tariff protection, arguing that "the Pittsburgh iron workers are shrewd men and are trying to get all they can for their services, but they are perfectly aware that their chances of obtaining good wages are lessened by every ton of iron which is imported into this country free of duty. In other words, being sensible men they not apt to embrace the idiotic Democratic free trade idea that a man can improve his face by biting off his own nose, a feat which would bear close resemblance to the effort of workingmen to ruin their employers because they happen to disagree with them."[16]

**The Steel Workers' Analysis**

Despite these efforts, the steel workers were only momentarily kept in the Republican fold. After the lockout at Homestead over the issue of the wage cut, the Amalgamated Association of Iron and Steel Workers began to critically evaluate the Republican's tariff policy on the products they made. Indeed, when a Congressional investigating committee came to Homestead, the city Burgess McLuckie testified that the McKinley bill not only reduced the tariff on steel billets on which wages were based, but also raised the tariff on all other articles of production. "It was," he said, "a gigantic conspiracy to rob the laboring man of a fair day's wages for a fair day's work."[17]

"Your idea, then, is that the Company, after having obtained a scale of wages based on the market value of steel billets, caused the duty on steel billets to be reduced?" questioned Congressman Oats. "Yes, Sir," answered McLuckie, "they shortly converted the Duquesne Works into a billet plant; increased the production; flooded the country so that prices may be reduced, and thus affect our wages. The Duquesne produces large amounts of billets."

The logic of McLuckie's argument was that Carnegie would stop at nothing to cut wages at Homestead. Today he would manipulate the tariff. Tomorrow he would shift billet production from Homestead to Duquesne and increase output so as to temporarily cut market prices. The next day he would lower the Homestead workers' wages an the sliding scale that pushed pay rates down with billet prices. Thus, for McLuckie, there were no hard and fast rules on the tariff question.

When Judge Taylor asked McLuckie, "The higher the tariff, the better your wages, then?" he could answer with candor: "Not always, Sir; simply because our wages were regulated by the selling price of billets, the value of which was reduced."

The Judge pressed: "I understand you to give it as your opinion as to the cause of the reduction on the selling price of billets, the reduction of duties thereon and the increased production by this company?"

McLuckie replied: "That is the idea exactly, Sir."

Judge Taylor: "How could this company make more money by selling their billets at a price less than before?" Unperplexed, McLuckie the skilled worker showed himself a brilliant economist as well:

> The Duquesne is not a union mill and the men have no defense there, they have no regular scale of wages. At any time the Company wishes to reduce these men's wages they have no defense. They can run billets down to almost any price so as to compel them [wages] to go down with them [prices] and then bring it [the lower billet prices and thus lower wages] to the Homestead plant also. We make some billets at Homestead, also, but while we may possibly make 100 tons of steel billets, we make 5,000 tons of beams, etc.[18]

Homestead wages were consequently based on the price of billets the workers made few of – a product made largely at Duquesne. Since Duquesne was unorganized, had low wages that could be cut with the tariff reduction on the billets it made, the purpose of the tariff cut was also to lower Homestead wages.

Not satisfied with this answer, Judge Taylor had persisted: "Why should the firm want this [tariff] reduction when they certainly lose by it?" McLuckie explained:

> Because every article produced is based on that particular item of steel billets. They [the managers of the Carnegie empire] could do it without the tariff, if they had our [Homestead wage] scale signed and our wages

> based on the price of these billets; they could have done much better than they have done. If they could get the minimum [billet-price] basis low enough it would pay them to take all the tariff off.[19]

In McLuckie's view, Carnegie had repeatedly tried to force the workers to take wage reductions below the minimum $25 per ton under the sliding scale. For this, he said, Pinkertons had been used in 1889, the tariff had been cut in 1890 when foreign steel was allegedly selling at a lower price, and the lower tariff was again used as an excuse that it was pressing down all U. S. steel prices in 1892. In the name of their ability to compete, the steel mill owners had then cut the $25 minimum wage, so that their profits were improved; and, in fact, the difference between the lower price and the lower wage to protect profits for Carnegie also resulted in an ever-decreasing fund for wages.

The crucial matter, then, was how the firm planned to lower U. S. wages to European levels, Judge Taylor and Burgess McLuckie engaging in the following exchange:

> Q: Do you know the difference between here and England in the price of billets.
>
> A: I understand $7 difference. There was a conspiracy by the Company to do by legislation [cutting the tariff on steel, to lower billet prices, so as to lower wages] what they did not succeed in doing with the deputies three years ago.
>
> Q: You think this conspiracy consisted of whom?
>
> A: I think the Pacific Railroad people [transporting iron ore and coke, as well as large-scale buyers of steel rails] were connected.
>
> Q: And whom else?
>
> A: H. C. Frick and the Carnegie Company interests. I think there was quite a large circle if you cared to gather them up.
>
> Q: Any members of Congress?
>
> A: The [tariff with its reduced rate on billets] was passed.
>
> Q: Then a majority of Congress must have been in it?
>
> A: I did not say so. I am not here for the purpose of insulting this Committee.[20]

But it required neither conspiracy nor majority in Congress to reduce tariffs and wages. The committee system in Congress assured that only a few people responding to Carnegie and the other mill owners needed to be influenced. These committees could draft legislation later presented to both houses of Congress. Agreement in the committee chambers therefore meant legislation cutting the tariff on steel billets could reach Congress without reason given. Since the Carnegie firm controlled the billet market, it could lower prices and wages as a seeming consequence of a lower tariff.

Burgess McLuckie had charged Carnegie Company and the Republican Party with duplicity: He exposed the tariff-wage connection, newspapers nationwide published his words, offering two ways at looking at the so-called doctrine of "protection to American labor."[21]

## The "Common Interests" of Labor and Capital

Just as Carnegie and other steel masters depended on tariff-protected markets, in the years preceding 1890, laborers depended on their employers for jobs in the tariff-protected steel industry. Republicans argued labor had an interest in maintaining the tariff to protect capital so jobs could be provided. That party and its newspapers attacked the view of the Democratic Party that the tariff did nothing for labor and had been the root cause of the Homestead strike[22] – we will see.

Four days after the battle, the Republican *Cleveland Leader* sarcastically polemicized:

> Insofar as protection has helped to build up a great iron and steel works in the United States it is responsible for strikes of iron and steel workers who, because of protection, are earning better wages than any class of workers in the United States. But for protection there would be no iron and steel mills in this country, and thousands of men who now find employment in these great industrial plants would have remained in England, Ireland, Germany and other countries from which they came and continued to work for the starvation wages which they came to America to escape. But for protection the vast army of industrial workers who now find employment in America would be turned adrift to seek employment of farmers engaged in raising wheat and other products for which they could not find a market. Without protection we might have no strikes of iron and steel workers, but the highways would be filled with tramps and vagabonds begging for money with which to buy food, and ready to commit crimes if they could not get it. Nobody but a fool, a demagogue or a knave would think of holding protection responsible for

a strike at an iron mill which, without protection, would never have been built.[23]

The idea that labor and capital had joint interests had become a common theme among both employers and trade unions in the steel industry in Europe and the United States. In mid-1892 the *Industrial World*, the official organ of the Welsh tin-plate workers, argued that it was "clearly in the interest of both employer and workmen to produce tin plates, tariff or no tariff, at a price that will drive all competitors from the field." Hence, it said, the workers should allow their wages to be cut to deal with the U. S. McKinley tariff: "the [potential] victory of the Republicans at the polls means the retention of the McKinley bill and means the rapidly running loss of 80 percent of the [Welsh] export American trade. Had there been no Democratic victory [reducing the metal tariff] in 1890 the spread of the tin-plate manufacture to the United States would have been both rapid and bona fide. It is not yet too late to do something to reduce the price of plates. Put them down to 11 s. per box of 100, 14 x 20, full weight basis. Let the workmen take half pay for a few months and turn out more. Then let the masters forego profits for the same time."[24]

The underlying idea was that the workers comprised a unit within each employer's mill, and the employer could conquer the market from American competitors only by an alliance with labor to reduce wages and/or profits for a time.[25]

In addition, they could jointly back Wales' own national tariff protection, copying the situation in America where high tariffs supposedly protected both employer profits and skilled labor's high wages.

They were blind to the fact that Carnegie had secured his profits by cutting the tariff on billets, demanding a corresponding cut in wages – and locking out the workers when they refused to accept the reduction.[26]

## The Alliance Breaks Down

Homestead workers fought back, organized, polemicized. "We were persuaded to vote the Republican ticket four years ago in order that our wages might be maintained," fulminated Homestead's town Burgess, McLuckie, in response to Carnegie's lockout of the workers at the Homestead mill and the use of arms against them:

As soon as the election was over a widespread feeling on the part of the manufacturers towards a reduction of wages was exhibited all over the land. As soon as the McKinley bill was passed, the article in the production of which we work was the only article that suffered a reduction. . . . The constitution of this country guarantees all men the right to live, but in order to live we must keep up a continuous struggle. This is the effect of legislation and nothing else. The McKinley bill reduced the tariff on the four-inch billet, and the reduction of our wages is the result. You men who voted the Republican ticket voted for high tariffs and you get [Frick's] high fences [surrounding the Homestead Works], Pinkerton detectives, thugs and militia![27]

Labor elsewhere joined the refrain, the radicals taking the initiative to call for a third party made up of workers and distinguished from the look-alike Republican and Democratic parties, both under the influence of manufacturers and financiers.

In truth, both Republicans and Democrats tried to use the tariff issue to garner votes. "Carnegie's locked out workers ought to begin to see the folly of voting with either of the old parties," the pro-labor *Milwaukee Daily News* said. "They should at least be consistent and vote the people's ticket."[28] Under the ticket of that party, *immediate legal authority* in Homestead and elsewhere would have been invested in the people, rather than in the hands of sheriffs and deputies.

At its Omaha convention in '92, the People's Party pointed out that:

The urban workmen are denied the right of organization for self-protection; imported pauperized labor beats down their wages; a hireling standing army, unrecognized by our laws, is established to shoot them down, and they are rapidly degenerating into European conditions. The fruits of the toil of millions are boldly stolen to build up colossal fortunes for a few, unprecedented in the history of mankind; and the possessors of these, in turn despise the republic and endanger liberty. From the same prolific womb of governmental injustice we breed the two great classes tramps and millionaires.

In overturning two-class society, the Party argued, the creators of wealth would become the owners and decision makers. "Wealth belongs to him who creates it," the Party proclaimed, "and every dollar taken from industry without an equivalency is robbery. 'If any will not work, neither shall he eat.' The interests of rural and civic labor are the same; their enemies are identical."

In the context of the tariff, company-hired Pinkertons, and scab laborers, the Party set up its principles:

Resolved . . . That we condemn the fallacy of protecting American

> labor under the present system, which opens our ports to the pauper and criminal classes of the world, and crowds out our wage earners; and we denounce the present ineffective laws against contract labor and demand the further restriction of undesirable emigration.
>
> Resolved . . . That we regard the maintenance of a large standing army of mercenaries, known as the Pinkerton system, as a menace to our liberties, and we demand its abolition . . . .[29]

## The Skilled and Production Relations

These were fine principles for a third party; but the skilled workers at Homestead were still largely Republicans at the polls and, while subscribing to similar beliefs about tariffs, Pinkertons and scab laborers, rejected the principles of socialism.

Regardless of tariff protection or lack of protection, these craft workers believed they were irreplaceable at Homestead and elsewhere; they thought the employers had to have their skills and ultimately would be forced to pay the wages they demanded. Eight days after the lockout began, Amalgamated official Cummings argued:

> The Company does not seriously hope to run the works with imported labor. Mr. Frick knows that if we provided carriages to bring black sheep from the depot to the mills and got the band to welcome them, he would not from all over the country get enough men for the Works. He depends upon breaking up the Amalgamated and upon taking back about two-thirds of the old men. This he can never do; and until he does do it the mills will stay idle and he will come to us. We can afford to wait. Not a cent in our treasury has been touched, and have powerful friends to fall back upon when we have spent our own money.[30]

The foremen, superintendents of divisions and boss mechanics of Homestead thought the same thing, declaring at a meeting that they would not operate the mill under armed guard, that this action would render the operation of the mill by non-union men impossible.[31]

The fact is, however, that these narrow beliefs of craft workers were unfounded. The entire American iron industry was deeply involved in the outcome of Homestead, whether or not they specifically were part of the inner sanctum that planned the tariff reduction and the ultimatum to Carnegie's workers. Even before the lockout, the Pittsburgh *Post*, a Democratic Party paper favoring free trade, had seen the writing on the wall.The impending conflict at Homestead, it wrote, was to be precipitated not in the interest of the Carnegie Company alone, but in that of all the iron and steel manufacturers of Western Pennsylvania and Eastern Ohio. Homestead, the *Post* said, was the chosen battleground because, first, the ease with which the mill property could be equipped for offensive and defensive purposes; second, because the ruin of the workers' town by a disastrous strike would be more sweeping than elsewhere; and, finally, because the Carnegie Company had the largest interest to serve and should, therefore, be willing to bear the brunt of the battle. If the industrial war was declared and the union lodges at Homestead were broken up, the other manufacturers were to follow the lead of the Carnegie Company, defy the Amalgamated Association, and reduce the wages of their employees to an extent varying from 20 to 60 percent.[32]

This was not an isolated opinion. In Congress, once the battle of Homestead began, Senator Voorhees insisted labor's riot had been due to the Republican policy of protection. The poor people who had laid down their lives at Homestead had been led to believe that the tariff was protecting them, though a greater lie had never been worked or woven into the nation's legislation. The Republican Party, Voorhees argued, had said that the great manufacturing barons had to be protected against foreign competitors, so that they might be able to pay higher wages. Mr. Carnegie had thereby received tariff protection of 55 percent on the price of iron and more than 70 percent on steel; but, instead of paying his workmen higher wages, he had notified them of a reduction from 12 to 40 percent. The real beneficiary of the tariff system had then responded with the employment of an armed mob of Pinkertons – mere mercenaries, a private corporation purchasing an armed squad to murder a direct consequence of the higher protective tariff.[33]

The political consequences for the Republicans during the presidential campaign of '92 were only too obvious. Every politician could see the voters were aware that tariffs, wage cuts and Pinkertons were a package at Homestead and perhaps elsewhere as well. But how could the employers now coordinate their tariff policy with the efforts to reduce wages and lower production costs? The answer was they could not without public scrutiny, the Baltimore *Sun* noting that in the U. S. House and Senate:

> The political aspects of the affair were discussed and it was commented on as a demonstration of the fallacy of the arguments that the protective tariff tends to increase the wages of workmen. It was pointed out that no interest had enjoyed such complete protection as the steel industry, and no single product of American manufacture was more carefully looked after by the framers of the McKinley bill than was structural steel, the

> principal product of the Carnegie Works. Yet, just on the eve of the presidential election, when the principal issue before the people is the approval or disapproval of this law, instead of wages being raised, as was promised, the country is treated to the spectacle of the chief beneficiary reducing wages from ten to forty percent, and forcing that reduction upon his workmen by shooting them down with Winchesters.[34]

Representatives of both parties now believed that the effect of the lockout and the conflict between the workers and the Pinkertons would see votes driven away from the Republican party in the fall elections. Republicans were particularly bitter in their denunciation of Carnegie for allowing the trouble to break out at Homestead and endanger the success of their ticket. After hearing accounts of the fighting, Representative Harner of Pennsylvania turned on his heels and remarked in a tone of disgust as he walked away: "If this thing goes on much longer Cleveland will carry Pennsylvania."[35]

**Wages, Technology and Skill**

Despite the political repercussions, wage reductions were viewed by the owners of Pennsylvania's steel mills as necessary security measures to tide them over hard times when the market fell away, when prices for iron products declined and profit margins eroded.[36] Their theory was that the good times to make profits at the workers' expense must be taken advantage of to offset hard times, one contemporary expert expounding:

> No industry is naturally so uncertain and consequently so competitive as the steel industry. The demand for the product is fitful and uncertain in the extreme because most of it goes into new constructions and new enterprises, and these are notorious for the spurts and depressions of demand which affect them. . . . The temptation of a fierce demand was always too much for some seller. He could not resist the temptation to ask $5 or $10 advance (clear profit) when he knew that to ask was to receive. Before the coming of the [steel industry] Trust the purchaser was always made to pay all that he would, whether it were $8 or $40 a ton. On the other hand, when demand slackened, no agreement [among competitors] could be made sufficiently binding to prevent the producer from cutting prices to secure orders. He had to do it or fail.[37]

The loss resulting from a slump often more than overbalanced the gain made during the sharp rise in demand and steel and iron prices, so any cut in prices of iron and steel was in part shifted to the iron workers in the form of reduced wages, speedups, and longer hours of work for the same pay.[38]

To enforce speed-ups and wage-cuts, technology had also been upgraded. H. C. Frick, as Chairman of Carnegie operations at Homestead, described needed changes in the sliding scale in terms of the new technology introduced:

> The skilled workmen in the Amalgamated Association work under what is known as the sliding scale. As the price of steel advances the earnings of the men advance; as the price falls their earnings decrease in proportion. While there is no limit to an advance of earnings on the scale, there is a point at which the decline stops. It is known as the minimum, the figure heretofore has been $25 per ton for 4 x 4 Bessemer billets. We believe that if earnings based on the selling price of steel can advance without limit, the workmen should be willing to follow the selling price down to a reasonable minimum, and so this figure was finally fixed by the Carnegie Company at the rate of $23 instead of $25. The reason for asking this on our part was that the Carnegie Company has spent large sums of money for the introduction of the new machinery in its Homestead plants, by means of which the workmen were enabled to increase the daily output [of steel], thereby increasing the amount of their earnings. We had originally asked a reduction to $22, but subsequently agreed to compromise the rate at $23. The Amalgamated Association was unwilling to consider a reduction below $24 on steel billets, notwithstanding the fact that the improved machinery would enable their members, even at $23, to earn more than is paid 36 in other Amalgamated mills.[39]

So far as Carnegie was concerned, the upgrading of his company's technology at Homestead meant the wage rate per ton would have to go down. Frick sought, he said,

> the reduction of tonnage rates in those departments in the mills where the improvements I have spoken of have been made and which enable the working men to increase the output and consequently their earnings. Where no such improvements had been made there was no request upon our part for a reduction of tonnage rates. In other words, we ask no reduction in any department at which the output has not been greatly increased by reason of our expensive improvements since the scale of 1889 went into effect. We are prepared to show that in nearly every department under our proposed reduction of the tonnage rates the skilled workmen would make more money than they did when the scale of 1889 went into effect.

As a rule the men who were making the greatest wages in the Homestead mill were the ones who most bitterly denounced the proposed revision of the scale, for out of 3,800 employed in every department only 325 were directly affected by this reduction.[40]

In fact, indirectly, every worker in the mill was threatened by Carnegie's lockout, so that the company's attempt to break the sliding scale minimum touched the entire work force. Not only would a lockout put the skilled workers' sliding scale and the minimum wage in jeopardy, but it was likely that it would be followed by an effort to cut the wages of semiskilled and unorganized, unskilled workers. If the skilled workers making high wages could be replaced by machines, moreover, the unskilled workers making low wages could be forced to tend those machines at even lower wages. True, the wages of Carnegie workers at the extreme foot of the scale were not reduced. But all others received wage cuts under Frick's plan for the new wage scale. No attempt was made for an equitable leveling of wages among those whose wages were reduced.[41] The only standard was to lower the cost of production, and the semiskilled and unskilled workers felt they were next in line for wage cuts.

To begin the wage reductions, Frick's goal was to first break the skilled workers' union at Homestead, to then extend nonunionism to all the Carnegie plants, and to thereafter spread it throughout the Pittsburgh area.[42]

Though one writer had gravely alleged that the Homestead men who went into the strike were getting as much as $30,000 a year in wages, in all the testimony Frick gave before the Congressional investigating committee, the highest wages quoted were the earnings of a roller who received in May, 1892, $279.30 for twenty-three days' work. Frick also admitted that May was above the average in tonnage; and at that rate this man would have only earned $3,280.50 in a year of 270 days, which Frick told the committee was the average number of working days. Yet this man was the highest paid. The lowest received $1.40 a day.[43]

When the workers had demanded wage participation in their increased output due to the introduction of new machinery into certain departments, Frick had refused. The sliding scale had already meant that the more technology each worker deployed, the greater was the output attributable to each, and, at the specified rate per ton of billets, the greater was the total pay – a payment the Carnegie firm wished not to make. "Everywhere throughout the works labor-saving appliances have been introduced at great cost," the *Sun* wrote the day after the Homestead battle began, "and within the last three years the necessity for human labor has very largely been reduced. The latter is one of the chief reasons why the firm insists on cutting the rates for work."[44] To do this, the union had first to be crushed.[45] And rather than strengthen its structure and expand its membership to the semiskilled and unskilled, the Amalgamated sought to secure the narrow craft orientation of its members to outmoded job positions.[46]

"The very life of the union," one writer summarizes, "rested upon its ability to control the supply of skilled labor. It would be able to maintain its position only as long as the skilled iron and steel workers which [sic] constituted its nucleus were in demand. This was the case in the iron industry" in the past, "for though limited steps were taken during the 1880s to introduce machinery into iron production, puddling, squeezing, and rolling operations by their very nature resisted mechanization. They remained almost entirely manual and required a good deal of skill. In contrast to iron production, however, steel manufacture was open to technological improvement, and opportunities for the use of machinery were present in almost every phase of the steel making process."[47]

In fact, steel-making at Homestead was subject to the introduction of technology that did away with the very skilled workers who had created the surplus capital used to invest in new technology. "When the Bessemer converter was introduced and steel proved superior to iron in many cases, particularly for rails," an expert explains, "the trend toward steel was accelerated, and this gradually resulted in the de-emphasis of the ironworker, as well as the Amalgamated Association. Technological methods incorporated into the production of steel eliminated the need for much of the highly skilled manual labor and with it the [Carnegie] company's dependence upon the union as its bargaining agent."[48]

Frederick Engels had once summarized a similar situation: "Thus it comes about," he said, "that machinery becomes the most powerful weapon in the war of capital against the working class; that the instruments of labor constantly tear the means of subsistence out of the hands of the laborer; that the very product of the worker is turned into an instrument of his subjugation."[49]

## Notes

[1] Livesay, 139-140.
[2] *The World* (New York, New York), November 2, 1892.

---

[3] *Ibid.*
[4] *Ibid.*
[5] Durant, 188, 191.
[6] *The Milwaukee Daily News* (Milwaukee, Wisconsin), July 5, 1892, 2.
[7] Percy Ashley, *Modern Tariff History* (New York: E.P. Dutton, 1905), 212-262; F. W. Taussig, *Tariff History of the United States* (New York: Putnam, 1888), 4th ed., 1898, 230-83.

[8] United States, Senate, 61st Congress, 2d Session, National Monetary Commission, *Statistics for Great Britain, Germany and France, 1867-1909*, Document No. 578 (Washington, D.C.: Government Printing Office, 1910),18, Table 8; 153, Table 3; 156, Table 6; 274, Table 2; 280, Table 8.
[9] W.T. Hogan, *Economic History of the Iron & Steel Industry* (Lexington, Mass.: D. C. Heath, 1971), Vol. I, 347.
[10] *Ibid.*; *The Sun* (Baltimore, Maryland), September 6, 1892
[11] See J. S. Jeans, ed., *American Industrial Conditions and Competition* (London: Report of the British Iron Trade Commission, 1902); Edward Stanwood, *American Tariff Controversies in the Nineteenth Century* (Boston: Houghton, Mifflin & Co., 1903), Vol. II.
[12] *The Phidelphia Press*, January 2, 1892; *The Sun* (Baltimore, Maryland), September 6, 1892.
[13] Burgoyne, 25-26.
[14] "The Iron Industry," *The San Francisco Chronicle*, July 1, 1892, 6, editorial.
[15] *The San Francisco Chronicle*, July 5, 1892, 6, editorial.
[16] *Ibid.*
[17] *The Sun* (Baltimore, Maryland), July 4, 1892.
[18] *Ibid.*
[19] *Ibid.*
[20] *Ibid.*
[21] Burgoyne, 25-26.
[22] *The Sun* (Baltimore, Maryland), October 26, 1892; *The Philadelphia Press*, January 2, 1892.
[23] *The Cleveland Leader*, July 10, 1892 .
[24] *The Industrial World* (Cardiff, Wales), June 10, 1892.
[25] E. J. Smith, *The New Trades Combination Movement* (London: Rivingtons, 1899), *passim*.
[26] Stanwood, Vol. II, passim; Wolff, 76.
[27] Burgoyne, 24-25
[28] *The Milwaukee Daily News*, July 5, 1892, 2.
[29] *Ibid.*, 4.
[30] *The Sun* (Baltimore, Maryland), July 14, 1892.
[31] *Ibid.*, July 15, 1892.
[32] Burgoyne, 27-28.
[33] *The Sun* (Baltimore, Maryland), July 8, 1892; *The Post* (Pittsburgh, Pennsylvania), July 8, 1892.
[34] *The Sun* (Baltimore, Maryland), July 7, 1892.

[35] *Ibid.*
[36] *The World* (New York, New York), July 3, 1892; *The Post* (Pittsburgh, Pennsylvania), July 3, 4, 5, 1892; *The Daily Dispatch*, July 4, 5, 1892; *The National Labor Tribune* (Pittsburgh, Pennsylvania), Thursdays, July 1892.
[37] J. Russell Smith, *The Story of Iron and Steel* (New York: D. Appelton and Company, 1908), 128-129.
[38] *Ibid.*,129.
[39] *The Sun* (Baltimore, Maryland), July 8, 1892.
[40] Ibid.
[41] Fitch, 128.
[42] *Ibid.*,132.
[43] *Ibid.*,129, fn.
[44] *The Sun* (Baltimore, Maryland), July 7, 1892.
[45] *Ibid.*, July 8 and November 22, 1892; *The World* (New York, New York), July 3, 7, August 20, 26, September 30,1892; *The Philadelphia Press*, January 20, 24, 1892.
[46] Hogan, Vol. I, 233.
[47] *Ibid.*
[48] *Ibid.*
[49] Frederick Engels, *Socialism: Utopian and Scientific* (New York: International Publishers, 1945), 34.

**Machines and Workers**
**(Credit: *Harvest Quarterly*)**

**The Battle Begins**
**(Credit: *The New York World*)**

# Part III

# Conflict Without End: Strike, Commune & Occupation

**The Great Battle of Homestead**
**(Credit: Amalgamated Association of Iron and Steelworkers, 1892)**

# Chapter 4

# The Great Homestead Strike

> My idea . . . is always to shut down and suffer. Let them decide by vote when they decide to go to work. Say kindly "All right, gentlemen, let's hear from you; no quarrel, not the least in the world. Until a majority vote (secret ballot) to go to work, have a good time; when a majority vote to start, start it is.
>
> – Andrew Carnegie, urging his negotiating stance on Frick when leaving for Scotland shortly before the contract with the Amalgamated Association expired[1]

The great Homestead lockout was planned by the Carnegies and expected by the workers. The skilled workers had sufficient reasons to know that Carnegie and his managers would ignore their demands. Andrew Carnegie already had experience in dealing with the Homestead workers on more than one occasion and knew the limits of labor's resources and power. Though previously the workers had been favorably situated, by 1892 the balance of power had shifted and, through his managers, Carnegie was determined to break the power of the workers' Amalgamated Association once, for all. He thus began making detailed military preparations to defeat the Homestead workers and to replace them with nonunion men.

**Fort Frick**

The planning was elaborate. "The Carnegie Company," one reporter noted, "has for some time past been making careful and exhaustive preparations to guard the property and the men it might find necessary to employ against any violence on the part of the workers. Around the entire plant was placed a palisading fence nine feet high, topped with three strands of barbed wire. There are applications by which these can be charged with electricity to prevent any possibility of invasion. All through the ground at various points are being erected high towers from which to view the surrounding country. Even every means of egress has been cut off, with one or two exceptions. Where the workmen and laborers entered hithertofore from the railroad tracks and other points, stout barricades of wood and iron have been put up and bridges placed in their stead as a means of getting from one portion of the Works to the other. Searchlights with strong reflectors have been erected on the building, so that a blaze of light can be thrown on the entire place to prevent surprise in case of a night attack. All approaches are firmly guarded by stout fortifications."[2] Searchlights placed in the skylights of the mill made objects over a mile distant plainly visible on the darkest night.[3]

Humor was not lost in the Homestead community, despite the dismal outlook. One joker became a poet, singing the absurdities of "The Fort that Frick Built":

> Twixt Homestead and Munhall,
> If you'll believe my word at all
> Where once a steel works noisy roar
> A thousand blessings did out-pour
> There stands today with great pretense
> Enclosed within a white-washed fence
> O'Wondrous change of great import
> The mills transformed into a fort.[4]

Fort Frick, surrounded by that tall, recently white-washed fence, enclosed every inch of mill property, shutting off all possible view of the Works from the outside. Looking out, the fences had two and three inch holes bored in them about five feet from the ground. "Portholes, three inches in diameter, were bored through this stockade, every twenty-five feet, commanding a full sweep of Munhall station by the company," recorded a witness. "Mr. Frick said these orifices were made to command a view of the outside, but failed to explain their extraordinary number."[5] Some believed they were portholes to shoot through at attackers, as frontier settlers had once emptied their weapons at Native Americans. Others claimed that water or steam pipes would be affixed to these holes to pour scalding contents on the workers. Whatever Frick actually planned, he did lay pipes along the fences. They were connected by foremen and bosses, who were then the only living beings on the premises. Their presence was attested to by a nearby water tower, picturing in chalk a hideous caricature of a man hanging by the neck, labeled in effigy "One of H. C. Frick's Scabs."[6]

These were only some of the preparations for battle. Halfway between the points of Carnegie's City Farm and Homestead's Munhall Station - the distance new arrivals in Homestead previously traveled to the Carnegie plant – Frick built a new railroad station for the accommodation of the Company. This facility was designed to enable a trainload of watchmen, workers and Pinkerton men to reach the center of the Works without interference from the locked-out workers. Inside, ample housing was guaranteed them in a new barracks-like frame building. Though the river front itself remained unprotected by any fence, the Company purchased and outfitted a steam launch to serve as a river patrol. The scow was huge, equipped with a powerful searchlight to protect the Works from an approach from the river under the cover of night. But the light was so bright it interfered with boat navigation, and the Company discontinued its use after a few nights.[7]

"A State of Siege: The Carnegie Works Ready to Fight the Workmen: Hot Water and Search Lights," headlined the Baltimore *Sun* on June 30th, 1892, indicating that the lines between Carnegie and his workers had been sharply drawn:

> . . . every union man in the employ of the Carnegie Works at Homestead will be discharged on July 1. The present scale expires on June 30, and it is asserted the firm has determined to run the mill nonunion thereafter. . . . [T]he undercurrent of feeling is very strong, and if nonunion men are put to work it is feared there will be trouble. The firm realizes this, and have taken every precaution to protect the plant. . .
>
> Something of a sensation was caused by the report that hot water will be thrown from the fire plugs in case the men showed an attempt to surround the mills or enter between the high board walls. Dangerous portholes look out on this narrow passage-way from the enclosure of the

> mills, and it was stated that in case of war it will be made very hot for any seeking to enter this pass. The men claim that detectives are already on the ground in considerable force, but this is denied by the firm.[8]

The workers had responded initially with dismay. But they were soon organized by the Amalgamated leadership, and on July 1 immediately began to search and seize newcomers in Homestead in preparation for a conflict they thought certain. All strangers in Homestead were suddenly viewed with distrust; and a number who appeared in town were unceremoniously hustled out of the city by hostile men. When one laborer arriving at Homestead had the temerity to say he had been hired to go to work July lst, he was taken to a ferryboat by a crowd of iron and steel workers, who sternly warned him not to return.[9]

This incident was part of the workers' overall preparation to confront the steel bosses in the Pittsburgh area. As Homestead workers tried to seal off the town to non-union labor, in Pittsburgh half the 35,000 to 40,000 members of the Amalgamated Association readied a strike against wage cuts by the steel masters of 60 leading mills. By the end of June, their numbers swelled to 50,000 and talk of a "General Strike" filled the air.[10]

Carnegie managers meanwhile moved to divide the work force by limiting the work stoppage to Homestead. Frick announced that after July lst, the Homestead Works would be operated as a nonunion mill. He posted warnings that after a certain date the places of those who failed to sign up for employment as individuals would be considered vacant. But at the same time, he signed the Amalgamated scale for the company's Thirty-third Street mill in Pittsburgh. "It was inferred by this move that the capitalists feared a long siege, and intended to fill contracts on hand for structural steel at this mill, in event of their being unable to complete them at Homestead," a contemporary wrote. "The latter plant was much larger, of course, but as the Pittsburgh mills were able to turn out the same class of commercial work, the inference was but natural."[11]

At Homestead the Amalgamated simultaneously prepared to close the plant and keep it closed. By July 3rd, most mill workers had taken up guard duty. Around the plant 235 workers stationed themselves in 11 posts. Another 66 men were on watch directly opposite the City Farm on the other side of the river, and 50 men occupied Munhall Station. The workers' army also prepared a general alarm system of rockets and whistles to be blown if necessary. In five minutes either day or night, 2,500 men could be summoned. This system was tested on Friday, July lst, when it was reported that a non-union gang was coming. The Advisory Committee had previously dispatched parties to nearby Altoona and Connessville to watch Carnegie Company efforts to recruit new men. When it was learned that the Company planned to bring 200 men to Homestead on that Friday, the workers went into action, dispersing the gathering of these nonunionists between Bethlehem and Scranton by sending in 100 Amalgamated members to join the new recruits and influence them from coming to Homestead. So successful were they that the trains had to be sidetracked between Tyrons and Altoona, and the men returned to their homes.[12]

To hold off the Company's import of armed nonunion labor, the workers also armed themselves. Later, puddler William Weihe, President of the Amalgamated Association, testified that the workers feared an armed confrontation at Homestead. "How," one House Representative asked – "How do the workmen feel about the employment of armed bodies of men for private purposes . . . in connection with the differences between workmen and employers?"

Weihe answered sharply: "They look upon that with great suspicion."[13]

**Pinkerton Preparations**

Pinkertonism had become an employer method to defeat workers organizing to improve wages and establish better working conditions, to say nothing as a method of protecting employers' factories. The employers hired the Pinkertons as a private army. This army had been mobilized by the City of Chicago's first detective Allen Pinkerton, a Scottish immigrant, who at mid-century investigated train robberies, then gained renown for uncovering a plot to assassinate President-elect Abraham Lincoln in 1861, and thereafter participated on the side of the employers in some seventy major labor disputes. Under its emblematic sign of crossed rifles and trademark "We Never Sleep," its two thousand active men and thirty thousand reserves totaled more than the standing army of the nation - though a mere 300 or so Pinkertons were to be deployed at Homestead. At Vanderbilt's New York Central Railway Strike in 1890, they had already displayed their skills by killing five people. At Homestead they ultimately would kill even more. The agency's preference was to hire those with military experience – ex-soldiers and policemen.[14] This hiring practice fit the tone of Allen Pinkerton's original plan as he expressed it in 1878: "Since the strikes of '77, my agencies have been busily employed by great railway, manufacturing and other corporations, for the purpose of bringing the leaders and instigators [of strikes and labor protest] to the punishment they so richly deserve. Hundreds have been punished. Hundreds more will be punished."[15]

When Allen Pinkerton's two sons, William and Robert, inherited his agency in 1884, they continued to arm their undercover operatives and uniformed guards. Almost never deputized, the guards operated beyond the pale of the law, save in a few states with relatively ineffective anti-Pinkerton statutes. And the agency evaded these by subterfuge or technicalities. The courts in some states may have established that the carrying of arms across state borders by groups of associated men was a criminal offense,[16] but the public believed otherwise and the newspapers often corroborated that belief.

"Apart from the very useful and perhaps necessary services of the Pinkerton agencies as private detectives," the Baltimore *Sun* editorialized, "their function as purveyors of fighting men, proprietors of a standing army, of rifles and men to handle them to be had for hire, is a frightful anomaly and a blot upon our nineteenth century Christian civilization. The law permits it, and, therefore, Mr. Pinkerton is not to be legally condemned. Circumstances have seemed to justify it and even encourage it, and, therefore, society has no right to complain."[17]

Carnegie and his managers, having for some years contested the adequacy of existing law and law enforcement officials, made the case that the use of the Pinkertons was a regrettably necessary final step to protect their mills and their right to undercut union workers with low wage nonunion laborers. When for example, Frick was asked if he had exhausted all means which the laws provided for the protection of Company property before applying to the Pinkerton Detective Agency for guards who might or might not be armed, Frick recalled that in July 1889, when Abbott was Chairman and he a mere manager, the Company tried to enforce an agreement with the Amalgamated Iron and Steel Association, failed to do so, called on Allegheny county Sheriff McCandless to outlay funds for special deputies and equipment to protect their Bessemer Steel Works at Duquesne, and, on the Sheriff's failure, refused to reimburse him,[18] and employed nonunion workers to take the Works. However, he continued: "The workmen were intercepted on the train going to the Works, were frightened, and they were unable at that time to introduce any of them into the Works, as the Works were surrounded by strikers." But, Frick lamented, Carnegie officials:

> then asked the sheriff for protection so they could introduce these nonunion workmen. . . . The sheriff sent to Homestead at their request

> about one hundred and twenty deputy sheriffs that he had selected in the county here. Immediately on the arrival of these deputy sheriffs the strikers took their maces away from them, took their coats away from them in many cases ,and their hats, and sent them all back to the city; didn't leave one of them on the ground, nor didn't allow them to enter the Works. The firm at that time made an arrangement with the Pinkerton Agency for guards. While the guards the Pinkertons had agreed to furnish were on their way to this city, the Amalgamated Association of Iron and Steel Workers asked for a conference with the firm and they were granted it, and they made an agreement for three years. It was an agreement that was very burdensome on the concern, and only agreed to by the firm because they could do no better. They saw it was impossible for the sheriff to give them protection and they concluded it was probably better to make the agreement than to go further in the matter, so that the guards employed through the Pinkerton Agency at that time did not reach the Works.[19]

In 1892, H. C. Frick still believed that, without the enforcement of laws to protect the Homestead Works and the bringing in of nonunion workers to replace the Amalgamated men, he was justified in bringing in the Pinkertons, whatever the price in money or human lives. Carnegie money both financed the unfunded sheriff deputies and paid for the oncoming invasion of Pinkertons, he told the Congressional subcommittee, giving the impression that the Pinkertons were a last-minute rescue force saving American industry from an illegal mob.[20] But in truth, the great Pinkerton raid on the Homestead workers was a premeditated, calculated move.

* * *

Frick's earlier arrangements for guards prepared for battle had been made in secret with Mr. Pinkerton himself. The House of Representative s reported that, "The testimony of Robert A. Pinkerton shows that the Carnegie steel works opened negotiations with his agency for armed men as early as June 15, or nineteen days before the strike actually occurred. It was a significant fact that while Mr. Frick had arranged for a conference with the men on the 24th of June, he was at the same time in communication, by long distance telephone from his office at Pittsburgh to their office in New York, with the Pinkerton Detective Agency to supply him armed men if needed."[21]

In response to the Pinkerton Agency's letter of June 22nd,, delineating the terms of employment of armed men, Frick had written on June 25th, "We will want 300 guards for service at our Homestead mill as a measure of precaution against the interference with our plan to start the operation of the Works on July 6, 1892," adding:

> The only trouble we anticipate is that an attempt will be made to prevent such of our men, with whom we will by that time have made satisfactory arrangements, from going to work and possibly some demonstration of violence upon the part of those whose places have been filled, or most likely by an element which usually is attracted to such scenes for the purpose of stirring up trouble.
>
> We are not desirous that the men you send shall be armed unless the occasion properly calls for such a measure later on for the protection of our employees or our property. We shall wish these guards to be placed upon our property and there to remain unless called into other service by the civil authorities to meet an emergency that is not likely to arise.
>
> The guards should be assembled at Ashtabula, Ohio, not later than the morning of July 5, when they may be taken by train to McKees Rocks, or some other point upon the Ohio River below Pittsburgh when they can be transferred to boats and landed within the enclosures of our premises at Homestead. We think absolute secrecy essential in the movement of these men so that no demonstration can be made while they are en route.
>
> Specific arrangements for movement of trains and connection with boats will be made as soon as we hear from you as to the certainty of having men at Ashtabula at the time indicated.
>
> As soon as your men are upon the premises we will notify the sheriff and ask that they be deputized either at once or immediately upon an outbreak of such a character as to render such a step desirable.[22]

Preparations for Homestead had meanwhile been extensive. Having engaged the Pinkertons, Mr. Frick instructed that they be brought to Homestead from Chicago by way of Youngstown and Bellevue (a short distance below Allegheny on the Fort Wayne road), where they were to take barges for their destination.[23]

As early as June 25th, Frick had contacted William B. Rogers, a steamboat business operator, to make detailed arrangements for the transport of 300 or more men at a future time to be decided upon by Carnegie personnel. Rogers said Frick told him, the men "were to act as watchmen in the Works and that they would be under the direction of the sheriff of the county. He said nothing about any apprehended trouble."[24]

In fact, however, as owner of the steamer *Little Bill*, Rogers had directly contracted with the Carnegie Steel Company to tow two barges filled with Pinkertons from Davis Island Dam to the Works at Homestead; and the extreme precautions taken were obvious to everyone involved.

Rogers and the Amalgamated Association were meanwhile watching each other's activities. The Amalgamated had kept close watch on the barges as they were fitted out at the Manchester docks, the preparations and transformations being so elaborate that not to notice them would have been difficult.

The barges had originally been heavy-duty, built to ship rails down the river from Carnegie's Edgar Thomson Works. They were now drastically altered, following Frick's long-range plan to use these scows to transport scab labor to the various Carnegie plants and hold off any attacks by unionists. Outside the barges appeared like any other river scows. But now both the *Iron Mountain* and the *Monongahela* had heavier wooden decking which completely covered them a few feet above the water lines – a form of construction that would have hampered any other barge in the amount of freight it could load. Midship were large, box-like hatches from which ladders led below. The back deck was steel plated. The insides were also converted for siege, lined with steel plate and equipped with bunks and cooking facilities. The *Iron Mountain* was a dormitory containing beds and tiers of bunks: cots, mattresses and blankets had been brought aboard. As well, the *Monongahela* had been transformed into a giant dinning hall, supplied by a kitchen aft, designed to carry a cook and twenty waiters, and loaded with groceries, hams and tinware.[25]

Once fitted out by the Company, and attracting much attention on account of the manner in which they were equipped, the barges lay for several days in the Ohio River at the foot of Beaver Avenue in the city of Allegheny. "They were supplied with bunks, cots and culinary arrangements to accommodate over a thousand men," an eyewitness reported. "No person seems to know for what use they were intended, although work on them was in progress several days prior to their fatal trip; and the mysterious atmosphere connected with the vessels caused no little newspaper comment."[26]

During the days preceding the embarking of the barges, moreover, several men had been actively engaged in loading them with boxes of various sizes, understood to contain not only bedding and food, but ammunition as well.[27] Thus supplied and protected, the occupants of the barges were prepared to withstand a protracted battle for the mills at Homestead.

"The men were better protected in the barges than most people imagine and could have stood a long siege in the barges except for the danger from dynamite or fire," said Carnegie Company Secretary Lovejoy – tormented as "Hatejoy" by the workers – after the battle had begun. "The barges were lined with sheet iron and could not be penetrated by bullets."[28]

With such sturdy means of transport awaiting them, in accordance with Frick's detailed plans, on Sunday, July 3rd, 125 men under the command of J. G. Hoffman, a clerk in the Pinkerton Agency, left Chicago on the Pennsylvania Railroad for Homestead. Included in his group were some 25 regular Pinkerton detectives, while the other men were temporary recruits, not in service except on demand for out-of-town purposes like Homestead.

Not regulars, these men were ordered to report at the Pinkerton Agency on Sunday evening for a trip to Pennsylvania. And, to remain unobserved, they were told to go to the railroad station in twos and threes and then board the fast train for Pittsburgh. So secret was their departure that even the railroad officials did not notice anything out of the ordinary.

The train arrived in Pittsburgh Monday morning and the men were quickly marched to Pinkerton headquarters. Then, on Monday night the scenario was repeated: Another carload of recruits left Chicago. There seemed no end to them, even after the battle of Homestead was fought and lost by the Pinkertons on the 6th. For on July 8th, still an additional 200 agents were sent from Chicago to Pittsburgh.[29]

Not all were aware of the role they were supposed to play. Some 316 men had been collected in Chicago and New York; most were unemployed, drifters sprinkled with students trying to earn summer money, a horde of Pinkerton regulars, a few hoodlums and criminals on the run.

The Pinkerton superintendent in the Chicago office had been vague in his explanation that the men were "hired to watch the property of a certain corporation, to prevent it from harm." He had said there was no danger, save a few brickbats might be thrown and names called, but there would be no reason to shoot. Though he refused to give the destination of the mission, the papers had been filled with stories about the Homestead lockout. A few of the recruits thus surmised they were being taken to Homestead and felt they were imprisoned by the Pinkerton regulars guarding the train exits. Some of the more perceptive volunteers wanted to know when they would be deputized to legitimize their roles; but the man to do it, Deputy Sheriff Gray, had not yet arrived.[30]

As many of the guards were hired under false pretenses and all of the new recruits were themselves guarded by regular Pinkertons, it is plausible they were "shipped to Homestead at gunpoint much against their wills," as one writer later claimed.[31] Most, however, came for money, and, as one Pinkerton explained after the surrender, "When I came here I did not understand the situation or I would never have come. I was told I was to meet and deal with foreigners. I had no idea I was to fight American citizens. I am a member of the Junior order [of

Pennsylvania fraternal societies] and I appeal to you for permission to leave and get myself out of this terrible affair."[32]

Nationalistic it may be, but this statement shows that many of the Pinkerton recruits were indeed deceived. The deceivers included not only the recruiting agents, but also the 50 Pinkerton regulars among the 300 men to come to Homestead. These "iron and coal police," as they were called, received their orders from above, while the new recruits had no legal standing and were never sworn in as deputies or anything else.[33]

They were nothing more than a privately paid army. Andrew Carnegie had paid the bills, and William Pinkerton supplied the mercenaries. Carnegie and his managers had wanted it to appear that the Pinkertons deployed were legally authorized. But the fact remains that the Pinkertons taken to Homestead were never sworn in as deputy sheriffs by either Deputy Gray who accompanied them from Pittsburgh (and admitted he had no authority to deputize anyone) or Sheriff McCleary, who authorized Gray to bring them.[34] For his part, William Pinkerton admitted on July 6th that, "We held off until the last moment on this business, but our company having done Carnegie's work for years, they [the Carnegie brothers and Frick] insisted that we supply the watchmen."[35]

* * *

The countdown began at the end of June, the Bellevue population having watched the reconstruction of the barge interiors for over a week. The Amalgamated grew more suspicious when Carnegie officials stated the barges were to transport laborers elsewhere and then on the night of July 5th heard by telegraph that several hundred strangers had gathered at the Bellevue shoreline.

About 10 P.M., the steamers *Little Bill* and *Tide*, each bearing a passenger license, moved their supplies down to Bellevue, where they were to take on their human freight. But the arrangements were unusual: the steamers had three extra engineers aboard, "as if trouble were anticipated with the crew when the latter learned of their true mission."[36]

The strikers now assumed they were faced by a naval invasion, immediately intensified their river patrol, sent alerts to their lookouts an the Smithfield Street Bridge, and moved their side-wheeler *Edna* downstream.

Meanwhile, the Pinkerton scows were loaded with nailed crates filled with 250 Winchester rifles and 300 pistols with ammunition.[37] The weapons had been carried on the train that came with the Pinkertons from Chicago, and the coaches carrying both ammunition and recruits were sidetracked, recoupled, and given a new engine at Ashtabula, Ohio, in accordance with Frick's previous arrangement with William Pinkerton.[38] It was now that the arrival of the rifles and as many maces boxed up in mysterious packages gave lie to the barge attendants' yarn about the vessels being prepared to aid in transporting men to make repairs at Beaver Dam. For one of the boxes containing maces was broken in loading them on the barges, several of the deadly weapons rolled out, and one was captured and taken to the officers of the Amalgamated.

So widely understood was the mission of the *Little Bill* on Tuesday night, July 5th, that Captain Rogers had great difficulty in obtaining a crew. After several attempts to engage a pilot, he finally obtained the temporary services of Captain Charles W. Wishart – just in time to leave the wharf to meet the Pinkertons. The Carnegie Company had given notice of the exact time of the expected arrival of the Pinkertons at Davis Island Dam and the *Little Bill* had to be there about 2 A.M. Here, five minutes down the river from Pittsburgh, the men derailed and marched to the barges.

Wishart later recalled:

> The Pinkertons arrived at the dam about 2 o'clock. There were 340 of them and from talks with them I learned that they came here from Youngstown. They are from Chicago and Cleveland and they have been collecting at Youngstown for several days. There was no delay at the dam. The men got off the train and immediately entered the model barges in waiting for them.
>
> Superintendent Potter, of the Carnegie Steel Company, and a number of assistants were there to receive them. Ex-Sheriff Gray was also there to take charge of the men. In addition to this there were a number of other Pittsburgh people who went along simply as spectators. We had a jolly crowd, and while we apprehended danger from bushwhackers along the river bank, everything was as merry as a wedding bell.[39]

Despite bungled steps at almost every juncture, the whole operation for Homestead had been planned clandestinely. Most of the Pinkertons, for example, were identified by numbers, and it only later turned out that 120 resided in Chicago, 75 in New York, 25 in Philadelphia, the remainder in the neighborhood of Brooklyn. William Pinkerton had denied press reports that he was recruiting an additional 500 men in Chicago to send East, asserting that – "We have enough in service to answer calls." But old Bill Pinkerton also issued stern orders to his agents – no man in the employ of his bureau was to talk about the Homestead strike.[40]

Once his men were routed at Homestead, however, he tried to pacify a nation of workers with their backsides raised: "No further effort will be made to force the strikers into order," he exclaimed. "It simply means a lockout that will last forever, so far as the union men are concerned. I guess Carnegie and his partners have money enough. I know what I am talking about when I say no union man will ever again set foot in the Homestead mill."[41]

## Sheriffs, Deputies and Political Power

Frick had hired the Pinkertons as a mercenary army because he found that he could not simply order the Governor of Pennsylvania to mobilize the Militia. Though Governor Pattison certainly had the power to mobilize the State Guard, he by no means led a government that was a carbon copy of the executive committee of the railroads and major industries of Pennsylvania.

In the inner sanctum of his office, he might promise the railroads and steel interests that certain measures would ultimately be taken; but he was also reliant on millions of workers' votes for his position. He could not possibly have justified using the Militia against Homestead's workers unless the county police powers were first used to squelch the uprising.

Pattison compromised by saying the dispute was a matter for local or county authorities, giving Frick the cue to play out the little act of seeking, in vain, help from Sheriff McCleary. After the fact, one reporter questioned: "Why did the Carnegie Company call upon the Pinkertons for watchmen to protect their property?" Frick answered sharply:

> We did not see how else we could have protection. We only wanted them for watchmen to protect our property and see that workmen we take to Homestead – and we have applications from many men to go there to work – were not interfered with.
>
> Q: Did you doubt the ability of the Sheriff to enforce order at Homestead and protect your property?
>
> A: Yes, sir; with local deputies.
>
> Q: Why?
>
> A: For the reason that three years ago our firm had an experience similar to this. We felt the necessity for a change at the Works; that a

> scale should be adopted based on the sliding scale of billets; and we asked the county authorities for protection. The workmen began tactics similar to those employed in the present troubles. The Sheriff assured the members of the firm that there would be no difficulty; that he would give them ample protection and see that men who were willing to work were not interfered with.
>
> What was the result? The posse taken up by the Sheriff – something over one hundred men – were not permitted to land on our property, were driven off with the threats of bodily harm, and it looked as if there was going to be great destruction of life and property. That frightened our people. Mr. Abbott was then in charge of the Carnegie, Phipps & Co. business, and was asked by the Amalgamated officials for a conference, which he agreed to, fearful if he did not do so there might be a loss of life and destruction of property. Under the stress, in fear of the Amalgamated Association, an agreement was made and work was resumed. We did not propose this time to be placed in that position.[42]

The 1892 scenario for the entrance of the Pinkertons and, on their defeat, the scabs, also began with the Carnegie Company deploying the legally constituted local authorities. Going beyond that, a reporter began: "the Pinkerton men . . . had been summoned and all arrangements with them to be on hand in case of failure by the Sheriff to afford protection" had been made. "Is that a fact or not?" Frick answered cautiously:

> The facts concerning the engagement of the Pinkerton men are these: From past experience, not only with the present Sheriff, but with all others, we have found that he has been unable to furnish us with sufficient number of deputies to guard our property and protect the men who were anxious to work on our terms. As the Amalgamated from the lst of July had surrounded our Works, placed guards at all the entrances and at all avenues or roads leading to our establishment, and for miles distant therefrom, we felt that for the safety of our property and in order to protect our workmen it was necessary for us to secure our own watchmen to assist the Sheriff, and we knew of no other source from which to obtain them than from Pinkerton agencies, and to them we applied.[43]

Frick had perfected this set-up to bring in the Pinkertons, and Sheriff McCleary's role was to make sure it appeared he could not occupy the mills due to the workers' control. Carnegie's attorneys had notified McCleary a week before the workers went out that there would be trouble, that the firm had arranged to send some 300 men to Homestead, and that they wanted them deputized on or before their arrival. The Pinkertons, due to arrive on Wednesday, July 6th, Frick had the Company attorneys, Knox and Reed, formally demand in writing that McCleary furnish 100 deputies to guard the Works.

This was not the first notice. On Independence Day, July 4th, Frick had publicly announced through his secretary that he would put the Homestead mills in the hands of the county authorities if necessary, repeating his request that one hundred deputy sheriffs be enlisted. On the same day he once more notified Sheriff McCleary of the presence of the locked out men on the premises of the Carnegie Company, demanding protection by the Sheriff as the executive official of Allegheny County.[44]

The next day, July 5th, the Company once more repeated the demands, that Sheriff McCleary protect its property - guard its mills and defend those men the Carnegie Company claimed were willing and anxious to go to work, except for their fear of bodily harm from the strikers.

"In years gone by the sheriff was an iron worker himself," noted the *Cleveland Leader*, "and his action in not swearing in the necessary deputies and taking them with him to the Works had accompanied considerable comment among the sympathizers."[45] In fact, except for two, all previous sheriffs in Allegheny County had openly sided with the steel workers; and, even in the case of these two supporters of "law and order," respectful, sympathetic role-playing had been expected by both workers and sheriffs. This time, however, Sheriff McCleary had every intention of siding with Frick, and not the workers.[46]

Frick continually used McCleary's inability to main order as a rationale for ordering in a private army:

> Our Homestead Steel Works were on July 1 taken possession of by a mob, which was immediately thereafter organized by local representatives of the Amalgamated Association of Iron and Steel Workers, and all our mechanics, millmen, and even foremen and superintendents of departments were forcibly denied admission thereto. We were also notified by as self-styled Advisory Committee that no fires would be permitted at the Works, lest the men become excited to certain unlawful acts. This continued until yesterday, when we called upon the sheriff of Allegheny County for protection and assistance in regaining possession of our property. The sheriff went to Homestead, and on his return sent deputies to the Works and posted a proclamation ordering the men to disperse. His deputies were routed and his proclamation torn down.[47]

When Frick had gone to McCleary's office on the 5th to ask the

Sheriff to take charge of the affairs of the plant, McCleary already knew what his role was to be. Frick no longer needed to state why the time had come for the county authorities to protect the interests of Carnegie Steel Company. Though Frick's request for 100 deputies could not be met by the Sheriff, nonetheless McCleary promised to render the Company all the help within his power and immediately went to Homestead accompanied by several deputies.

Thus on July 5th, Sheriff McCleary, accompanied by ex-Sheriffs Gray and Cluly, came from Pittsburgh at noon and headed straight for the strikers' headquarters. There they announced that they would assume control of the Carnegie properties and would send some deputies to guard the Works. "We are deputy sheriffs, and our instructions are to proceed to the Homestead Steel Works with all possible speed," Cluly announced to the crowd on their arrival. A voice answered from the assembled strikers amidst wild cheers: "You fellows will never get to the gates alive." But the crowd was calmed by the representative of the Amalgamated Executive Committee: "Order, boys, order. These gentlemen are now in our care and you must protect them from the unthinking mob." Quieting the unruly, the representative then ordered a path cleared and said to the deputies: "Follow me, or I will not be responsible for what may happen."

"Move on. We will follow you," replied Cluly.

* * *

Off they marched to Amalgamated Headquarters, narrowly escaping a critical confrontation, though the union Headquarters was immediately surrounded by a massive crowd – a reporter catching the moment:

> once there it halted and spread out until the neighborhood streets and lanes were filled to overflowing. Slowly the deputies and their escorts made their way to the side entrance of the hall. When they had climbed the stairway and found themselves in the big smoke-scented room they breathed sighs of genuine relief.

Then:

> When the deputies and their guards were safely housed the doorkeeper was instructed to close and bolt the outer doors and allow no one to enter . . . . While all this was going on the captured deputies were invited into a private office. Six of the principal leaders of the men followed them and a star chamber session ensued. The conference lasted for fully an hour and when it was over it was announced that the deputies were to return to

> Pittsburgh by the first train down.

But:

> Then to the surprise of the leaders, the crowd outside began to show unmistakable signs of irritation and it was plain to the close observer that if the crowd gained another glimpse of the deputies there would be a serious if not fatal collision.

To avoid it, the Executive Committee ordered the crowd to disperse, and decided to quickly get the deputies out of Homestead on board the steamer *Edna*.[48]

Despite their narrow escape, the next day the Sheriff audaciously returned with Cluly to the Amalgamated headquarters. But this time, fearing it might be breaking the law, the Advisory Committee followed a legal script, offering to assist them in "preserving peace and protecting property hereabouts."[49]

The Sheriff then asked to make a general inspection of the grounds and interior of the mill. The Advisory Committee magnanimously gave them permission and offered a 10-man Amalgamated bodyguard to accompany the party into the mill to inspect the premises and see that, in fact, the workers surrounding the property had damaged nothing.[50] Satisfying himself that all was in order, the Sheriff prepared to leave.

Before departing, however, he had to listen to the grave statement of the Chairman of the Advisory Committee now representing the workers.

"Mr. Sheriff," he implored:

> The Homestead people both before and after the trouble came and since its precipitation, have expressed their intention of not only guarding their own property, but the property of their employers, the Carnegie Steel Company, as well. The property is not endangered, but the bringing into Homestead of deputies, ostensibly for the purpose of guarding property which is not in danger, is unnecessary. Now we, the people of Homestead, wish to make an offer. We will place from fifty to five hundred of our best citizens inside of the mill to guard that property if the firm thinks it is in danger, and these men will protect the property with their own lives if necessary, and further that, a bond of $10,000 will be given for the faithful vigilances of each man posted. We do not want deputy sheriffs here, as it will precipitate trouble, and therefore we make this offer.

The Sheriff replied he was unable to accept, although he acknowledged the common sense and fairness of the offer, adding that he had been instructed to bring in 50 deputy sheriffs. Though McCleary explicitly admitted there was "no sign of disorder anywhere" in the mill and that there was "no necessity of having deputies here," he repeated that he had to perform his duty and clear the mill of strikers. And to this the strikers finally agreed.[51]

The Advisory Committee then went into private session for over an hour, called in the Sheriff, and advised him that the Committee had dissolved and was no longer "responsible for any disorder or an lawless act perpetrated either in Homestead Borough [where most of the population lived] or Mifflin Township [where the mill was located]." Absolving themselves, all Committee members then tossed their badges on the committee room table, the fifty bits of crimson ribbon piling into a mound. Thereafter, all official Amalgamated documents which might signify legal violations were laid on the open grate of the fireplace and burned in McCleary's presence. The Sheriff tugged his silver mustache, said he had nothing to say, and left with a low bow, backing out of the room, and heading out of Homestead for Pittsburgh.[52]

Since the Sheriff did not have the fifty deputies the Amalgamated agreed to allow into the mill, but only ten, before he left he put them on duty within the barricaded plant while McCleary returned to Pittsburgh. Yet, scoffed one reporter, "Within a comparatively short time, the ten deputies were convinced their lives were safer elsewhere, and every mother's son of them got out of the mill and Homestead as quickly as their legs could carry them."[53]

Unable to raise a large posse (though technically under the law he could deputize any male citizen of voting age he wished) and knowing in advance that the Pinkertons were due on Wednesday, McCleary assiduously neglected to take advantage of the Advisory Committee's earlier agreement to allow the Sheriff a *posse comitatus* of fifty deputies to take over the Works. He rationalized that he hadn't sufficient time to do so, though he must have known that the Pinkertons had long since left Ashtabula by train and were perhaps already afloat for Homestead.[54]

By the evening of the 5th, the situation was so polarized that neither the workers' Advisory Committee nor Frick wanted responsibility placed on them for what seemed certain: a violent confrontation. To avoid legal complications, the Advisory Committee had not only formally dissolved itself and refused to accept responsibility for any ensuing events, but had also sent a request for an injunction to Judge W. J. Brenner in Pittsburgh to restrain the Sheriff from bringing in deputies – a request later denied on the basis that no judge could prevent the Sheriff from protecting the Company's property.

In Homestead, meanwhile, the word quickly spread that the Advisory Committee had dissolved and that the Sheriff would again attempt to place his men in the mill. In half an hour, the town's population poured into the streets. Many workers armed themselves. Worried, the town's businessmen and merchants asked Governor Pattison to come to Homestead immediately to avert conflict and to prevent possible bloodshed.[55] Tension among the population then increased when neither Judge Brenner nor Governor Pattison replied promptly.

Meanwhile, in the lull before the storm its agents had planned, the Carnegie firm took final measures to prepare. By turning the mills over to the care of the county authorities, Frick freed Manager Potter and twenty of his superintendents to leave the city and scout elsewhere for 260 skilled laborers to act as replacements for those strikers without whom it would be impossible to run the mill. Other nonunion men were also lined up. Then the firm lit the final fuse: The Pinkerton rifles were ordered in.

* * *

After 4:00 P.M. on the 5th, rumors floated wildly in Homestead. A few minutes after the hour, word was received that 75 blue-coated Pinkerton men had just landed at the Union Station in Pittsburgh, on their way to Homestead and 2,000 men immediately assembled along the railroad track near the station. Then the word came that acting-Sheriff Cluly had arrived with 11 deputies at Homestead's City-Farm Station, and 3,000 people began to congregate around the depot, though no one came. When the deputies finally arrived at Munhall Station a half mile away, a crowd of 1,500 swarmed in. But here they found a mere twelve men who, under Deputy Sheriff Cluly, had gotten off the train and attempted to stage an entrance into the mill property – an entrance denied by the hundreds of determined strikers on guard, packed against the gates in a solid body, making it futile for the "invaders" to enter by force.[56]

"The scene that followed is indescribable," reported an on-the-spot witness:

> The crowds at the City Farm station were compelled to come alongside of the fence about 300 yards before getting onto Eighth Avenue, leading to the Munhall station, and as every man wanted to be the first to get to Munhall station, the narrow passageway was crowded to suffocation by the struggling mill workers, all tumbling over each other in their great haste.

> When the crowd reached Eighth Avenue they were joined by a larger crowd, all pressing toward Munhall, and the sixty-foot street was hardly able to hold them. In the meantime the deputies in charge of ex-Sheriff Cluly, were having an interesting time at the Munhall gate. As soon as the train stopped the deputies, headed by Cluly, started toward the gate, against which stood several hundred brawny mill men. The deputies came close up to the determined mill workers, who looked them square in the face without flinching or moving an inch. Ex-Sheriff Cluly advanced and in a clear voice said: "Men, we command you to fall back and give us entrance to the property of the Carnegie Steel Company, which we have been detailed to guard."
>
> There was silence for a moment, when one of the workers stepped forward and said, "We cannot permit you to enter the mill. The mill property is not in danger, nor does it require your guardianship, and we think that should we admit you to this mill it would be against our interests, so we will not allow you to enter. Should you wish to return to Pittsburgh we will see that you get safely out of town."
>
> Acting Sheriff Cluly, after a few moments' consultation with his aides, said: "We will go back to Pittsburgh and will depend on you for a safe escort out of the town."
>
> A large guard of workingmen formed about the deputies and started toward Amalgamated Headquarters, where the men were to be kept until the steamer *Edna* could be summoned to transport the deputies to Glenwood.
>
> There was but little demonstration. Half an hour later the march to the *Edna*, a quarter mile distance, was begun. Each deputy was escorted by a workman, and none were permitted to follow in the rear. Not a loud word was spoken. Acting Sheriff Cluly brought up the rear of the procession with Burgess McLuckie.[57]

When the river bank was reached, each deputy shook hands in turn with his escort, thanked him, and stepped on board. As Sheriff Cluly stepped gingerly upon the gangplank, he turned sharply and admitted defeat: "I am much obliged, boys, for your kindness, and now I will walk the plank."

Once Cluly was escorted across the river by members of the Advisory Committee, the workers diligently guarded the mill – remaining wide awake at their posts throughout the night. They reasoned, as one prominent worker explained, that: "The company knows that no Homestead men will work at the mill, and they have tried to get their guards into the mill to protect all nonunion labor which they hope to get into the mill. We resisted them today; and we will do so tomorrow and the next day if necessary, as we don't propose to give the company any advantage."[58]

Thus, on three separate days, despite arrogant persistence and tenacity, the Sheriff and his deputies had not been able to take charge of the mills. If they had, they might well have acted as the unseen occupants of a Trojan Horse, emerging when all was calm to open the gates for the Pinkertons – asking for admittance, one reporter said, as a cover-up "in clandestinely introducing a body of Pinkerton detectives into the mill enclosure."[59]

The implications were clear: Carnegie officials had planned in advance to mobilize the means of violent conflict – to bring in their privately-paid mercenary army to defeat unionized labor. They had gained the *ex-officio* aid of the constituted judicial authorities. And they were now trying to make their actions appear completely legal from the start, though the measures they took might have struck the population as laughably absurd were they not carried forward with such deadly seriousness.

Thus on the afternoon of July 5th, McCleary posted notices all over Homestead and Munhall calling attention of the general public to the fact that those interfering with the operation of the Homestead steel workers were liable to arrest, fine and imprisonment, and commanding all to abstain from congregating or assembling about the Company's property, under penalty of law.

The Company meanwhile announced in the newspapers that they anticipated no trouble, that they expected to begin the traditional repair work on the mill the following week with 500 or more men at work, and that they anticipated the return of the full body of 2,000 strikers – though Homestead was still in a fever and the narrow streets were congested with citizens awaiting serious trouble.[60]

**Embarking for the Homestead Mill**

To work for the Pinkerton Agency, one had to take orders without question. Even Captain Hein, a seasoned director of Pinkerton raids and leader of the detective forces at Homestead did not know much about the strike, but admitted that he had been ordered there to take charge of the mob and protect Carnegie's property. He said that he did this the best he knew how, though the odds were against him and "it was hard to do anything."

Other, lesser Pinkertons agreed: They were ordered to do the duty they were assigned to and that was all they knew or cared about. New recruits were in a worse position, as one of the injured Pinkertons explained: "I was never in this business before, and did not read much about it . . . ."[61] That was the consensus of many. They were about to enter a struggle "which for bloodthirstiness and boldness of execution has not been excelled in actual warfare . . . ," one reporter said.[62] Yet they did not know why they were fighting. Their role in the gunfight at the Homestead mill was that of mercenaries.

* * *

On the evening of July 5th, the towboats *Little Bill* and *Tide* had moored at the Davis island dam on the Ohio River, having in tow the two barges, well fortified and provisioned. At 1 o'clock, the special train from Ashtabula bearing the Pinkertons had arrived at Bellevue and they were marched over to the barges, which were then propelled up the Ohio River to the Point at Pittsburgh and then down the Monongahela to Homestead.[63]

The tug towed the barges up the river slowly, trying to make as little noise as possible. But the *Tide*, which had towed one of the barges thus far met with an accident at Lock No. 1, calling attention to the operation and leaving the *Little Bill* to pull both barges the balance of the way to Homestead.

After leaving Lock No. 1, the passengers knew their movements were being watched.[64] The Advisory Committee had created so perfect a system of patrolling the river that no sooner had the barges started than word was sent to Homestead. At a little after 2 o'clock in the morning, information was telephoned to the office of a morning newspaper that the *Little Bill*, towing two barges filled with armed men, had just passed Lock No. 1 on the Monongahela.[65] The Amalgamated scouts stationed there meanwhile sent word ahead – the men on the advancing flotilla were each armed with Winchester rifles. An instant later a horseman crossed the upper bridge headed for the South side of the river, breaking into a wild gallop to Homestead. Within a half hour he had reached the little city, given the alarm and set off an uproar as the word spread like wildfire.[66] The steam whistle at the general light plant blasted the general alarm. Flashing arc lights and skyrockets directed the Homestead people where to assemble. And the workers' little battleship, *Edna*, was quickly dispatched down the river to act as an escort, shrieking its whistle at 2:30 A.M. as a continual alarm. A silent approach to the Works was now impossible.[67]

On reaching Brown's landing, near the Salt Works approaching the city, the workers sounded another alarm of the Pinkerton arrival. It was now 4 A.M. and the barges were only a mile below Homestead.

As soon as this fleet was sighted, three messengers, each mounted on fresh horses, started for the city, shouting Paul Revere-style as they rode in advance of the dreaded barges. Someone again notified the engineer at the electric light works and he sounded another shrill whistle of general alarm.

"The preconcerted signals, blasts from the electric light plant whistle, filled the air with hoarse, ominous shrieks," said a witness. "Humanity began to pour from the houses and buildings all over the town." Out came the old and young of Homestead, women and men, mothers with infants in their arms and an array of children. "Men, women and children who but an instant before had been in sound sleep," he continued, "thronged into the streets like panic-stricken sheep. The men began to shout: 'On to the river!' 'To the river!' 'The scabs are coming!' 'Don't let the black sheep land!'"[68]

Within ten minutes not a soul was indoors. The streets were one surging, congested mass of human beings headed for the river bank, shouting, cursing, screaming and laughing. Many carried guns, rifles, or revolvers. Some had clubs which they picked up on the street; others tore pickets from the fences as they passed along; and still others moved forward empty-handed.

A throng of people headed towards the river, the first of several thousand reaching the dark bank under a dense fog hanging over the valley and almost obscuring the lights of the blast furnaces on the opposite side of the Monongahela at Braddock. Reaching the bank, the men turned up the river towards the mills, running along the rough, uneven surface of the bank, passing the word that 500 non-union men were coming up the river on boats to be landed at the Works. This news accelerated their already rapid pace; they were soon at a dead run to the infamous barricade Frick had built right up to the river's edge, just above the old City Farm.[69]

As dawn reddened the midsummer sky, the first rays of light confronted the thick, gray vapor hanging over the water. Thousands of eyes were now strained to catch a first glimpse of the invading armada, a mighty shout from the lower end of the town announcing that the headlight of the approaching steamer had been sighted. Those above the bend in the stream could not yet see it, but word passed from mouth to mouth that the *Little Bill* had arrived. The lights of the steamer then broke through the shadows, and a few moments later the red lights of the boat shone through the dense mist.[70]

"The moment the people on the bank caught a glimpse of the flotilla," said a witness, "they made a break down the road towards the Works." Then:

> When the head of the line reached the City Farm fence, which had been built out several feet into the water, fifty men or more used their shoulders as battering rams and almost in a twinkling had battered down

the barrier.

Once through the fence the crowd poured over the steep embankment skirting the river fence of the Carnegie Company, and despite the mud and uneven character of the ground they made rapid headway. It was a strangely silent crowd. Few words were spoken and these sprang from the lips of women, who were apparently as anxious to participate in the conflict as their husbands, brothers and sons.

Enduring the march from Homestead to the mill, whole families turned out to take on the Pinkertons:

From Homestead to the pump house is a good two miles, but, all things considered, these half-dressed men and women made the journey in record-breaking time. Once there they gathered in one solid mass, the women, or at least most of them, on the outskirts. The front rank of the solid wall of humanity stood at the water's very edge and every moment the crowd grew in numbers and determination.[71]

Despite the heavy fog upon the river, and with poor visibility, many Homestead skiffs plunged into the water. Though the men armed with revolvers were a poor match for the protected barges, gun-fire filled the air as the *Little Bill* plowed forward. The barges were headed above the Pemicky Railroad Bridge at the foot of the Carnegie pumping station where a siding ran down to the water and a steep slag and cinder bed rose above. The skiffs and barges arriving about the same time, now the main battle was the race for the mills.[72]

They had company waiting for them: with a whole hour's advance notice, thousands had gathered on the river bank and broken through Frick's fence. Some had entrenched themselves behind piles of steel billets in the mill preparing to resist.[73] Slowly the *Little Bill* brought the barges down the channel, and just before the Pemicky Bridge, turned its prow toward shore. Shrieks of steam whistles, angry groans and hisses broke the air from the bank. The occupants of the boats stripped for action, each one of them taking up a Winchester rifle, every weapon loaded to kill.

Slowly the crafts edged toward the landing, the men rushing to the water's edge as the flotilla advanced. The *Little Bill*, now firmly secured between the two barges, proceeded to a point directly in front of the mill yard, running the barges closely against the shore where they grounded with a soft crunching sound of gravel under the keels.[74] As the keels grated on the river sands, the grim faces of the invaders were brought within a few feet of their opponents gathered on the shore.

New York World 7-92

## The Carnage Begins

Here at the shoreline a number of men were waiting. They told the Pinkertons that their services were not required, since everything in and about the mill was quiet and peaceful; and they also warned the Pinkertons that if they came ashore, they would be powerfully resisted.

The strikers were ready to back up their warning. Some at the top of the bank were armed with clubs. Others held revolvers, though no rifles were yet in view. Against this formidable grouping the Pinkertons were intended to enter and take charge of the mills at all hazards. Then a deafening fusillade issued from the barges, cutting down William Foy, outstretched arms still imploring the Pinkertons to go back.

Another eyewitness viewing the conflict from another position gave a different account of the initial exchange:

> The first shot of the engagement came from the barge. It was aimed at a big Hungarian who stood at the water's edge. The ball went wide of the Hungarian target, but it was the signal to the Pinkerton men to begin, and for a full ten minutes they continued to fire. The first man to fall was Martin Merry, a heater in one of the mills. He was shot in the left side and fell downward in a pile of ashes. Close besides Merry stood a big Hungarian. He stooped over Merry's prostrate body and as he was in the act of raising him, he was also shot. This bloody spectacle aroused the drooping spirits of the crowd, and with a hoarse cheer, half a dozen men rushed to the place where Merry and the Hungarian lay. They picked up the bodies and carried them behind the trestle. One of the rescuers, a Welshman, who refused to give his name was shot in the left leg as he raised Merry's head from the ground[74]

"Directly forty or fifty men stepped out from a door in the end of the nearer barge to the small deck on the bow," still another observer wrote. "Each man carried a 45-70 Winchester repeater, and on every face was stamped an expression of dogged defiance, almost desperation. Through the small doorway at their backs could be seen many more faces wearing similar expressions, their owners carrying those ugly looking magazine guns."

A gangplank was thrown out, the men started toward it, then hesitated at the sight of the thousands on shore, and half-turned as if to go back.

"Don't step off that boat!" bellowed a warning voice from the shore. But an authoritative Pinkerton responded in stern tones of command: "Forward." The little column then moved toward the gangplank, while a shot fired from a porthole in the side of one of the barges was followed by a volley from the guns of a score of onlookers on land. The invaders lined up, raised forty rifles and hurled forty bullets into the ranks of the crowd, so densely packed that the men could scarcely turn around. A wild scramble for shelter followed.[76]

The attack on, the Pinkertons attempted to land, crowding out of the rear of the barges, keeping up a rattling fire into the groups of men as they scurried in retreat up the steep embankment and along the river toward the Water Works, a few falling under the fire as they tried to escape. But the gunfire was briskly returned by some 200 mill workers, some of whom stood their ground while others retreated to the mill yard at the top of the bank to gain protection behind piles of steel billets.[77] Then, as the news spread of armed invaders trying to force their way into the Works, throughout the borough whole families quickly gathered weapons and ammunition, hurrying them to the workmen assembled at the river near the mill.[78]

After the initial exchange left two strikers dead and one wounded, the Pinkertons retreated into the covered barges, firing ceased on both sides, and a brief exchange took place between a middle-aged Pinkerton on the boat and a millworker approaching the water's edge. A witness recorded the words of the millworker, pleading:

> On behalf of 5,000 men I beg you to leave here at once. I don't know who you are nor from whence you came, but I do know that you have no business here, and if you remain there will be more bloodshed. We, the workers in these mills, are peaceably inclined. We have not damaged any property nor do we intend to. If you will send a committee with us we will take them through the Works, carefully explain to them all the details of this trouble, and promise them a safe return to their boats. But in the name of God and humanity don't attempt to land! Don't attempt to enter these Works by force!

The leader on the boat stepped to the front, resting his rifle across his left arm, and elevating his voice for all on the river bank to hear:

> Men, we are Pinkerton detectives. We were sent here to take possession of this property and to guard it for the company. We don't wish to shed blood, but we are determined to go up there and shall do so. If you men don't withdraw, we shall mow every one of you down and enter in spite of you. You had better disperse, for land we will!

There was silence for a moment, followed by the millworker's reply: "Before you enter those mills, you will trample over the dead bodies of 3,000 honest workingmen."[79]

For the next two hours an ominous silence reigned. Then at 6 A.M., the leader of the Pinkertons again stepped out and commanded the strikers to disperse, saying that an hour later he would take his men into the mills against all obstacles. This provided the strikers with additional time to erect substantial breastworks of structural steel to protect themselves as they crouched with loaded guns. Then forty-five minutes after the appointed time, at 7:45 A.M., the Pinkertons stepped out on the forward deck to prepare for a landing. The Pinkerton officer in charge then ordered his men to advance, but as they left the barge they were met by a shower of lead from the workers' pistols and rifles. The officers fell back for an instant as if repulsed, then rallied and, marching eight abreast European-style, attempted to get ashore. But they were driven back again, the rattling volley from the mill yard putting them in hasty retreat . . . as four men dropped in their tracks.

The Pinkertons returned the fire from the portholes and ends of their boats, wounding a number of workers who were in exposed positions.[80] "Already," wrote an on-the-spot reporter, "four of the locked out men have been killed. John Morris was killed in the last effort to land the forces. Six more workers were shot in the second fight. . . . The killing of Morris seems to have crazed the people, and men, women and children are running through the street crying for revenge and for blood."[81]

* * *

The exchange continued till high noon. All the Homestead victims were shot through the head, evidence that the legs and torso of the workers were well hidden and the Pinkertons, using high-powered Winchester rifles, took deliberate and murderous aim. On their part, the workers took their bead from behind beams and piles of billets – and their fire also took a heavy toll on the Pinkertons.[82]

Disabled under this fire were Pinkerton Superintendent Klein and Captain Hein, a seasoned leader of many large anti-labor actions. So fierce was the exchange of shot and so easily could the workers use their refuge of steel and billet breastworks as an advantage to pick off detectives venturing on the decks that the worried crew of the tugboat hastily sought an escape. Cutting loose from the barges, they steamed up-river and carried as many of the wounded Pinkertons as they could to Braddock, from which point the wounded were sent down to the hospitals for treatment at Pittsburgh. Seven of the Pinkerton force were thus cared for. Back at Homestead, meanwhile, the strikers that fell were carried to their homes, while the dead were taken to the morgue and the undertakers.[83]

The other Pinkertons felled in the early gunfight and picked from the water and decks by the *Little Bill* were carried away with Superintendent Potter to Pittsburgh. "There was a large number of men on the boat," a reporter recorded the pick-up. "The moment it reached the shore a regular fusillade took place. It continued for ten minutes and was mixed with hoarse, derisive cheering from the men in the mill."

Leaving under fire, the *Little Bill* was back in Pittsburgh shortly after 12 noon, Captain Rogers exonerating himself and asserting: "I never saw such a cowardly attack as those the strikers made. They had a fortification of pig iron, and the minute the boat arrived there they commenced firing, and not a shot was fired by Pinkerton men until three of their comrades were shot down like dogs."[84]

Then the *Little Bill* again returned, steaming in under heavy fire towards the moored barges she had left behind. Mocking hoots and yells went up from the 150 men behind the barricade and the 2,000 still unarmed who were back in the mill, on the trestles, and at other points out of range.

A hot volley from the sharpshooters and the millworkers raked the steamer fore and aft once she turned her broadside toward the shore, and a dozen bullets struck the pilot house. The occupants dropped to the floor as the workers cheered and the men on the boat returned the fire. But instead of landing, the *Little Bill* floated past the Works, "running the fiercest blockade that has been witnessed on this continent since the days of 1865," a reporter compared the event with those of the closing year of the Civil War. "There was a perfect shower of lead from the boat, but it was returned with an energy to which her perforated sides attested for weeks and this attack was kept up as long as the craft was in range."[85]

Captain Wishart on the *Little Bill* corroborated:

> When we started down between 10 and 11 o'clock, the entire population seemed to be laying for me along the bank, and bullets rained through all parts of the boat. . . . The window glass was riddled. . . . It got so

> uncomfortably warm that I was compelled to abandon the wheel and steer the boat from the engine room, where we were better protected by the machinery. I only remained in the engine room for a short distance, when I again returned to the cabin. They shot at us from every direction, and at the Homestead bridge men were in wait for us on the abutments. They fired down upon us through the roof, and kept peppering away at us after we had passed.[86]

Such risks were worth the money, so far as Captain Rogers, owner of the *Little Bill*, was concerned. And when his contract ended, he started home, insisting he "was simply doing what I was paid for doing, and what anyone would do."[87]

But there were also those who fought, not for money, but for justice that day. Some strike supporters had volunteered from Pittsburgh, when the news reached there at 6 A. M. Thousand of idled mill workers had congregated on Pittsburgh's main roads and streets, pending their own negotiations on the wage scale. Homestead was on everyone's lips, and hundreds of men, armed with guns and revolvers and well supplied with ammunition, took up a line of march to come to the aid of the Homestead strikers.

Meanwhile, the war at Homestead intensified. As soon as day broke, the strikers had secured a small ten-pound cannon and planted it within a steel billet enclosure, so as to command the barges still moored at the river bank. On the opposite side of the river, another thousand men had positioned themselves on the hillside where, concealed by shrubbery, they planted a small brass cannon of the *Griffin Post 207*, and loaded it with iron scrap, nails, slugs and "punches" – the iron centers punched out in the making of nuts. Though the first three shots fired at the barges fell short, the workers then fired a fourth that struck inside the workers, own fortifications on the opposite shore, where a glancing piece hit striker Silas Wain in the neck, severing his jugular vein, almost tearing off his head and instantly killing him. To avoid repetition of the tragedy, shortly before 9 A.M. the cannonading party was signaled by a secret code and then brought their cannon to the mill side of the river and, from behind a big armor plate 25 yards from the boats, they fired shots directly into the barges.

The slugs easily tore through its upper ramparts, driving the Pinkertons into the hold. This prevented them from shooting through the loopholes as before, except when the *Little Bill* returned and gave them protection.[88]

So now there was a double rain of fire. For simultaneous with the small arms shot hitting the barges along their sides, the workers trained their cannon on them, then kept up a steady, awful bombardment for several hours. The stout oak timbers making up the sides of the barges splintered, but the cannon balls were not able to penetrate the heavy steel plates fastened to the interior. Shot after shot of small arms fire filled the portholes in the bow, inflicting terrible injury on the imprisoned men.

But the barges continued to float, even though the workers were more determined than ever to destroy them and compel the Pinkertons to surrender. The bursting of their largely ineffective cannon was temporary setback. "When it was found that little impression could be made by the cannon an the boats," said one eyewitness, "an effort was made to fire the barges, and thus compel the detectives to leave the vessel or suffer the terrible fate of being burned alive."[89]

By loading a car with lumber and empty oil barrels, then igniting them and pushing them down the siding toward the barges, they hoped to set the vessels on fire. But the car jumped the track before it got near enough. The workers then forced crude oil through the hose of an old fire hand-pump in an effort to drench the barges. The stream of oil did not quite reach the barges, so the workers threw flaming torches upon them, yet again failed to set them on fire."[90]

An on-the-spot reporter described the scene: "At this time a body of men is rolling barrels of oil to the river. It is their intention to set fire to and burn the Pinkertons from their position, which they have held since daylight. The excitement and confusion last night have developed into a desperate storm. The Pinkertons seem about to make another move to land. Their boat is almost riddled with bullets."[91]

At 11 A.M. a reporter for the *Toronto Globe* watched the grisly events unfold: "The strikers have fired a car of oil standing near the Works on the P. V. & C. Railway, for the purpose of firing the boat in the river." The cannonading had not ceased, he added: "The men have placed a cannon on the north side of the river and are firing into the boats. Balls and pieces of iron are being used, and at every shot the boat's body is penetrated. . . . On every hand men are gathering ammunition of all kinds. Every form of firearm is being gathered and shotguns are being loaded with buckshot."[92]

Still more hose was procured at 12:10 P.M., when a wheeled carriage belonging to the city and half a dozen barrels of lubricating oil were brought to the water tanks, together with a fire engine. But there was difficulty in getting the engine to work,[93]though now the oil was trained on the decks and sides of the barges. While this was being done, barrel after barrel of oil was emptied into the river above the mooring place, the object being to allow it to float against the boats and then to ignite it. This terrible deed was attempted several times, but still the boats did not burn.[94]

"All sorts of plans were tried to fire the boats," eyewitness Stowell wrote:

> . . . the stores with overstocked Fourth of July fireworks were drawn upon. Rockets, Roman candles and the like were used, but without avail. The oil was of the lubricating kind and was not as flammable as other grades. But if the mill men had succeeded, an appalling fate would have overtaken the unhappy men in the barges. To save themselves from death in fire they would have been forced to face the bullets from the watchers, and escape of any of them alive would almost have been beyond hope.[95]

By 12:30, the strikers finally succeeded in setting the Pinkerton barges on fire by breaking open an oil car standing nearby and letting the oil run down the river. "The fire from the burning oil is enveloping the barges and those on board are fighting it off," another observer reported. "In the meanwhile the strikers on both shores keep up a constant rain of bullets upon the detectives in the boats."[96]

By 1:15 in the afternoon, flames were spreading up the riverfront, the mill was threatened, and several explosions frightened the population. "The people," a witness wrote, "are taking to the hills to get away from the fire."[97]

Dynamite had become the principal means of destruction after 1 o'clock, when a tall brawny workman waved two sticks high above his head at the new railroad station. Displaying a basket full of the deadly explosives, he railed onlookers that those murdered by hired thugs must be avenged: "We must kill them! No one must escape alive!"[98]

Five hundred voices responded - "aye, aye, aye!" And a score of men within the range of Pinkerton rifles followed him, using their penknives to make holes in the cartridges for fuses with short wicks, lighting them, holding the "messengers of death until they burned closely," and, with arms drawn until every muscle showed like a whipcord, letting them fly to the cheers of the excited crowd as the explosions raked the roofs of the barges.

An enlarging circle of millmen then began to hurl dynamite at the vessels. The cartridges, which were tossed toward the boats, exploded with great violence, though few reached their mark. It was difficult to hurl them where they would stay long enough to take effect, but several did considerable damage.[99] The barges heaved, a hole was blown in the roof of one, the sides of the barges were torn into, and a door in the bow was blown off another.

Fire spreading at 1 o'clock the millmen refused "quarter" to the Pinkertons who raised a white flag. The workers now expanded their defenses - to every pile of iron and stack of coal, to the water tank, the pump house, the gas house, the old shearing machine under the Pemicky Bridge, and the piers of the bridge on the Braddock side of the river. Every target the Pinkertons offered drew their fire, and from positions of safety all along the trestled Pemicky road and station thousands of men and women cheered the sharpshooters.

The bleachers included the now deserted offices of the Carnegie plate mill and the hills on both sides of the river where crowds loudly saluted the blood running from every fallen Pinkerton. Shouts – "Don't let one escape alive!" – followed each report of a rifle and boom of the cannon. "Then another shot and another cheer told that somebody has been hit," a reporter noted the cracking of Homestead's veneer of gentility. Fresh ammunition and arms had arrived from Pittsburgh for the strikers. The men moved in closer to the river to get a better shot, and, as Pinkerton officers were revealed with every blast of the cannon, a volley of shots from the defenders followed.[100] More dynamite tore off portions of the planks on the barges; and the battle continued as dynamite combined with renewed attempts to burn the boats with oil.

By 1:35 the Pinkertons were on the bow picking up their dead and wounded, when half a dozen shots felled two more, bringing additional cheers for victory and curses for Carnegie.[101]

An eyewitness diary sums up the events of late afternoon:

> 3:45 the steamer "Little Bill," trying to check the strikers across the river with their cannon, was driven off with her pilot killed. . . .
>
> 4:00 The men have secured another cannon and three of them are now bombarding the barges. Other workmen are trying to blow up the barges with dynamite cartridges. Another Pinkerton has been shot while trying to raise the white flag.[102]

"The explosions were terrific," still another witness reported. "When the dynamite sticks struck the roof of the boats they demonstrated the character of the roof coverings. Not a mark was left where the explosion occurred. The violence of the explosive stuff could be heard for miles around, and it was kept up, now and then varied by the report of cannon or the sharp crack of the rifles, until 4 o'clock in the evening."[103]

**No Exit**

Cooped up in the barges, even the stoutest heart was dismayed by the maddened army of strikers. "A flag of truce was displayed by the Pinkerton men and was shot down," a reporter documented. "It was hoisted the second time with the same result. The third time the flag was riddled with bullets and hardly enough of it was left to hoist again. The mill workers have heard that the Militia have been ordered here, and they now seem determined to rid the place of every Pinkerton man before the troops get here."[104]

As the Pinkerton's ammunition supply dwindled and their exhaustion increased, they discontinued the fight. For nearly two hours before the end of the struggle was reached, not a shot was fired.

The Pinkertons had little choice at this point. They were unprotected. Many had sought crevices in the side of the boat for security from which to fire at the strikers. But, according to three Pinkertons who later violated the orders of their supervisors not to talk, they "fought at a disadvantage, for the men on shore could see into the open ends of the barge and so pick off any men who approached too closely toward the opening; and in addition the bullets from the rifles of the Homestead strikers were continually piercing the boilerplate sides of the barges."

In any case, when the ammunition began to grow low towards 5 P.M., all the men save one decided to run up another white flag – the sole dissenter wanting to hold out to the end. Outvoted, he picked himself off the floor, walked to the end of the boat, raised his Colt revolver to his head, pulled the trigger – fell back on the deck dead, with his brains oozing out on the already blood-soaked boards.[105]

With the steady rise in the temperature and the close confinement of the Pinkertons under fire from the shore, conditions within the barges also grew unbearable. The sun was beating down on the low roofs of the barges, the air within was stifling, and the Pinkertons were sweltering. Yet they dared not seek an opening at the portholes or the hatches, as it would only draw a storm of bullets from the angry men outside.

As the sun grew hotter and the suffering of the inhabitants and wounded grew, sounds of an ax at work within the boat told the crowd that the Pinkerton guards were taking desperate chances to prevent suffocation. Soon a hole was cut through, and a moment later it was made twice as large by bullets from the shore, wounding the axman and preventing further attempts to secure ventilation. Death in a stifling atmosphere was better, the Pinkertons decided, than from the guns of the crowd on the bank.[106] As one big, tough Pinkerton from Chicago later said, the experience in the boats was the worst he had ever had, though "he had been in some very warm places." Some of the men, he continued, even cried from fear, and not many expected to get away with whole skins.[107]

After attempting to surrender several times, the Pinkertons had become desperate. "The groans and curses of the wounded mingled with the prayers and pleadings of the dying, as they lay at the bottom of the barges; the whistle and ring of the bullets, the report of the guns, and the shouts of caution, formed a scene of indescribable horror," said Pinkerton A. L. Wells, who was on the barge. "It was a terrible experience, and one that blanched the cheek of the most fearless."

"We were caged like rats in a trap," he continued:

> The chances were one hundred to one that not a man of us would get out alive. After we had run up a flag of truce for the third time, and each time it had been shot down, we gave up all hope. The dynamite bombs thrown at us blew out one of the sides of a barge as though it was paper, and I saw men in the party who were contemplating suicide in preference to enduring the terrible strain, which would undoubtedly end up only in death if no mercy had been shown us. If we had known the awful treatment which we were subsequently led to undergo, it is a question whether the men would have capitulated when they did.[108]

**Making the Peace**

With the Pinkertons trapped in the barges, Sheriff McCleary's office became the seat of negotiations between the Amalgamated and Carnegie Steel. A two hour consultation began between Messrs. Weihe and Wakenight of the Amalgamated Association, attorney Knox representing the Carnegie Steel Company, and Sheriff McCleary and his attorney R. B. Petty. Knox and Petty then left and went to consult with H. C. Frick, while the Amalgamated President Weihe remained behind, hoping to settle the battle. "I am here," he told a reporter, "to see if this matter cannot be arranged, and I am hopeful this day will see the trouble end. I believe the whole question can be peaceably arranged and no more bloodshed follow. The attorneys for the Carnegies and the Sheriff are now in consultation, and I hope for good results."[109]

But the conference failed and Weihe was discouraged. It was only in the late afternoon that he began promoting Sheriff McCleary's peace proposal to send a boat to tow away the barges, if the men would cease firing.[110] Thereby a peace faction among the workers was initiated and began to gain ground.

The grand *finale* came towards dusk when the new efforts by the strikers to burn the boats with their living freight were openly opposed by the leading officers of the Amalgamated Association, who had come to the scene in the afternoon. They argued that the workers should allow the Pinkertons to surrender. Many workers strongly objected, having lost friends and acquaintances during the day. Thousands of others attracted to Homestead, with no allegiance to any local organization or community, were on hand to advocate a continuation of the battle. "They wanted to see the carnage go on," stated one reporter, "and it was not until some of the strikers pointed their guns at the outsiders that a hearing was obtained."[111]

Not only outsiders wanted the battle to continue, however. Many strikers wanted to see the Pinkertons ground under. Yet only the leadership of the Amalgamated was staunchly in favor of accepting the Pinkerton surrender and organized to do so. For this the Amalgamated leaders had met; then called a 4 o'clock meeting of nearly 1,000 men, where these leaders exhorted them to raise the olive branch.

President Weihe promised that the barges containing the deputies would immediately be removed and President-elect Garland, towering from a furnace top, pleaded: "Men, for God's sake and your families' sake, and for your own sakes, listen to the pleadings of cool-headed men. We have positive assurances that these deputies will be sent away, and all we want is the statement that you will not do any more firing."

But Garland was booed for his appeal to the thousand workers to put down their arms. The crowd was determined to burn the boats, annihilate the Pinkertons, and give no quarter to the murderers of their comrades. Though Amalgamated vice-president G. H. McEvoy warned that, if the Pinkertons were not released, the Militia would be sent in, his words were soon drowned out by a new blast of dynamite thrown at the barges. Many rank-and-file had already jilted their leaders.

McEvoy's warning was heard by some, though: "I know full well that if you continue doing as you are the State Militia will be ordered out before dark and you will lose ten men in every instance where you have so far but lost one. You have gained just achievement over Frick and now aren't you willing to allow these boats to return?"

Excited, some men took this warning seriously, agreeing to let the Pinkertons return if they again showed the white flag. But others continued to pump oil on the barges, waiting for nightfall to burn the men with them. Hugh O'Donnell, head of the Homestead Amalgamated, became frantic, grabbed a small American flag, mounted a pile of pig iron and pleaded for mercy for the Pinkertons lying like "caged wolves" in the barges. Though a few cried out "Let's burn them," they were silenced as O'Donnell continued his plea to negotiate with the enemy.[112]

Finding that he had won the crowd's attention, O'Donnell suggested that a truce be arranged until the arrival of the Sheriff. He insisted that a white flag should be carried to the bank; but a thousand voices defiantly replied, "Never!" And someone yelled that they had shot at one flag on the barges in the morning and "if there was any white flag to be shown, it must fly from the boats!"

"What will we do, then?" asked O'Donnell, to which one man replied loudly: "We will hold them in the boats till the Sheriff comes, and we will then swear out warrants for every man on a charge of murder. The Sheriff will then have to take them in charge."[113] Shouts of approval gave O'Donnell his cue: "Let us turn them, over to the Sheriff and make information for murder against them," he projected to cheers from many of the 2,000 armed strikers and maddened men who had been shouting for blood. O'Donnell then stepped down and went to work to hold the crowd to his proposal to prevent further conflict if possible.[114]

Thereafter, on a one-to-one level, the strike leaders began to reason with the workers, building up a peace faction to offset the deadly demands laid down by some strikers, outraged families, and outsiders from Pittsburgh and elsewhere.

Through manipulation, the peace faction began to grow, operating out of proportion to its actual strength. Then about 5:30 a striker, waving his hands and asking the Pinkertons to surrender, received an immediate response from the terror-stricken guards on the barges.[115] Suddenly two Pinkertons emerged from the interior of the barges, stepped to the front and threw up their hands. When a Pinkerton draped a white handkerchief from a porthole, cheers greeted it, and cries rang out: "They surrender!" "Victory!" "We have them now!"

O'Donnell seized the advantage, and started down the steep embankment accompanied by two members of the old Advisory Committee, moving towards the outer barge with a white flag – sternly being warned by his own men to drop it or be shot!

Going forward the rest of the way down the embankment with his companions, O'Donnell entered the barges for a short parlay, promised the Pinkertons protection if they surrendered, and emerged with a little party displaying white handkerchiefs above their heads. Though a multitude of men were still howling for the blood of those who had killed their comrades,[116] a round of cheers followed, 100 or more spectators from the shore rushing forward from all sides bent on capturing rifles and looting everything.[117]

**Victory Without Grace**

Inexperienced in the arts of military victory, the steelworkers ordered the first Pinkerton out with his Winchester in hand. "Disarm them!" cried the crowd on the bank. And now as each man came out, his gun was taken away and became the property of the taker. Then, as general looting of the barges with bunks and provisions began, everything of the slightest value that was portable was carried away by the crowd.[118] In possession of the barges, the workers also took charge of the large remaining store of Winchester rifles, some say with unlimited ammunition. These rifles, made of the best patterns, were a great acquisition to the strikers, many of whom were still unarmed.[119]

But again, disparate plans of the strike leaders and the strikers emerged. The leaders had no way to force the workers to observe the guarantee of safe conduct these leaders had promised to the Pinkertons, provided they left their weapons and ammunition behind and agreed to leave the place under guard. The Pinkertons had no alternative and promptly accepted the terms – some men saying that it was the first time they had ever submitted to such a humiliating surrender. But the rank-and-file were determined to make it more humiliating still.[120]

Despite the surrender, O'Donnell tried unsuccessfully to negotiate two concessions from the rank-and-file: that no violence would come to the surrendering Pinkertons; and that the Pinkertons' Winchester rifles would be boxed and carried to the railroad depot. The barges were then occupied, the Pinkerton arms laid aside, and the decks were cleared. The dead and wounded were brought out on stretchers and directed to a place of safety.

Immediately after the surrender, the Pinkertons had put on their uniforms and slouch hats and nailed up some Winchesters in crates. But now each was taken charge of by a striker. Their barge, the Iron Mountain, was swarming with a hundred armed strikers, who disarmed each Pinkerton, threw some of their jackets into the river, gave others to the striking Hungarian workers, and confiscated the crates of rifles.

At least 209 of the original 300 Pinkertons were then herded together and marched out, single file, while the barges were looted for food, tools, crockery and mattresses.[121]

Down the gangplank, one by one, they came, all forced to uncover their heads to be clearly distinguished from the men on the bank and to prevent an escape.

The strikers had now increased their cache of arms. At least 24 sixteen-shot Winchester repeaters were appropriated in the presence of one reporter. "Ammunition was also saved, and there were never better plans for a stubborn fight," this witness observed, adding that the workers then had about 960 Winchesters and repeating rifles in all.

"It looks much more like warfare than on the previous arrival of the Pinkertons," he continued. "The men are so well armed that it will take a force of several thousand men to drive them from their fortifications. It will also be foolish for the sheriff to come with a few hundred men as the strikers will not give up without a desperate struggle. Conservative people think that nothing short of the military can quell the disturbance."[122]

Then the Homesteaders began to taste revenge: "They carried oil into the holds, poured it over the [remaining] bedding and furniture, and then set it on fire, first securing . . . [the barges] so they could not float down the river and cause damage at points below. When the flames broke through the decks the cheers which rent the air were deafening, and the noise could be heard miles away. The hills on either side of the river were literally crowded with people, who could witness from this high point all that was transpiring on the battlefield and be out of range of the deadly bullets."[123] The barges burned to their waterlines, and were then cut adrift – chased for miles along the banks by camera fiends seeking to capture the events for posterity.

Hugh O'Donnell, always the negotiator, had gone among the workers announcing the Pinkertons' unconditional surrender, exhorting all the workmen he could assemble to lay down their weapons and allow the Pinkerton men to come out of the barges and pass through the Works down to the town, where they would depart on the train. Though O'Donnell gained the pledges of some that they would not assault the vanquished, the workmen who heard him were comparatively a small number who could not bind the others. Nor did O'Donnell's appeal reach the crowd that rushed in a few minutes later. His pledge to the surrendering Pinkertons that they would not be harmed was not known or agreed to by the population along the line of march through the town. In fact, the crowd had the impression that the Pinkertons had been driven out of the barges fighting, rather than surrendering, and they steeled themselves for a confrontation.[124]

## The Rout of the Pinkertons

The barges held at least seven dead Pinkertons and twenty or thirty wounded, many so badly maimed they were expected to die. "As they were brought from the boat they presented a terrible appearance," an eyewitness recoiled. "Many were besmeared with blood, while all of them showed signs of exhaustion from the long confinement in the close quarters between decks."[125]

When the crowd first rushed upon the barges after the surrender, the Pinkertons had hidden themselves under the bunks, behind boxes – anything that offered concealment. They were then dragged out of their hiding, but not maltreated. They shivered though it was hot, and their lips trembled when they spoke. "For God's sake, don't ask me anything," one of them painfully moaned, "only let me get out of here!" Tears rolled down his cheeks, marking furrows through the perspiration and dust that gathered there.[126]

The crowd's fear then turned to rage. Throwing himself on his knees on leaving the gangplank, one young fellow burst into tears and begged to be spared – only to be kicked to his feet, then felled to earth by a blood-stained club.

An Irish Pinkerton, by his looks and behavior, was subject to special harassment. Once on shore, making his way through the crowd, an Irish steelworker caught sight of his face. "Luk at the Oirish blackguard!" he shouted, and a hundred sons of Erin showered blows upon his head, crying hotly: "You're a disgrace to your counthry!" Only when he got away did he shout back: "Oim as good an Oirishman as any of ye, and oim as good a man, too." Rushing after him to deliver a few extra cracks, one of the assailants shouted what only an Irish reader may understand: "Be jabbers, and he's a Tip!"[127]

The landing of the vanquished lasted for over an hour, another witness wrote: "They came ashore singly and in parties, and as they reached the top of the bank, they were pounced upon by the crowd, beaten with clubs and rifle butts, kicked and knocked down and tramped upon. . . . Many with ashen countenances, blood-streaked faces, clothes almost torn from their backs, cried for mercy."[128]

Yet the worst was to come, for all the streets of Homestead were filling with thousands of weeping and wailing children, women and men. The narrow roads leading to the mill were overflowing, as the crowd of some 3,000 came down from the hills back of the mills, where they had viewed the battle without risk to themselves.[129]

From all directions a tremendous throng meanwhile gathered along the shoreline, hooting and yelling, most carrying weapons of some kind to greet and direct the disembarked.

Then as if by a preconcerted plan, the crowd on the shore formed itself into a 600 yard gauntlet – two lines between which the men from the barges had to pass.[130]

Ordered up the long slope into the heart of the crowd and again to remove their protective hats, the Pinkertons blanched, but were forced along. Stumbling up the embankment, they were pushed and pulled from one side to the other by spectators, those responding with anger taking even more abuse.

Halfway up the hill, the fury of the mob exploded. Though men began by slapping and striking detectives with open hands and others followed their example, before long men took to clubs, children to rocks, women to anything to beat or maim.

The Pinkertons shrieked, "Murder!" and pleaded for their lives; but their hats, satchels and coats were taken from them, torn into shreds and carried away, while the patrol guards composed of the steel workers and the Advisory Board were helpless to save them. "When a man reached the top of the bank," wrote an eyewitness, "he would receive a blow from a club that would fell him to earth. When he got up he always ran, and blows from fists and clubs were rained upon his unprotected head."[131]

Yet the bitter end had still not come, for in the mill yard several hundred people were gathering. Among them were strangers who had flocked into Homestead for the great events, though most had kept their distance from the fighting. Also mingling among the crowd were Hungarian workers, who had been badly treated in the mills and were on the lowest wage level in and outside of Homestead. At last they had a way to vent their anger without being held personally accountable. And reports circulated that many Hungarians in the crowd began to handle the passing prisoners roughly. Pinkertons were slapped and buffeted on all sides, as well as knocked down, often with clubs.

Clubs became the weapon of choice in the mill yard near the end of the first of two gauntlets. The leaders were unable to do anything to control the crowd or stop mob-rule: "When you see an uncovered head, hit it," was the prevailing thought. And while the carnage went on, the Pinkertons' valises were also knocked out of their hands and made footballs, until the curious contents were scattered all around.[132]

**The First Gauntlet Run**

Entering the first of two terrible gauntlet runs, the vanquished were rushed forward at top speed. The Amalgamated men tried to guard them from abuse,[133] begging the people to restrain their anger, trying to rescue the more unfortunate, and guarding them through the lines as best they could.[134] In this concerted way, the Amalgamated men acted as policemen to regulate the actions of the people at large, though they could not constrain the many thousand now present.

"Women, converted for the nonce into veritable furies, belabored Mr. Frick's janizaries with bludgeons, stoned them, kicked them and spat upon them," a witness commented of the sisters, mothers and grandmothers who had seen their men shot down and killed. "Cries for mercy were treated with derision."[135]

Repeatedly ordered to uncover their heads for identification, the Pinkertons offered the half-craze assailants an easy mark. "Men, for the love of God, have mercy on me," shrieked a gray-headed man, wounded and bleeding – bringing forward new beatings as he was struck down before reaching the end of the terrible lines.

Though the guards bravely tried to shield the Pinkertons from blows on the way to the Homestead lockup – the guards themselves becoming the objects of clubbings and curses for their acts of humanity – the Pinkertons were unceasingly assaulted by the maddened crowds,[136] a reporter painting an ugly scene:

> An angry mob lined the streets on both sides. As the deputies passed by, each in charge of two [Pinkerton] deputies, the mill men and their friends kicked them and threw them down. The unfortunate detectives begged for mercy. Some of them had pistol-shot wounds in their heads, and three were seen that had their eyes shot out. Several were shot in the shoulders, arms and legs and could scarcely limp along. Blood was running in streams down their shirts, and they fairly yelled with pain.[137]

But the blood-letting continued: one Pinkerton had his eye punched out by a woman wielding an umbrella. Sand was thrown intro the Pinkertons' eyes; they were hit time again with clubs, trampled upon. Some were already too weak to walk when the trek to the Town Hall began. But mill men armed with rifles still swung the stocks to strike detectives over the head and shoulders, inflicting serious and sometimes fatal injuries. Disdaining or lacking weapons, the Slavs grabbed the conquered by the back of the neck, punching their faces with bare fists. Forty victims felled under the pounding were then dragged toward the lockup at the skating rink and its adjacent theater. A striker shot one of the captives, then clubbed him; another bashed in the head of a wounded Pinkerton with the butt end of a musket - both these victims dying that evening.[138]

As the wretched staggered on by the Amalgamated Association building, the crowd forced the detectives to remove the shreds of their few remaining hats and salute the flag – only to again be beaten over the head with sticks and umbrellas, and abused in every imaginable way. "There seemed a determination to kill the prisoners," a reporter observed, "and it was with the greatest difficulty that the demon-like crowd could be restrained."[139] Kicked and pounded with stones by 20 or 30 men, Pinkertons John Holway was thus pulverized, barely escaping with his life . . . .[140]

**The Ordeal of the Second Gauntlet**

A second gauntlet formed and the brutality intensified as the Pinkertons emerged from the Works at Munhall Station and had to pass along the township road to the Borough of Homestead.

At this point the Pinkertons were confronted by a veritable wall of people. Hatless, clothes torn, many of them bloodied, all of them frightened out of their wits, they were hurried onward through the crowds of people who all along the line showered abuse upon them.

The Police Committee of the Amalgamated, with the entire Borough police force under David Lynch, could do little more than push the crowd aside and threaten broom-or club-brandishing women with arrest.[141]

Thousands of women, children and a few straggling steelworkers crowded the streets. While the workmen still congregated at the mills expecting another invasion, the women left to deal with the captured Pinkertons armed themselves with clubs and stones, hooting and urging the remaining millmen on to take revenge. As the wounded guards then ran by, mud, stones and other missiles struck them, and hoots and yells rent the air – while, still following them, was a crowd of men whose duty it was to see that none got away.[142]

"One woman, who appeared to be the queen of the battle," said a witness, "raised her broom, and in a shrill voice said: 'Where are the dirty black sheep? Let's have them, boys.' "

At this juncture, the leader of the workers appointed to guard the Pinkertons shouted: "Why, my good woman, we want our shirts laundered, and we are going to make these tramps do the job at cut rates."

The angry mob was transformed into a cheering and good humored one for the moment. "Make way for us," commanded the joker, the women obeying and the crowd moving up against Frick's high whitewashed fence, opening a narrow lane to let the column of Pinkertons advance. Heads bowed, laggard steps, the Pinkertons went forward, not daring even to glance at the stern faces and gleaming eyes of the crowd.

Insatiable, the strikers, the women folk, and their allies soon sought still more revenge for their losses. Men like J. E. Morris had been slaughtered, many felt, and his death from a Pinkerton bullet through the forehead was seen by many as he fell forty feet after being hit, his nose smashed from the fall and his chin crushed. A popular leader, Morris was to draw a great crowd of 2,000 to his funeral from his relatives, members of his Amalgamated Lodge, the Odd Fellows and the Knights of Pythias – his death enraging many a friend and acquaintance who wanted revenge. And this was but one case among many that aroused the hatred of Homesteaders for the Pinkerton invaders and murders.

Even at Munhall Station where the wounded were carried, the mob was unrelenting. One man, lying dying, feebly called for water, while the mob scoffed. When a man finally brought some to him, a woman knocked it from his hand, crying he should die. "Kill him!" the crowd screamed, though, despite protest, another drink was offered.[143] "We relied implicitly upon the word of the leaders to give us protection," young Pinkerton recruit A. L. Wells, a medical student, was to say later, "but instead all but a few of the wounded were tortured worse than if they had fallen into the hands of savages. Even some of the men who were injured were kicked and beaten into insensibility."[144]

* * *

Crossing the intersection of Heisel Street and Eighth Avenue, the Pinkerton column reached the crest of the hill, at the foot of which was the workers' Headquarters, and, in between, a veritable sea of enraged people. Looming ominously below, one witness reported more than a thousand determined men and pale-faced women lined the avenue. This human gauntlet extended from the brow of the hill to McClure Street, a distance of several city blocks.

For fully an hour these men and women had stood and waited for the captives, cat calls announcing their arrival. Speedily the Pinkertons then tried to pass, the first of the prisoners moving unharmed behind the worker-guards, the rest trailing behind unprotected, being bloodied by the crowd.

A tall, handsome woman began the fray by throwing a handful of dust into the eyes of a 300 pound prisoner, blinding him, then heaving a sharp stone through his right cheek, shattering four teeth. "Serves you right, you dirty cur," she muttered as he fell face downward on the road.

The Homestead guards pleaded, a few conservative workers protested, but the mob continued to knock the prisoners down, pound them with clubs and stones. Women spat in their faces and tore their clothes amid screams, cheers and pandemonium. Big-boned, muscular Pinkertons became childlike at the hands of their frenzied assailants, pleading for mercy of which there was none.

For three-quarters of an hour the Pinkertons passed. Then, when the last man straggled through and the mob had dispersed, the column moved off Eighth Street to a small side street leading into the town's Skating Rink. Here the great door was swung shut, locked, and the men were temporarily secured.[145]

**The Violent and the Damned**

Although it was the *skilled* men of Homestead who first lost their jobs, it was the *foreigners* and *women* who were accused of the most vicious brutality against the captured Pinkertons.

As Frick planned to replace the Hungarian and other laborers who joined the union men with unskilled scab labor, they were fighting not only for their jobs, but their existence. The women too had a way of life at stake. "The women were the most virulent and savage after the surrender," wrote one reporter, "and it was due largely to their acts and to their goading of the men that the leaders were unable to restrain the mob. Tales innumerable are told of the scenes along the line of the gauntlet. The Madame Dufarge of the movement was a woman who stood near the headquarters and outdid all the men."[146]

While reporters from outside Homestead made these charges of violence against Homestead women, the local papers portrayed them as ordinary looking - not handsome; genteel, not violent; and at their *apogée* as home-bodies whose men folk would not permit them to do factory work, thus giving them the opportunity to become well-educated helpmates of their men. They maintained the home, did all domestic chores, and held together the strong family units supported by their men's wages.[147]

Such a division of family labor made the women painfully aware that, should the breadwinners no longer be able to work, the whole family would be placed in jeopardy. That they were so militant had a good deal to do with their knowledge of daily needs: if non-union workers took their husbands', fathers', sons' or brothers' jobs, it would cut away the only source of money for necessities, threatening what remained of their traditional family stability. Valuing their way of life under threat from Frick and the Pinkerton invaders, the women of Homestead fought heartily for what they knew, and the Pinkertons who earned their wrath were seen as murderers of their family breadwinners and Carnegie agents attempting to take away the jobs of their men. Their violence reflected their struggle to sustain family life itself.

Few Homesteaders lost the meaning of an analysis by the *Local News* on "An Umbrella as a Weapon," for it had been reported that the women of Homestead were adept at tormenting the Pinkertons with such implements. Artists also took pains to depict them with umbrellas in charge of the invaders. And in an analogous quip, this paper noted that mining prospectors in the Rockies without arms had fended off mountain lions with an umbrella, and that in India a lady had driven off a large tiger about to attack her party by suddenly opening, almost in his face, a cherry red parasol. "For the benefit our readers," mused the *News*, "we think it is well to give these incidents and suggest that an umbrella may be sometimes almost as good as a gun."[148]

**The Release of the Pinkertons**

The march to the town lockup had taken its toll. Nearly every Pinkerton was maimed running the gauntlets, and 20 had been shot earlier on the barges, resulting in at least seven deaths.[149] By the end of the day on the 6th, seven Pinkertons had been killed, 11 wounded,, several thrown from the *Little Bill* into the river, and at least 234 placed in the lockup or incarcerated at the hastily arranged prison at the city Skating Rink.[150]

The newspaper accounts of the condition of the defeated Pinkertons turned many a staunch defender of the workers into a skeptical critic. Nor did the Amalgamated's O'Donnell change this impression much. "When I went out toward the barges with a white flag," he told the Congressional Committee, "fully forty rifles were turned upon me by our men and I heard one of them call out, 'We don't want to hurt you, Hugh, but you must put down that white flag or we will kill you.' We forced the men to accept the surrender just the same, at a peril we knew well. The mobbing of the Pinkertons we could not prevent. It was done, not by the men who had been in the fight, but by the outsiders who were safely stowed away while there was danger in the front."[151]

O'Donnell knew this was not the whole truth: The community had been enraged to the point where the captured Pinkertons were seen as hired killers, who had come into the mills armed and invaded Homestead from the barges. Their fears were fanned by rumors and the complaints and hatreds of the supporting workers from Pittsburgh and other mill towns, as well as sympathizers from the outside, and a mass of sensation-seekers. The self-appointed leaders of the strikers had no way of making promises for the many they could not speak for, as only a democratic and self-disciplined community with its own committee on public safety or other democratic body might have retained flexibility to care for an injured enemy in captivity.

"This day," surmised one reporter, "was one that will be remembered with horror by the people of the borough, as well as by the citizens of the entire county, who, for the second time will be forced to pay the enormous amounts of money entailed in the shape of riot losses."[152]

* * *

Though accounts vary as to the exact location of the make-shift lockup, the Pinkertons were finally lodged in both the Skating Rink and Opera House, where they were to be kept for the night. Thousands of Homesteaders meanwhile surrounded the buildings, keeping the wounded men in a constant state of terror.[153]

The crowd had mounted to 2,000, but was prevented from setting fire to the temporary prison and shooting the Pinkerton men as they escaped only through the persistent efforts of the strike leaders. The Pinkertons, a reporter wrote, "huddled . . . absolutely terror-stricken, being afraid of a lynching by the immense crowd gathered outside."[154] While armed strikers kept the crowd at bay, the lack of food and water for the Pinkertons was the least of their worries, for most of the Slavs were in favor of murdering them all![155]

Though the Amalgamated leaders' initial plan had been to remove the Pinkertons from the Skating Rink and get them out of town on an evening train, now they deemed it best to keep them safe until the situation cooled down. Filling the Skating Rink and Borough lockup with Pinkertons, they put the rest in the Opera House. The Pinkertons then had several hours to rest, receive medical attention and food, water and coffee.[156]

Meanwhile the workers sent a message to Judge W. J. Brenner asking him to come to Homestead and advise the victors as to the disposal of the prisoners-of-war.[157] The Advisory Committee then went to the county Courthouse to confer with Sheriff McCleary to make arrangements to hospitalize the most severe casualties and to plan for these Pinkertons to be escorted by the Sheriff and twelve unarmed deputies to the West Penn Hospital in Pittsburgh. Yet McCleary was unable to recruit a single deputy for this task![158] To get the Pinkertons out of Homestead would require the cooperation of the Amalgamated, the Carnegie people, and the legal authorities.

"Last evening," the *San Francisco Chronicle* wrote on July 7th, "a lengthy conference was held by Chris L. Magee, H. C. Frick, Sheriff McCleary, W. J. Brennan and others." Magee represented Carnegie, as did Frick, while McCleary was the highest enforcement officer in the county and Attorney Brennan was the judicial officer in charge. "At its conclusion," the *Chronicle* continued, "a special train bore the party, with the exception of Frick, to Homestead."[159]

Meanwhile, the situation outside the temporary jail had grown more tense. Although there were fewer than 300 strikers in front of the Opera House where the Pinkertons were caged at 9 p.m., the Amalgamated leaders wanted the Pinkertons removed from Homestead to Pittsburgh as soon as possible. President Weihe and President-elect Garland, fearing that another day would bring dangerous trouble, anxiously awaited the six-coach special train. When the Pittsburgh, Virginia and Charlestown (P. V. & C.) finally arrived at midnight, about 100 men gathered around the temporary jail, some 40 of the sturdiest steel workers entered the hall, and the apprehensive Pinkertons feared for their lives – a lynching was still a possibility!

Quiet, in ill humor, the crowd outside had not agreed to let the Pinkertons out of Homestead. But a leader of the Amalgamated quickly came out: "Boys," he pleaded, "we are going to send the Pinkertons away. Many claim they were deceived in coming here. They are all hurt and crippled, many dangerously; do not repeat the scenes of this afternoon. I want to hit from the shoulder here; none of the men who went to the front hit unarmed people from the rear as cowards did this afternoon."

"That's right," affirmed a voice from the crowd, "we'll protect them."

Asking assistance for the crippled Pinkertons, President Weihe made a similar speech, and he and Hugh O'Donnell both counseled moderation. Some men now offered to help.[160]

The Pinkertons were then marched out, pale and apprehensive; but not one was molested on the way to the train where Sheriff McCleary, President Weihe and others assisted. Marching two-by-two, the Pinkertons quickly walked to the cars, escorted by the workmen. They boarded the train. And when the train pulled out a ringing cheer went up all around,[161] and the crowd dispersed.[162]

Leaving Homestead, accompanying and watching over the Pinkertons, were also a troika for enforcement – Sheriff McCleary, Amalgamated President Weihe and Attorney Brennan.[163] And, as they planned, two hours later, McCleary arrived with his Pinkerton charges in Pittsburgh.[164]

Though reports differ, it appears that many workers had no idea that the Pinkertons were to be released under the Sheriff's control. The strikers had agreed to spare the Pinkertons with the understanding that these mercenaries would be locked up in the county jail until indicted for murder. The first burst of anger over their release was soon calmed by the realization that if the strikers or others appeared as witnesses, they themselves would be liable to prosecution for riot. Thus, several of the leaders expressed relief that the Pinkertons were released, believing there would be no further legal complications.[165] But many strikers retained their belief that, had the Pinkertons been kept hostage and not released by the self-appointed leaders, the bargaining power of the emerging city commune would have been greatly enlarged.

The evidence to bring murder charges was contained in a roll book, which the strikers found before burning the barges. It contained the names of 266 Pinkertons, divided into 13 squads of twenty each, commanded by a lieutenant and regulars. The book showed that some of the men had been on guard at Wallston near Nuxsutawney, Cleveland, and at other points where labor troubles had occurred. Such notes as "strike ends," "transferred to Wallston," "sent to Cleveland," etc., told the story. "The workers intend using the names in prosecuting the Pinkertons for murder," said a reporter. And once the rumor circulated that the Pinkertons were being taken to the Ohio line and turned loose with only a warning to keep out of Pennsylvania thereafter, the strikers swore to avenge themselves against the Sheriff if he showed up in Homestead.[166]

By evening of the 7th, reports of workers' discontent with the release of the Pinkertons proliferated. One alleged that a party of Homestead strikers were making their way to Pittsburgh to capture Captain Hein, who was in the Homeopathic Hospital. Pittsburgh Superintendent of Police Robert O'Meara responded – "I don't think those men are foolish enough to attempt such a thing. However, we will be prepared for them. We will not interfere at Homestead, but it is our duty to protect anyone within the limits of the city, and we will give any of them a warm reception . . . we can marshal our entire forces within fifteen minutes, and that will enable us to stop any trouble here."[167]

Police protection of the Pinkertons was the policy in the steel city of Pittsburgh. Indeed, as soon as Captain Hein, who was wounded at Homestead, had his wounds dressed at the Pittsburgh Homeopathic Hospital, he sent a message to Superintendent of Police O'Meara, arranging a private meeting.[168] Few doubt what they planned in the light of the fact that no Pinkertons were subsequently jailed or arrested. For when the defeated Pinkertons arrived in Pittsburgh in the wee hours of the 7th a number of city detectives were immediately assigned to the vicinity of the railroad cars all morning, "not to make any arrest of Pinkertons," sallied a reporter, but, as Superintendent O'Meara said, "to see that no one interfered with them."[169] Everywhere detectives protected their own kind; and everywhere Carnegie operated there seemed to be detectives.

In a momentary time warp, after the last lot of 25 seriously injured Pinkertons had been taken to the West Penn Hospital, the remainder lay under the Seventeenth Street incline in the Union Station, securely locked in their cars, with curtains drawn tightly. Then at 9:30 A.M. an engine and express car secretly steamed up to Twenty-Eighth Street, where provisions for the Pinkertons were obtained. Only after the special train left the yard on the Pennsylvania Road and they were fed, did the reporter from the *Sun* gain an inside view of their condition and their ordeal:

> The men inside the car presented a very sorry sight with their bandaged heads and arms. Nothing could exceed their impatience to get away, and they were crazed for papers to get an idea of the situation at Homestead.
>
> "Why can It we get out of this?" said one Pinkerton man, as he poked his head out of the window. "Will we ever get away?" The expression was only one of many.
>
> John Matthews, of New York, who was one among the party, said: "I never want to go back to Homestead and I hope we may get out of Pittsburgh soon. No description could fit our condition inside the barges. It was a veritable hell. I laid on my back for hours in momentary fear of getting a bullet through my head. We expected to be killed if we had to stay inside until midnight. The abuse we got on the street was the most inhuman I ever experienced or witnessed, and I have been a detective for twenty years."[170]

Embarking for their homes at 10 A.M., the Pinkertons were battered and torn, two-thirds of them having been brutally beaten.[171] The forty who left the train at Germantown Junction at 11:30 P.M. on July 7th on the way to Philadelphia "were an unshaven, grim, bedraggled, weary, bruised and blood-stained lot of men. Many of them had managed to save their gripsacks, but there were few among them who could boast a whole coat on his back. The countenances of the majority were adorned with black eyes, and bruised and swollen noses."[172]

On the 8th, the Pinkertons living in the New York area arrived at 2 A.M. While most of the detectives departed for their homes, the dozen or so men in charge were ordered to report directly to Robert Pinkerton in his private office at Exchange Place and New Street to assess their tactical expedition.

But the rank-and-file Pinkertons arriving at New Jersey City at 2:30 A.M. were overwrought with emotion over their near-death experience then and their lucky freedom now. "Some of the men wept for joy at the idea of being home again. Many showed in their wounds the terrible ordeal through which they had passed. The men in charge of the crowds forbade any of them from talking to reporters, but notwithstanding some of them told harrowing stories. They said the workmen on the shore fired first." And one Pinkerton complained that, in addition to clubbing them unmercifully after the surrender, the Homestead people took their watches, chains and traveling bags.[173] Only by a stroke of fortune had their lives had been spared.

Back in Pittsburgh, though, the Pinkerton agency made sure all survivors of the battle were hushed up. In the morning on the July 8th, two Chicago chiefs of the Pinkerton agency went to Pittsburgh's Penn Hospital to arrange for the removal of the injured detectives to their homes that night. Infuriated because the names of their wounded agents had been given to the newspapers, they hotly denounced the hospital authorities.

The injuries were now public knowledge. While most of the Pinkertons were in fair condition and would recover, there were serious cases. One detective from Brooklyn had a terribly shattered right arm and thigh; another from Chicago had a bad wound in the abdomen; a third from Philadelphia, a dangerously injured skull – and all three were in critical condition. Two dead Pinkertons were boxed and unceremoniously shipped off to Chicago.[174]

There were others from both sides on the casualty list, and deaths resulting from the battle continued for many weeks and months. Nine strikers and seven Pinkertons are said to have died. At least forty strikers and twenty invaders were shot, and practically every one of the captured detectives was beaten bloody.[175] Another source documented 16 killed, 62 wounded, six of whom were expected to die – "making the total much larger than in the riots of '77."[176]

Yet, some of the Pinkertons were still not accounted for. On July 8th, Captain Sloan, representing the Pinkerton agency, went down to the river at Homestead looking for some of the missing men. Reporters were on hand to question him. Though he refused to talk about the origin of the detectives or the number missing, he did note the costly significance of deploying Pinkertons.

"It means that the mill at Homestead will be closed down until non-union men are permitted to go to work," he surmised. "The owners will never again permit a union man in the mill. All damage resulting from the present or future fights will be demanded from the county."[177] This had already been legally confirmed in the 1889 steel strike.[178] Thus a private army employed by Carnegie could lose a military struggle and impose the burden of joblessness and damages on the very working population that won by arms.

* * *

Only later was the House of Representatives Committee to write:

> Your Committee is of the opinion that the employment of private armed guards at Homestead was unnecessary. There is no evidence to show that the slightest damage was done or attempted to be done, to the property on the part of the strikers. True, there was apparently an unlawful assemblage which refused to obey the authority of a weak and irresolute sheriff, and committed acts that cannot be defended; but their submission to the governing authority of the State showed that the power resided in the State to execute the laws after the sheriff had failed to invoke the power of his authority, and that order would have been restored without the interference of armed men from another state, brought here clandestinely.[179]

Burgermeister McLuckie had given the initial order to arrest the Pinkertons, feeling it was justified since the detectives were an armed force invading his legal bailiwick – giving him the right to arrest and disarm them.[180] He had set in motion a whole sequence of events that brought the Pinkertons to their knees – and some to the grave. The Pinkertons had lost the battle of Homestead hands down.[181] It was now up to Frick and his master Carnegie to win the war.

## Notes

[1] Livesay, 140.
[2] *The Sun* (Baltimore, Maryland), July 7, 1892.
[3] *Ibid.*, June 30, 1892; *The New York Times*, June 30, July 1, 1892.
[4] *The Local News* (Homestead, Pennsylvania), July 2, 1892
[5] Stowell, 29.
[6] *The Local News*, July 2, 1892; *The Sun* (New York, New York), July 2, 3, 4, 1892

---

[7] *The Times* (Philadelphia, Pennsylvania), July 6, 7, 1892; *The Sun* (Baltimore, Maryland), July 7, 1892; Stowell, 30
[8] *The Sun* (Baltimore, Maryland), June 30, 1892.
[9] *Ibid.*; *The World* (Baltimore, Maryland), June 30, 1892
[10] *National Labor Tribune* (Pittsburgh, Pennsylvania), weekly, issued Thursday for last week of June 1892; *The Sun* (Baltimore, Maryland), June 28, 30, 1892; *The Post* (Pittsburgh, Pennsylvania), June 30, July 1, 2, 1892.
[11] Stowell, 30.
[12] *The Cleveland Leader* (Cleveland, Ohio), July 4, 1892); *The Cleveland Press* (Cleveland, Ohio), July 4, 5, 1892.
[13] U.S. Congress, House of Representatives, 52nd Congress, 2nd Session, *Report 2447: Employment of Pinkerton Detectives*, 197.
[14] Richard W. Rowan, *The Pinkertons* (Boston: Little, Brown, 1931). See also Barnaby J. Feder, "Securitas of Sweden to Acquire Pinkerton's in $384 Million Deal," *The New York Times*, February 23, 1999.
[15] Wolff, 69-70.
[16] *Ibid.*, 70.
[17] *The Sun* (Baltimore, Maryland), July 8, 1892.
[18] See McCandless v. Allegheney Bessemer Steel Company, Supreme Court of Pennsylvania, *Pittsburgh Legal Journal*, N. S. Vol. XXIII, O. S. Vol. XI, No. 25 (January 18, 1893), 223-228.
[19] U.S. Congress, House of Representatives, 52nd Congress, 2nd Session, *Report 2447*, 162.
[20] *The Sun* (Baltimore, Maryland), November 24, 1892; *The New York Times*, November 23, 24, 25, 26, 1892.
[21] U.S. Congress, House of Representatives, 52nd Congress, 2nd Session, *Report 2447*, XIV.
[22] *Ibid.*, 161-162.
[23] Stowell, 31.
[24] U. S. Congress, Senate, 52nd Congress, 2nd Session, *Report 1280: Investigation of Labor Troubles*, 186.
[25] Wolff,101-102; *Local News*, July 9, 1892, 2; *Sun* (Baltimore), July 7, 1892.
[26] Stowell, 31.
[27] *Ibid.*, 31-32.
[28] Wolff, 101-102; *The Cleveland Leader*, July 7, 1892; *The Daily Gazette* (Pittsburgh, Pennsylvania), Julv 7, 8, 1892.
[29] *The Sun* (Baltimore, Maryland), Julv 8, 1892; *The Local News*, July 7, 8, 9, 10, 1892.
[30] Wolff, 100-101, 103-104.
[31] Jeremy Brecher, *Strike!* (San Francisco: Straight Arrow Books, 1979), 58.
[32] *The Cleveland Leader*, July 7, 1892.
[33] *The Local News*, July 9, 1892.
[34] Wolff, 104; *The Sun* (Baltimore, Maryland), July 7, 1892; *The Mail and Express* (New York, N.Y.), July 7, 1892.
[35] *The Sun* (Baltimore, Maryland), July 7, 1892.

[36] Stowell, 31.
[37] *The Local News*, July 9, 1892; Wolff, 102-103; *The New York World*, July 6, 7, 8, 9, 1892.
[38] Wolff, 101-102.
[39] *The Cleveland Leader*, July 7, 1892.
[40] *Ibid.*; *The New York World* (New York, N.Y.), July 7, 8, 1892; *The Sun* (New York, N.Y.), July 7, 8, 1892.
[41] *The Cleveland Leader*, July 7, 8, 9, 1892; *The Sun* (Baltimore, Maryland), July 9, 1892.
[42] *The Sun* (Baltimore, Maryland), July 8, 1892.
[43] *Ibid.*
[44] Stowell, 32.
[45] *The Cleveland Leader*, July 6, 1892.
[46] *Associated Press Report* (Homestead, Pennsylvania: Postal Telegraph Company), July 6, 1892.
[47] *The Philadelphia Press*, July 6, 1892.
[48] *Ibid.*
[49] *The Cleveland Leader*, July 6, 1892; *The Chicago Tribune*, (Chicago, Illinois), July 6, 7, 1892; *The Times* (Pittsburgh, Pennsylvania), July 6, 7, 1892.
[50] *The Chronicle Telegraph* (Pittsburgh, Pennsylvania), July 6, 7, 1892; *The Daily Dispatch* (Pittsburgh, Pennsylvania), July 6, 7, 8, 1892; *The Cleveland Leader*, July 6, 7, 1892.
[51] *The Philadelphia Press*, July 6, 1892.
[52] *The Cleveland Leader*, July 6, 1892.
[53] *Ibid.*; Stowell, 33.
[54] Wolff, 96-98.
[55] Stowell, 33-34.
[56] *The Cleveland Leader*, July 6, 1892.
[57] *The Sun* (Baltimore, Maryland), July 6, 1892.
[58] *Ibid.*
[59] Stowell, 34.
[60] *The Cleveland Leader*, July 7, 1892.
[61] *Ibid.*
[62] *Ibid.*; *The Times* (Pittsburgh, Pennsylvania), July 7, 8, 1892.
[63] Stowell, 38-39.
[64] *The New York World*, July 7, 1892; *The Chicago Herald* (Chicago, Illinois), July 7, 1892; *The Cleveland Leader*, July 7, 1892.
[65] Stowell, 38.
[66] *Ibid.*, 39.
[67] *The Local News*, July 9, 1892.
[68] Stowell, 39-40.
[69] *Ibid.*, 40-41.
[70] *Ibid.*, 41.
[71] *The Cleveland Leader*, July 7, 1892.

[72] *The Local News*, July 9, 1892.
[73] *The World* (Baltimore, Maryland), July 7, 8, 1892; *The Sun* (Baltimore, Maryland), July 7, 1892..
[74] Stowell, 44.
[75] *The Globe* (Toronto, Canada), July 7, 1892.
[76] Stowell, 42-43.
[77] *Ibid.*, 43.
[78] *The Local News*, July 9, 1892; *The Chronicle Telegraph*, July 7, 8, 9, 1892.
[79] Stowell, 45-6.
[80] *The Local News*, July 9, 1892; *The Cleveland Leader*, July 7, 1892; Stowell, 46-74; *The New York Times*, July 7, 8, 1892; *The Sun* (New York, N.Y.). July 7, 1892.
[81] *The Globe* (Toronto, Canada), July 7, 1892.
[82] *The Local News*, July 9. 1892; *Freiheits Freund* (Pittburgh, Pennsylvania), July 7, 8, 9, 1892.
[83] *The Philadelphia Press*, July 7, 1892; *The New York Times*, July 7, 1992.
[84] *The Globe* (Toronto, Canada), July 7, 1892.
[85] Stowell, 48.
[86] *The Cleveland Leader*, July 7, 1892.
[87] *Ibid.*
[88]*The Local News*, July 9, 1892; *The Transit Journal* (Pittsburgh, Pennsylvania), July 8, 9, 1892; Stowell, 48-49.
[89] *The Sun* (Baltimore, Maryland), July 7, 1992.
[90]*The Chronicle Teleqraph*, July 7, 8, 1892; *The Post* (Pittsburgh, Pennsylvania), July 9, 1892; *The Local News*, July 9, 1892.
[91] *The Globe* (Toronto, Canada), July 7, 1892.
[92] *Ibid.*
[93] Stowell, 56-57.
[94] *The Times* (Pittsburgh, Pennsylvania), July 7, 1892; *The Transit Journal* , July 7, 1892; *The Sun* (New York, N.Y.), July 7, 1892; *The Sun* (Baltimore, Maryland), July 7, 1992.
[95] Stowell, 59.
[96] The *Milwaukee Daily News*, July 6, 1892.
[97] *The Globe*, July 7, 1892.
[98] Stowell, 52.
[99] *The Local News*, July 9, 1892; The Dailv Gazette, July 7, 1892.
[100] Stowell, 50-52.
[101] *Ibid.*, 49.
[102] *The Milwaukee Daily News*, July 6, 1892. See also *The Globe* (Toronto, Canada), July 7, 1892.
[103] Stowell, 49.
[104] *The Globe* (Toronto, Canada), July 7, 1892.
[105] *The New York Times*, July 8, 1892; *The Sun* (Baltimore, Maryland), July 8, 1892.
[106] Stowell, 58-59.

---

[107] *Ibid.*, 61.
[108] *The Cleveland Leader*, July 8, 1892.
[109] *The Globe* (Toronto, Canada), July 7, 1892.
[110] *The San Francisco Chronicle*, July 7, 1892.
[111] *The Sun* (Baltimore, Maryland), July 7, 1892.
[112] *The Cleveland Leader*, July 7, 1892; *The Times* (Philadelphia, Pennsylvania), July 7, 1892.
[113] Stowell, 59-60.
[114] *Ibid.*, 60.
[115] *The Cleveland Leader*, July 7, 1892; *The National Labor Tribune*, issued Thursday covering July 7, 1892; *The Globe (Toronto*, Canada), July 7, 1892.
[116] Stowell, 60-61.
[117] *The Cleveland Press*, July 7, 1892; *The Local News*, July 9, 1892.
[118] Stowell, 61.
[119] *The Herald*, July 7, 1892; *The Cleveland Leader*, July 7, 1892.
[120] *The Post*, July 7, 1892; *The Sun* (Baltimore, Maryland) July 7, 1892.
[121] Wolff, 122-129; *The Cleveland Leader*, July 7, 1892.
[122] *The Milwaukee Daily News*, July 7, 1892.
[123] *The Sun* (Baltimore, Maryland), July 7, 1892.
[124] *The Local News*, July 9, 1892; *The Mail and Express*, July 7, 8, 1892.
[125] *The Sun* (Baltimore, Maryland), July 7, 1892.
[126] Stowell, 65-66.
[127] *Ibid.*, 66.
[128] *The Milwaukee Dailv News*, July 7, 1892.
[129] *The San Francisco Chronicle*, July 7, 1892.
[130] Wolff, 127; Burgoyne, 83.
[131] Stowell, 63.
[132] *The Local News, July 9, 1892; The Times* (Pittsburgh , Pennsylvania), July 7, 8 , 1892.
[133] *The Local News*, July 9, 1892; *The Daily Gazette*, July 7, 892.
[134] Burgoyne, 84.
[135] *Ibid.*, 84-85.
[136] *The Sun* (Baltimore, Maryland), July 7, 1892.
[137] *Ibid.*; also see also *The Post*, July 7, 1892.
[138] Wolff, 128-129.
[139] *Ibid.*; *The Local News*, July 9, 1892.
[140] Wolff, 127-128.
[141] *Ibid.*; *The Local News*, July 9, 1892.
[142] Stowell, 64.
[143] *The Cleveland Leader*, July 8, 1892.
[144] *Ibid.*; see also *The Local News*, July 8, 1892.
[145] *The Cleveland Leader*, July 8, 1892; *The Local News*, July 8, 1892.
[146] *The Cleveland Leader*, July 8, 1892.
[147] *The Local News*, July 23, 1892.

[148] *Ibid.*; see also Stowell, 64ff.
[149] *The Local News*, July 30, 1892.
[150] Wolff, 130; see also *The Cleveland Leader*, July 7, 1892; *The Times* (Pittsburgh, Pennsylvania) July 7, 8, 1892.
[151] *The Sun* (Baltimore, Maryland), July 14, 1892; Stowell, 64.
[152] *The Sun* (Baltimore, Maryland), July 7, 1892.
[153] *Ibid.*; *The New York World*, July 7, 8, 1892.
[154] *The San Francisco Chronicle*, July 7, 1892.
[155] Wolff, 129-130.
[156] *The Local News*, July 9, 1892; *The Daily Gazette*, July 7, 8, 1892.
[157] Stowell. 64.
[158] Wolff, 129-130.
[159] *The San Francisco Chronicle*, July 7, 1892.
[160] *Ibid.*
[161] *The Local News*, July 9, 1892.
[162] *The Cleveland Leader*, July 8, 1892.
[163] Wolff, 129-130.
[164] *The Cleveland Leader*, July 8, 1892; *The Post*, July 7, 1892.
[165] *The Sun* (Baltimore, Maryland), July 8, 1892; *The World* (Baltimore, Maryland), July 8, 1892.
[166] *The Milwaukee Daily News*, July 7, 1892.
[167] *The Cleveland Leader*, July 7, 1892.
[168] *Ibid.*
[169] *The Sun* (Baltimore, Maryland), July 8, 1892.
[170] *Ibid.*
[171] *The Cleveland Leader*, July 7, 1892; *The Press* (Cleveland, Ohio), July 7, 1892.
[172] *The Sun* (Baltimore, Maryland), July 8, 1892; Wolff, 130.
[173] *The Sun* (Baltimore, Maryland), July 9, 1892.
[174] *Ibid.*; *The Chicago Tribune*, July 8, 9, 10, 1892.
[175] Wolff, 130.
[176] *The Milwaukee Daily News*, July 6, 1892.
[177] *Ibid.*, July 8, 1892.
[178] See McCandless v. the Allegheny Bessemer Steel Company, Court of Common Pleas No. 1. *Pittsburgh Legal Journal*, No. 32, March 9, 1992, 299-302.
[179] U.S. Congress, House of Representatives, *Report 2447*, XIII.
[180] U.S. Congress, Senate, *Report 1280*, 186 ff.
[181] *The New York World*, July 7, 8, 9, 1892; *The National Labor Tribune*, issued Thursdays covering July 8-12, 1892.

# Chapter 5

# Origins of the Homestead Commune

> It is part of the program of the strikers that the great Carnegie plant must be preserved intact and free from outside intermeddling or trespass. Their labor has built up this mammoth plant, they say, and their care must preserve it for the future support of their families when the battle shall have been won.
>
> – Report from Homestead, July 1, 1892[1]

> The proletarians of Paris, in the midst of the defeats and betrayals of the ruling class, have come to understand that they must save the situation by taking the conduct of public affairs into their own hands. . . . They have realized that it is their highest duty and absolute right to make themselves master of their own fate and to seize the power of the government.
>
> – March 18, 1871, Proclamation of the Central Committee of the National Guard, Defending the Paris Commune

The commune at Homestead was initially organized around the wage question, was polarized by preparations for a military confrontation, and was then regimented in the name of defense, self-sufficiency, democracy, and survival of the city's population.

Before the lockout, the strike and the battle with the Pinkertons, the central question for the Amalgamated at Homestead had been whether or not skilled labor would receive in added wages a portion of the increased output due to Frick's introduction of new machinery and the speedup of work. By 1892 the manufacture of steel had been revolutionized by the Open Hearth process and other inventions. Bessemer and rolling mill machinery hitherto employed had become obsolete, and the firm had spent several million dollars reconstructing and enlarging the Works.

Carnegie argued that since the new machinery turned out about 60 percent more steel than the old, the 200 to 350 highly skilled men who were paid by the ton of steel had increased their earnings almost 60 percent before the contract expired on June 30th. Carnegie also claimed he had offered to divide this 60 percent increased output with them in the new wage contract, so that the earnings of the men would be 30 percent greater than under the old scale. But he wanted the other 30 percent to go to the firm as compensation for its outlay. As well, Carnegie asserted that the men's work would not be much harder than it had been, because the improved machinery did most of the work. He argued his was a generous offer, and, under ordinary circumstances would have been accepted by the men with thanks.[2]

But the men were not grateful, feared Carnegie would seek to replace them with non-union workers, and sought means to impose their demands.

## The Commune and the Economic Question

Over the question of secure jobs, minimum wages on the sliding scale, and right of the Amalgamated to represent the workers, an armed commune was organized in Homestead on July 1st, 1892 – the very day Carnegie managers had rejected the workers' demands.

For some time Carnegie had planned to cut wage rates and eliminate the current minimum. He was determined to meet his competitors who had cut the market price of steel by lowering wages in any way possible. So when the workers at Homestead refused to allow reductions in both wage rates and the minimum, Carnegie managers locked them out. At that point, the workers declared a general strike and, fearing an invasion of Pinkertons and scabs, organized themselves into a commune under trade union leadership.

A militarized city directed by centralized workers' authority at first seemed a strange way to solve a wage matter. But fears of an armed encounter heavily overlaid the initial economic demands – knowledge of past Carnegie strike-breaking antics fueling steps towards organizing the commune. There was something deeper than wage cuts, too, as Carnegie typified employers interested in maximizing their profits by underselling competitors, made possible by introducing the latest machinery - displacing some workers, cutting wage rates of others, and eliminating the skilled workers' union.

According to one prominent Pittsburgh steel manufacturer, the employers had been "confronted with conditions which never before menaced Pittsburgh," explaining: "The conditions governing the trade have entirely changed. We have competitors now which we never had before, and they are selling iron cheaper than it has ever been sold."

To compete, he reasoned, wages had to be kept in line, for the main question weighing upon Pittsburgh mill owners was comparative wage rates: "How can we continue to pay $3.50 for puddling when in the Eastern mills the puddlers work for $3.60 per ton and [from this fund] pay their helpers? They also pay their finishing men less than we do by 25 percent. It is simply out of the question to pay the Amalgamated scale and compete with the East."[3] Since the employers were not about to reduce their profits, and the workers refused the pay cut, the situation was polarized. Labor began preparing for long strikes while, on their side, the employers planned protracted lockouts.

At Homestead, manager Frick issued his ultimatum that either the union agree to a new scale lowering the men's wages by about 18 per cent, or he would only deal with the workers as individuals, not with the Amalgamated representing the workers. It was clear that Frick wanted to dictate wages, and to destroy the union as well. Thus, on June 28th, he proceeded to lock out 800 workers by closing down some departments; and on July 2nd he locked out the rest.[4]

The Amalgamated Association responded immediately by organizing the entire Homestead population. By July lst, a subcommittee from the Amalgamated lodges had already taken steps to control any unplanned workers' outbursts against the Carnegie Company and its managers. Even greater control was exercised through an Advisory Committee acting as the central body of the Amalgamated Association. This Committee, initially formed to settle the question of wages with the Company, continued its existence after Frick refused to confer with any representatives of the Amalgamated.

Unable to negotiate a settlement, the Advisory Committee had initially selected three (then raised to five) delegates from each of its eight Homestead lodges for the purpose of recruiting those outside the union to support the strike. This group then became the new central Committee and immediately went into action.

Even before the last of the workers were fired, this Committee had met to make arrangements for protection of their interests during the "wage war." Their organization was thorough, consisting of committees and subcommittees for the preservation of order, regulation of the sale of liquor in saloons, control of the press, and protection against outside, nonunion labor.[5] The regulation subcommittee began by visiting all saloons to request the proprietors not to allow excessive drinking and to use their best efforts to keep activities in bounds. All saloon keepers promised to keep things quiet, some expressing a willingness to close entirely if needed.[6]

Frick in turn expressed the Company viewpoint: "Finding that it was impossible to arrive at any agreement with the Amalgamated officials, we decided to close our Works at Homestead. Immediately the town was taken possession by the workmen. An Advisory Committee of fifty took upon itself the direction of the affairs of the place; the streets were patrolled by men appointed by the Committee and every stranger entering the town became an object of surveillance, was closely questioned, and if there was the slightest reason to suspect him he was ordered to leave the place instantly under threat of bodily harm. Guards were stationed at every approach to Homestead by this self-appointed local government. Our employees were prohibited from going to the mills and we, as owners of the property, were compelled to stand by powerless to conduct the affairs of our business or direct its management."[7]

We may disbelieve Frick on at least one major point: The regular Homestead workers and the population no longer desired to go into the mills to work for Mr. Carnegie. Rather, the workers had begun to fraternize and express solidarity in a way that had never before taken place. Homestead's Steel Works were so large that, in the past, men working in one department scarcely ever met those in other departments. But by July 2nd they were introducing themselves all around, and improving their acquaintances with each other.[8]

The workers were brought together not only by their opposition to Frick's firing and lockout policies, but also by the Advisory Committee's resolution demanding that all nonunion employees at Homestead join the Amalgamated men by refusing to return to work by July lst. This resolution passed overwhelmingly at a packed meeting of some 3,000 union and nonunion men in the Opera House, binding all to strike together and withhold their combined labor from the mill.

Typically, Frick responded with wrath and fired all Homestead workers, thereby pressuring the Advisory Committee to make immediate decisions to protect its members. Before long, the Committee had extended its power, acting for laborers citywide, and providing protection to all of the city's workers and their families.

In fact, the Committee now functioned as the central committee of the Homestead commune, while within this central committee the Amalgamated's Hugh O'Donnell emerged as the principal leader, empowered to issue mandates that became the laws regulating the city. Thus, against Frick's rigidities, the usually even-tempered O'Donnell declared the Advisory Committee had organized to deal with an economic question that the Company had transformed into a potential source of armed struggle:

> The Committee has . . . decided to organize their forces on a truly military basis. The force of four thousand men has been divided into three divisions or watches, each of these divisions is to devote eight of the twenty-four hours to the task of watching the plant. The Commanders of these divisions are to have as assistants eight captains composed of one trusted man from each of the eight local lodges. These Captains will report to the Division Commanders, who in turn will receive the orders of the Advisory Committee.
>
> During their hours of duty these Captains will have personal charge of the most important posts, i.e., the river front, the water gates and pumps, the railroad stations, and the main gates of the plant. The girdle of pickets will file reports to the main headquarters every half hour, and so complete and detailed is the plan of the campaign that in ten minutes' time the Committee can communicate with the men at any given point within a radius of five miles.
>
> In addition to all this, there will be held in reserve a force of 800 Slavs and Hungarians. The brigade of foreigners will be under the command of two Hungarians and two interpreters.[9]

This was a fine way to run a military body, but a poor one to operate a commune, for no democratic body had ever given the Advisory Committee of Amalgamated men, or its leader O'Donnell, authority to dictate maneuvers.

Inexperienced in such matters, few people in Homestead had thought much about the implications of organizing themselves in armed divisions simply to secure wages and jobs. After all, Frick had built his fort around the mills and prepared for a military campaign, so it was perfectly logical that the Advisory Committee should respond to Fort Frick in kind. Whatever military maneuvers the central committee planned at this point, the workers simply went along.

The Committee directed that the rivers be patrolled and watched day and night by some forty skiffs, and that the steam yacht *Edna* be used to guard both sides of the river. The Committee also called upon experts, with the river patrol directed by one man with considerable experience, assisted by a large number of others.

The Committee then decided that Headquarters should be secured by moving to Eighth Avenue. This office was to be fitted to suit the needs of the Association, with thirty incandescent lights and a private telegraph wire, so as to put the Association in communication with every city, town and hamlet in the United States.

In addition, the Committee decided that 20 members of the Amalgamated Association were to act as special officers to police the workers themselves. Whereas previously the lockout had led workers to hang effigies of Frick and Potter from the Works, now, under the Committee's orders, all effigies were to be cut down, with the special officers empowered to prevent the hanging of others. The Committee also imposed more stringent rules on the saloons: keepers were ordered to regulate their sales, keep all mobs out, and close up entirely if circumstances made it necessary.[10]

Following the strike day of July lst, the Committee forced the saloons to close at 8:00 P.M. and to remain closed on Sunday, the 3rd. Since all the employees were ordered by the Works to pick up their last paycheck with their notices of discharge Saturday, the Committee prevented them from spending their "wad" at the usual royal celebration Saturday night. "That there has been no serious trouble since the struggle was inaugurated is due to the splendid organization of the workers," the *Cleveland Leader* concluded.[11] This was the work of the leaders of the Amalgamated Association, who remained at the center of control over the Advisory Committee, which in turn controlled the military line organization headed by the Amalgamated lodges in maintaining control over all workers, and, through them, the Homestead family structure and the city's population.

By July 3rd, "Quiet reigned supreme in Homestead and nothing occurred to disturb the usual Sunday serenity," one reporter noted. "The workers' organization in Homestead is the most thorough ever known here. They have divided the watches into three relieves of eight hours each and their guards so completely surround the big steel plant that it would be impossible for anyone to get inside without their knowledge."[12]

In fact, following the orders of this leadership pyramid after the Pinkerton raid on July 6th, the people held the city in a state of siege. According to one paper:

> The town is picketed and no man can enter without his presence being noted. A cordon of watchful and intensely suspicious workmen are around the entire city. Every road is guarded. Along the river, above and below, on the bank and on the opposite bank, a ceaseless patrol is maintained.
>
> No boat, no party of men can come along without being followed, and if it is thought worthwhile, questioned. Railroads are watched.
>
> Two lines of roads, the Pennsylvania and the Pittsburgh and Charleston, enter the place, and as dusk fell, a tour along the tracks showed that at nearly every street crossing were from half-dozen to two score of men up and down on the track; others were going and coming. The men intend that the Works shall not be taken by surprise. They are stronger now in number and death-dealing equipment.[13]

## The Armed Commune

The workers of Homestead, armed and determined to settle their dispute with Carnegie without outside interference, made their own laws and regulations and recognized no authority higher than themselves. For the moment the town was under their own martial law, not that of the law of the land.

This commune was armed for two central purposes: first, to protect the Amalgamated in ultimately resolving its economic disagreement with the Carnegie Company; and, second, to counteract Frick's attempts to turn the mill into a fort and bring in Pinkertons to protect nonunion workers. When the threat of the Pinkerton invasion materialized and, later, when a series of new invasions seemed likely, Amalgamated leaders expanded and rigidified their control apparatus and, out of the seeming necessity of defense, began to police the liberties of the very population they were supposed to protect – justifying their actions by reiterating, and making a shibboleth of, the original economic goals of the commune.

From July 2nd to the 10th, commune authority held sway. The Sheriff was powerless, the Governor had not yet intervened, and the armed guards of Homestead patrolled the streets night and day. Every approach to the town and the Carnegie Works was watched. Every stranger was compelled to give a satisfactory explanation of his presence and business in Homestead or was escorted to the next outgoing train and warned it would be dangerous to return.

The strikers were also prepared for further battle, being drilled and equipped with Winchesters, revolvers and shotguns. Guns were carried openly in the streets, and the organization of more than 4,000 workers by the Committee left every man and woman determined to fight to survive. Desperate after the first Pinkerton attack, they would give no quarter to another force of Pinkertons. "The mill men," a reporter depicted their state of mind, "will die or win if there is another battle, and there is every indication there will be one and soon."[14]

The traditional, local legal authorities were powerless to restrain the strikers and, in fact, sided with the mill men. McLuckie, as Burgess and the highest officer of Homestead, was openly in sympathy with them. Also supporting the armed commune were the chief and two patrolmen of the regular police force. The twenty additional policemen sworn in by McLuckie were not only with the mill men, but many of them were themselves mill men thrown out of work by Frick's lockout. Some members of the Town Council were also strike leaders, and those who served on important city committees distributed ammunition on the night of July 9th when the Pinkertons were again expected.

"Our men are well armed and know how to use their arms," said a prominent citizen while distributing weapons. "We have plenty of ammunition and our men are desperate. Some are aching for an encounter. There are plenty of relatives and friends of the boys who fell last Wednesday who are thirsting for revenge. If the Pinkertons come or scabs are put in the Works they will get it."[15]

The workers at Homestead were, not alone in building a popular, armed body to protect themselves against an avaricious employer and his hired gunmen. As a result of the events at Homestead, several state legislatures immediately moved to protect labor against the Pinkertons. By July 8th, two days after the Homestead invasion, workers in many places in the nation began to discuss union organizing of military companies to oppose the Pinkertons in future fights. In Chicago, some 500 men enrolled in an open military organization. On July 9th, the Kentucky Legislature passed resolutions expressing sympathy with the Homestead strikers, denouncing the Pinkertons and introducing three anti-Pinkerton bills. On the same day, in Denver, Colorado, Governor Routt, now a labor supporter, refused to renew the license of the Pinkertons sent during the strike at Crested Butte and at Burlington.[16]

Chicago's Typographical Local No. 16 offered "hearty endorsement of the recent suggestions that workingmen organize military companies to protect themselves against the invasion of the . . . Pinkerton detectives and the disastrous effects of the presence of these unscrupulous and unpatriotic mercenaries in the times of labor troubles."[17]

Workingmen everywhere were now being enlisted in armed battalions against Pinkertonism. "There appears to be a determination on the part of organized labor to form military companies within their ranks to combat Pinkertonism when the next clash between capital and labor occurs," penned the *Sun*, adding that in Chicago: "The leaders of the various labor organizations . . . when questioned in regard to the matter all confirmed the rumor that the working people intend to organize themselves."[18]

Said Ben Garland of Chicago's Gas Fitters, Union:

> We propose enlisting and arming 1,000 men. We want Jas. O'Connell, president of the Building Trades Council, as colonel. I can raise a company easily enough. So can A. C. Boyd, business agent of the Building Trades Council. We will open our enlistment papers tomorrow. We have already circulated papers and a number have signed the papers. We will organize companies of 100 men. They will be armed, drilled and equipped. We will join the State Militia if necessary, but the prime object of our organization will be to fight the Pinkertons whenever they are called in to fight workmen. We will send a petition tomorrow to the Secretary of State asking for a charter or license to organize and arm as an independent military body. As soon as we get that, we will arm ourselves and begin drilling.[19]

From Boston came the same response. A secret organization of workmen called *Our Own Club* was formed to give union men military training, their oath of allegiance stating that the time had come when the working people of the country had to prepare to overthrow the Pinkertons or any similar body of what they termed "hired assassins and mercenary desperadoes." "It is the opinion of these men," reported the *Boston Herald*, "that the large capitalists of this country have concluded to destroy all labor organizations in order that reductions of wages may be enforced without protest from the wage earners. Thus the workers claimed that the New England Granite manufacturers' Association was in sympathy with Carnegie, Frick and Pinkerton, and its members would take similar action to that at Homestead in the granite centers if they had the courage."

Members of *Our Own Club* asserted that the legislation enacted against the Pinkertons by Pennsylvania and other states was of no practical value to union workers. A Pinkerton army, they maintained, could be placed in any state by the capitalists regardless of any law. So, as a defense, the *Club* established subcommittees to organize branches of the society in every industrial center of the Commonwealth of Massachusetts, with membership to be comprised only of wage-workers who could answer relevant questions.[20]

What happened at Homestead, Chicago and Boston, then, were similar responses leading to popular organization of militia, though limitations inhered in the goals sought under the union leadership groups of each city. For it had become clear that trade union leaders seeking to militarily protect the wages and jobs of skilled workers against employers' attempts to break unionism with Pinkertons could only lead to armed conflicts over these narrow issues.

## Commune Structure, Trade Union Goals

Could the commune at Homestead go beyond this singular vision?

Certainly the struggle over the issues of wage levels and jobs had led to the establishment of the armed, citywide commune and made it a popular organization, with the support of almost every working class family in the city. Nonetheless, from the very start it was not democratically organized, and the craft-orientation of its leadership fettered both a wider imagination of the possible and the storming of tradition.

The leadership of the commune had emerged from the upper ranks of the highly skilled workers in the Amalgamated Association. These were craft union leaders seeking tradition goals of secure jobs, control over the immediate labor process, surveillance of the introduction of labor-displacing technology, and high pay for their members. They accepted the wage system as well as Andrew Carnegie's legal right to own and control the mills of Homestead, conditioned only by their firm belief that they had an irreplaceable role in production and thus a vested interest in the continued operation of the mill.

But how would they assert that interest? And would the executive committee of the union, rather than a democratic, grass-roots body, best represent the workers' demands?

The structure of the commune reflected the organizational structure of the Amalgamated Association. In the beginning, the Advisory Committee thus consisted of the president and two members of each of the eight Amalgamated lodges at Homestead. Later, additional lodge members were added. But their function was to direct the workers both inside and outside the union, assigning them duties and seeing that they kept sober and orderly, acted as a unit, and served the general interests of the workers.[21]

By controlling these male mill workers in Homestead, the Advisory Committee also gained influence over their families. By and large, the males in each family over the age of 12 or 14 were the only wage earners, and they frowned on women holding jobs in the mill because women were likely to undercut the men's high wages. Dependence on male wage-earners made both women and children reliant on the men and more or less compliant to their decisions within each family.[22] And this reflected their deeper fears – that, without the jobs ands wages of their bread-winners, there would be neither bread nor any other means to buy necessities.

These dependent and fear-driven relationships also provided the Amalgamated leaders making up the Advisory Committee with a powerful chain of command: skilled union men at the top; nonunion unskilled and mechanics on a second tier; foreign day-laborers on a third; and members of working class families on a fourth level. This organization enabled the Committee to give general direction to the flow of events that affected the entire Homestead population.

In using this structure, the Committee's major goals were to secure jobs and high wages for the skilled workers and, toward this end in defending the city, to act as a center of public safety. The Committee's organization of the commune along military lines, as well as its direction of militant and sometimes brutal methods against Frick's manipulations and his use of Pinkertons, were limited to these objectives. So the prevailing mentality and consciousness was that of craft unionists with special self-serving goals, reflecting the stage of development of the skilled work force and its elevated position in the production process.

Immersed in the traditions of American patriotism and democratic forms which these skilled believed helped to secure their rights, other goals were kept from becoming issues in the struggle simply because the Advisory Committee was organized around limited trade union demands.

**Accepting the Wage System**

The emerging commune thus set its sights on future negotiations with Carnegie over wage rates, secure jobs and the setting of production rules that would allow workers to control immediate production tasks.

The Amalgamated and Advisory Committee were neither revolutionaries seeking government power beyond the local community nor anarchists attempting to destroy the state and assassinate its representatives. They had no goal to use the strike or commune organization to take charge of the ownership of Carnegie's mills. They granted the future power of Frick and his managers to direct the overall production and marketing of steel. And they completely accepted the position of their members as Carnegie employees, entitled to jobs and good wages under the American

Federation of Labor standard of a fair day's pay for a fair day's work. Yet they sought to maintain the workers' power to set the pace of production and contested Carnegie's right to cut wages and eliminate the workers' skilled jobs by introducing technology without their consent.

Like the Amalgamated leaders, most of the mill workers had a firm belief that wage system benefited them and that their employer was entitled to ownership of the Works.[23] Though after repelling the Pinkerton raid, the workers at Homestead accepted with some pride the support of Socialist workers from Chicago for their heroism in holding off the invaders, they did not share the Socialist's hope that they would "continue to hold the fort until they own it."[24]

Yet, hold the fort that Frick built they did. On July 10th, the strikers permitted the few clerks and other employees still working at the mill to pass in and out unmolested, but they consciously held the Works for eventual return to and benefit of the owner. By refraining from placing guards within the Works, the strikers affirmed Carnegie's ownership and thus claimed that there was no excuse for intervention by State troops. Though they had no qualms about defeating the Pinkertons by any means in the civil war with Carnegie, they would preserve his Works at all costs, unwilling as they were to apply their military might with a political vision that they themselves could operate, own and defend the plant without the direction of Frick and his administrators.

"If they decide to blow up the Carnegie bridge from the Works to the Monongahela," mused a reporter, "they will do it even if a trainload of Pinkertons should happen to be crossing at the time. If they decide to demolish the Works it is in their power to do it, but there are too many sensible men among the strikers and they have too much interest in the preservation of the Works for any such step."[25]

Indeed, the strikers believed it was in their interest to preserve the plant which offered them jobs, enabling them to earn their livelihood, maintain their homes and support their families. They knew no wages would be paid if they alone operated the mills, which they had fought to keep closed to bring Carnegie to their terms over rates. "It is part of the program of the strikers that the great Carnegie plant must be preserved intact and free from outside intermeddling or trespass," an observer wrote. "Their labor has built up this mammoth plant, they say, and their care must preserve it for the future support of their families when the battle shall have been won."[26] They believed, then, that Carnegie would again employ them at wage levels they imposed by military force.

Many Homestead workers also locked their vision into the patterns of the past, having purchased their homes near the Works with the understanding that they were assured of their positions at the going wage. They felt they had no choice but to stay and try to maintain these wages by sheer force. Otherwise, one paper pointed out, they would have to accept Frick's terms simply because their homes were in Homestead.[27]

Taking a slightly different perspective, Pennsylvania Governor Pattison believed these factors would make the workers compliant. "The residents of Homestead are a law-abiding, intelligent and orderly people, and owners of their own homes," said he. "In such a community there ought to be no difficulty in adjusting their differences and certainly none in keeping the peace and protecting property. I believe that wise and conservative counsel will prevail and the trouble will be settled."[28]

Fearing the Militia, Homestead's local papers asked which would benefit the locked-out men the most - to have the troops come into the city and take possession of the Works, or to have the strikers hold the Works themselves? They raised the possibility that if the troops had possession of the mills, not a single man would be able to enter until a settlement was made between the Company and the strikers. They also reported they had reliable information that the troops would not only protect the Company property, but would ensure the safety of nonunion workers inside the mill. It now became clear to the people of Homestead that the Militia could play the same role as the Pinkertons in supporting the Company against the workers. This realization is what eventually led to popular hostility against the presence of Pinkertons, deputies, soldiers and newsmen alike.[29]

**The Commune and Freedom of Criticism**

The militarized Advisory Committee, which took over the powers of government in Homestead in the name of the workers and the population at large, proceeded to regulate all public affairs and concentrated on the defense of the city against outside dangers and inside subterfuge. In fulfilling this mission, the right of correspondents to freely document, discuss and criticize the commune was sharply curtailed through the Advisory Committee's Subcommittee on the Press. Early freedoms taken by correspondents were closely regulated, only to be later liberalized after the press castigated the restrictive policies of the commune under trade unionist O'Donnell.

When the Militia later replaced the central committee and its subcommittees and established the autocratic *régime* of General

Snowden, the press was given the authority to say anything it wished about Homestead and its plight. Thus, though the commune had sought to protect itself by regulating freedom of criticism, it ended by being exposed to the most virulent condemnation.

The initial reaction of the leaders of the commune to the press was one of healthy distrust of a media steeped in opposition to any aspect of popular control over American institutions. In past strikes and dealings with the employers, the views of the officials of the Amalgamated Association had been distorted by reporters and courts alike.[30] Determined this time to control what was written about the Homestead strike and lockout, on July lst the Advisory Committee set up a subcommittee to keep watch over all reporters entering Homestead as well as what they wrote and the political purposes their writing served. What the Committee leaders feared most of all was that their views would be distorted, and the resulting negative public opinion would permit Carnegie, the politicians of both major parties, and the military to crush the commune ideologically as well as with arms.[31]

Even before the Pinkerton raid, the workers had become suspicious of outsiders. They had expected the Pinkerton mobilization for several months, realizing that Frick was preparing to break the Amalgamated[32] And so every stranger, whether a real reporter or otherwise, was viewed with extreme reserve.

Reporters quickly learned that in Homestead a press badge issued by the relevant subcommittee of the commune was essential for mobility in the city. Detention of reporters occurred frequently, and not everyone was freely issued a press badge. Then, on July 6th, the day of open warfare, press badges were given out only after an examination of each reporter. Even the day after the Pinkerton raid, reported the *Cleveland Leader*: "The newspaper men could get but little news. Each one was suspected of being other than he was. Some of them took the precaution to establish their identity and secure credentials in the shape of a stamp of the order. These were very useful documents to those who had secured them before the last night's fight, for it was necessary then to show what right you had on the premises or 'git.'"[33]

The next day, on July 7th, the leaders of the commune also threw the "Anarchists" out of the city, thereby attempting to prove their desire for orderly negotiations with Frick, loyalty to the wage system, and efforts to protect the Carnegie mills.[34] Although the strikers themselves had used destructive means against the Pinkertons, they now had no intention of allowing the "Anarchists" to distribute a call for resistance to capital, for making the Carnegie mills the property of the workers and people of Homestead, and for socialism in one city. "Although the strikers had been liberal in their use of dynamite in dislodging the enemies who had carried death and destruction into their ranks," eyewitness Stowell wrote, "they at once rejected the overtures of the Anarchists, whose profession teaches the use of this destructive agency, and with considerable difficulty the cooler heads saved [Anarchist John] Powers and his companion from the wrath of the indignant strikers after they had realized the full import of the 'address.'"

Powers and his unnamed "assistant" had suddenly appeared in church-filled Homestead with their literature, announcing that "Resistance to tyrants is obedience to God!" But their distribution of this tract calling for capital's overthrow "almost proved their ruin, for the excited crowd demanded that they suffer the penalty of death," Stowell reported from Homestead. They were taken by "wiser counsel" to the lock-up, where Powers renounced his role, saying he "had found the bunches of circulars in Homestead, and supposing that they belonged to the strikers, he had proceeded to scatter them among their original owners."[35] These "owners" either never saw the literature or, if they read it, did not accept it, or were influenced by the commune leaders to reject it – the leaders of the commune seeking above all to strictly control the central thoughts regulating the life of the city.

This was just the beginning of Homestead's control over the distribution of literature and the press. Four days after the Pinkerton raid, restrictions on the press assumed major proportions. The commune *Subcommittee on the Press* based its censorship on a close accounting of what reporters had sent back to their newspapers. Every hour the mails brought in the latest editions of the prominent newspapers of the country, and members of the *Subcommittee* poured over them to see how events at Homestead were-being reported.

The town's hostility to correspondents meanwhile increased. With sentinels everywhere guarding against surprise, the workers were ready to rise at the slightest provocation. and reporters were viewed as agents of espionage.[36] Newly arriving newspapermen were marched off by guards to Amalgamated Headquarters to identify themselves to the satisfaction of the men in charge, an ordeal often taking as long as two hours.[37]

Since the workers continued to look on the pressmen with suspicion and antagonized them, a few days after the battle with the Pinkertons some reporters began to complain to the Advisory Committee that their rights were being infringed. The *Subcommittee on the Press* then had large white badges printed, invited the press to meet at Headquarters and handed out the credentials. Hugh O'Donnell himself handed each reporter a badge after taking down his name and the paper he represented.[38]

But this did not prevent harassment. Though the metropolitan press of the entire country was thoroughly represented, though these gentlemen allegedly did not manufacture news, but faithfully chronicled every event in its true light, the Homestead residents remained suspicious, subjecting every reporter to unfavorable criticism and uncomplimentary comments as he walked the streets, attending to his business.

Several reporters were arrested at different times. one was run out of town at midnight and compelled to walk to Pittsburgh. Too, the commune's bureau of press censorship subjected the copy of every correspondent to the closest scrutiny, before he was allowed to put it on the wire. "These gentlemen," wrote an eyewitness, "were on the *qui vive* during the last day of control of the Advisory Board, the generally perturbed condition of the town and people combining to keep them very busy."[39]

The procedure was clear-cut: All newspaper men were either issued press badges by the *Subcommittee* or picked up by street guards and expelled from Homestead. Immediately after arriving, "unbadged" correspondents were first taken to Headquarters. Well-dressed commune policemen met the trains, asked reporters their business in town, and marched them to the *Subcommittee* offices for identification and to obtain badges. "The whole thing was done so quietly and with such a complete air of good humor that it was some time before the reporters imagined that there was anything like duress intended," noted one reporter coming to Homestead. "Even then it only dawned upon them gradually."

The head of the central committee, Hugh O'Donnell, was also the key member of the *Subcommittee on the Press* and was the only one empowered to issue press badges. When one little party of pressmen was brought to Headquarters on July 10th, they round O'Donnell absent, resting from his activities of the night before. Yet they had to wait for his personal approval of their presence in Homestead. "Nobody else could issue the badges," mused one reporter, "and the gentlemen were advised that it might be imprudent to go away from Headquarters without their little squares of white calico."

"'Of course, you can go if you wish,' said Mr. Crawford, one of the committee, 'but you may be subject to annoyance, and I would advise you to stay here until O'Donnell has seen you.'"

Tongue in cheek, this reporter added:

> The gentlemen all agreed to stay, and a few moments later some of the old reporters gathered in and warmly recommended the most exact obedience to any advice tendered by the committee. The duress lasted about two hours before O'Donnell appeared, and even then it took more than half an hour for the badges to be awarded, the committee being in anything but an amiable mood; one of them suggesting that the move of identification was so loose that if the Pinkertons were not fools they would be well represented among the so-called reporters.[40]

The *Committee* conceded that there were at least a dozen Pinkertons in Homestead and thus explained the stringency of their press regulations.[41]

But there was more to it than that. Under O'Donnell's press regulations the Advisory Committee was able to prevent criticism of their rule in Homestead. Only reporters with badges were allowed to remain in the city, and O'Donnell personally handed out no more than 48 badges for identification during the early days of the commune.[42]

**Leadership Vacillations; Rank-and-File Goals**

This proved unworkable, though, culminating in a split between O'Donnell and the rank-and-file, in turn leading to the initiation of a new policy of rigidity towards the press.

Thus the first policy of caution begun before July continued after the battle of the 6th, though the press had favored the workers' position, was largely opposed to the Pinkertons and called Frick's methods lawless. During this period the commune under O'Donnell's cumbersome subcommittee placed the press under more or less regulations. Thereafter, the second period began when some newspapers turned against the militant and radical methods of the commune; and during this period the rank-and-file itself took charge of trying to keep out certain reporters and papers.

July 9th provided the transition to another period, when the representatives of the Pittsburgh *Dispatch* were notified by the strikers to leave Homestead at once. Reports printed in the *Dispatch* had been unfavorable to the strikers and were the object of sharp criticism by the rank-and-file. One of the *Dispatch* reporters, Robert Herbert, had already been under the workers' suspicion for several days for sending critical reports; and on the morning of the 9th he was escorted to the Borough limits and told in no uncertain terms not to return. "The men in possession of the Homestead mill are not allowing anyone to enter today, not even the firm's representatives," wrote a *Milwaukee Daily News* stringer, "and it is said it is intended to exclude the newspaper reporters, as has already been done in the case of the Pittsburgh *Dispatch* men."[43]

This exclusionary policy did not make liberal newsmen feel any better than those who did not like the commune, and by mid-July heavy criticism was levied by many hitherto sympathetic reporters. When O'Donnell finally realized that his efforts to censor the press dispatches had led to sharp criticism of the Advisory Committee and his *Subcommittee on the Press*, he attempted to ease up the requirements and stop the workers from harassing reporters. "Everything that can be done by Hugh O'Donnell to protect the newspaper men from harassing interference has been done," a reporter noted the about-face, "and it is safe to say that so long as he is the dominant spirit no more reporters will be ordered out of town."

But O'Donnell's new liberality with the press was opposed by the rank-and-file, who were now determined to keep all outsiders under control. "The spirit of tolerance which he evinces is not shared by many of the rank-and-file," this reporter explained. "The great majority of the strikers are still disposed to regard the newspaper men as interlopers. . . ."[44]

The split widened as the days passed and a countdown to occupation by the Militia began. Additional press badges were given out by O'Donnell, mounting to a total of 110 during the entire lockout. And, after the arrival of the Militia, at least 25 pressmen who came to Homestead functioned without badges under the new rulers of the city.[45] But in the interim, the rank-and-file workers called for O'Donnell and his *Subcommittee* to tighten control on the press in the name of survival of the commune and its goals.

Against them traditional libertarian arguments were put forward by national pressmen and publishers trying to further ease restrictions by claiming the commune leaders were agents of their own kind of espionage on establishing autocratic government. They should, these conservatives said, control Homestead workers who censored the rights of correspondents. The *Leader* of Cleveland, Republican to the end, attested to O'Donnell's loss of control over the commune:

> A half dozen new arrivals today were promptly approached by a self-constituted committee and marched off to the strikers' Headquarters, where they were forced to remain an hour or two until they satisfied the men as to their identity. Of course all of this is irritating to the newspaper men, and when the strikers make their habitual assertion that peace

reigns in Homestead and perfect freedom maintains on every hand, the victims of their martial law methods wonder how, if that be true, they should have been forcibly marched away by the strikers' committee and deprived of their liberty for an hour in violation of the laws of Pennsylvania and in defiance of the constitutional liberty of the press.[46]

With their main focus on defense, freedom of criticism was never fully allowed by the rank-and-file, their arrest committees, or the subcommittee in charge of the press. Many representatives of the press in Homestead became cautious about what they wrote for fear of being kept out of the city or having to reckon with an irate committee with a different set of goals and ethics. The commune members, one reporter said, "while clamoring to be released from the oppression of capital, do not show such a fine appreciation of liberty when another's rights are to be observed."[47]

The Associated Press went on record saying that it was only when temerity outweighed discretion that correspondents in Homestead could exercise that perfect freedom of news expression which would characterize their dispatches from any other point, and which even the gravity of the situation in Homestead should not obscure.[48]

"The strikers assume not only to run the whole town and to control the movement of boats on the river," the Cleveland *Leader* editorialized, "but they have now driven several newspaper correspondents from the place. A censorship of all dispatches sent out of Homestead may be the next step. Still these people claim that they are perfectly law abiding."[49]

This barrage of criticism began to have its effects. O'Donnell, hoping to appease public opinion about the justice of the struggle at Homestead, had already relented and loosened restrictions on the press, driving a wedge between himself and the rank-and-file, whose members became ever more suspicious that outsiders would undermine their goals.[50]

And once the Militia marched into Homestead, General Snowden voided the cotton badges issued by the advisory Committee. His first general order was to give special passes good at all hours to all reporters bearing proper credentials, an eyewitness noting "the correspondents heartily welcoming the new reign of the state."[51]

**Preliminary Lessons of the Commune**

Towards those who questioned the wage system and Carnegie's

ownership of the mills, the Advisory Committee acted as a regulatory body, policing and repressing them and their views. But such regulations had limits, and, later, when the rank-and-file to refused to be bound by the Committee's promises to the Sheriff and the Governor to bring in the Militia, the Committee would lose control, refuse to be responsible for rank-and-file actions, and become increasingly less responsive to the needs of the population.

In fact, by the time the Militia entered Homestead, general disorganization and confusion prevailed among the populace. The self-appointed Advisory Committee spoke for no one save itself. The rank-and-file were leaderless and so disoriented that they could be controlled by the soldiers who, in contrast, were organized under strict discipline. Among the various groups of workers comprising the skilled, the mechanics, the unskilled, and the, day laborers, individuals already had begun to look to their own family needs for future security, thereby undermining the solidarity that previously had taken priority over their individual interests.

Two basic weaknesses in the organization were to become evident as events unfolded during the ensuing weeks. First, the Advisory Committee became isolated as the elite central committee of the commune with its self-appointed craft union leaders, who, accepting the tenets of the wage system, reaffirmed Carnegie's ownership rights in the mill. Given their beliefs, they limited the issues in the struggle to jobs, wage levels and recognition of the Amalgamated as the representative of the skilled workers. And second, hinged on military defense in pursuit of these limited goals, they played an undemocratic, dictatorial role, separating themselves from the rank-and-file workers and the people of Homestead. They failed to broaden the struggle to encompass the needs of other sections of the population or to include political goals that questioned the underlying inequities of wage system itself. These weaknesses were eventually to lead to the breakdown of the commune.

## Notes

[1] *The Cleveland Leader*, July 10, 1892.

[2] Andrew Carnegie, *Autobiography* (New York: Houghton Mifflin, 1920), 223.

[3] *The Cleveland Leader*, July 2, 1892.

[4] Burgoyne, 27-39.
[5] *Ibid.*, 39-40; *The Cleveland Leader*, July 2, !892.
[6] *The Local News*, July 2, 1892; *The National Labor Tribune*, July 1, 2, 1892.
[7] *The Sun* (Baltimore, Maryland), July 8, 1892.
[8] *The Local News*, July 2, 1892; *The National Labor Tribune*, July 1, 2, 1892.
[9] Wolff, 90.
[10] *The Cleveland Leader*, July 2, 1892; see also *The New York Times*, July 2, 3, 1892; *The Herald* (New York, N..Y.), July 2, 1892.
[11] *The Cleveland Leader*, July 2, 1892.
[12] *Ibid.*, July 4, 1892.
[13] *The Sun* (Baltimore, Maryland), July 8, 1892.
[14] *The Cleveland Leader*, July 11, 1892.
[15] *Ibid.*
[16] *The Milwaukee Daily News*, July 9, 1892.
[17] *The Sun* (Baltimore, Maryland),, July 12, 1892.
[18] *Ibid.*
[19] *Ibid.*
[20] *The Herald* (Boston, Massachusetts), July 11, 1892.
[21] *The Cleveland Leader,* July 11, 1892.
[22] *The Local News*, July 9, 15, 23, 30, 1892; *The National Labor Tribune*, July 10, 11, 12, l892; *Freiheits Freund*,, July 9, 10, 12, 14, 1892.
[23] *The Cleveland Leader*, July 11, 1892; see Burgoyne, 1-26.
[24] *The Cleveland Leader*, July 11, 1892.
[25] *Ibid.*
[26] *Ibid.*, July 10, 1892.
[27] *The Chicago Tribune,* July 7, 1892.
[28] *The Post* (Pittsburgh, Pennsylvania), July 7, 1892.
[29] *The Local News*, July 9, 1892.
[30] See McCandless v. the Allegheny Bessemer Steel Company, Supreme Court of Pennsylvania, *Pittsburgh Legal Journal*, No. 25, January 18, 1893, 223-228.
[31] See Roe Wellington, *Juggernaut* (New York: Lippincott and Co., 1948); Carroll D. Wright, "The Amalgamated Association of Iron and Steel Workers," *Quarterly Journal of Economics*, July, 1893; Ferdinand Lundberg, *America's 60 Families* (New York: Vanguard Press, 1937).
[32] *The Local News*, July 9, 1892; *The Cleveland Leader*, July 11, 1892; see also *The New York Times*, July 10, 11, 1892.
[33] *The Cleveland Leader*, July 8, 1892.
[34] *Ibid.*; see Alexander Berkman, *Prison Memoirs of an Anarchist* (New York: Mother Earth Press, 1912), *passim*; Emma Goldman, *Living My Life* (New York: Alfred Knopf, 1934), 84-91; Richard Drinnon, *Rebel in Paradise* (Chicago: University of Chicago Press, 1961), 41-47.
[35] Stowell, 79-80.
[36] *The Cleveland Leader*, July 10, 1892; *The Daily Dispatch*, July 9, 10, 1892.
[37] *The Sun* (Baltimore, Maryland), July 11, 1892; *The Press* (Pittsburgh, Pennsylvania, July 9, 10, 11, 12, 13, 1892.

38 *The Local News*, July 23, 1892.
39 Stowell, 124-25.
40 *The Cleveland Leader*, July 11, 1892.
41 *Ibid.* See also *The New York Times*, July 8, 9, 10, 11, 1892 .
42 *The Local News*, July 23, 1892.
43 *The Milwaukee Daily News*, July 9, 1892.
44 *The Cleveland Leader*, July 11,1892.
45 *The Local News*, July 23, 1892.
46 *The Cleveland Leader*, July 11,1892.
47 *Ibid.*
48 *Ibid.*; see also Burgoyne, 91-94 for another view.
49 *The Cleveland Leader*, July 11, 1892.
50 *Ibid.*; see also *The National Labor Tribune*, July 11, 13, 16, 1892.
51 Stowell, 128.

**Hugh O'Donnell**
**(Circa 1892)**

# Chapter 6

# Homestead's New Elite

People who pronounce themselves in favor of the method of legislative reform in place of and in contradistinction to the conquest of political power and social revolution, do not really choose a more tranquil, calmer and slower road to the same goal, but a different goal. Instead of taking a stand for the establishment of a new society they take a stand for surface modification of the old society. . . . Our [Social Democratic Party] program [thereby foresees and] becomes not the realization of socialism, but the reform of capitalism: not the suppression of the system of wage labor but the diminution of exploitation, that is, the suppression of the abuses of capitalism instead of the suppression of capitalism itself.

- Rosa Luxemburg[1]

The capitalist has his money, the laboring man has his muscle – the one is capital as much as the other. If the government protects capital why not protect muscle. Capitalists all over the country combine and set the price of labor and no one seems to object or raise an outcry against it. Why then should not labor combine and have something to say in fixing the scale of wages? Shortsighted conservatism says that if men are not satisfied at the wages offered at Homestead, let them go elsewhere. But suppose they do. They will find capitalists combined to lower their wages wherever they go.

The only alternative then, for the laboring man is to contend for his rights where he is – fight his battle on his own ground. He must also fight mental and moral battles as well as physical. He must strive for equal laws, which will compel the rich man as he grows richer to bear more of the public burdens. We can learn a lesson from ancient Athens, where the rich men by law were compelled to bear the chief burdens of the state.

- Rev. W. T. Galloway, Pastor of the First Baptist Church of Homestead, 1892[2]

Homestead had come under the authority of a new elite and an expanded central committee which was enlarged to include, besides Amalgamated leaders, shop keepers, owners of small factories and other middling and tiny cottage businesses and commercial undertakings. With the commune under pressure from Pennsylvania's constituted authorities, these leaders were particularly vulnerable to appeals for law and order, measures which were to be implemented by both State officials and the Militia.

At first, the commune leaders claimed they were the legally-constituted government of the city. But that claim was quickly undermines when the rank-and-file pursued their own objectives and prepared militarily for another Pinkerton raid. To reassert their position, the Amalgamated leaders then made plans to negotiate with State officials and to convince the people of Homestead that the military would protect them from the Pinkerton invaders. For these leaders had concluded that there was no way in which the commune could win its demands by military means, and that the State authorities would be satisfied with their recognition of Carnegie's legal ownership of the mills, exemplified by their efforts to allow the Militia to enter Homestead unopposed.

They argued that the State authorities and the Militia would not protect nonunion workers, though there was rank-and-file apprehension that the Militia would bring in both Pinkertons and scab laborers to replace the Homestead work force.

## The New Elite and Control Over the Rank-and-File

After July 5th, the form of the commune was altered. The old Advisory Committee had formally disbanded before the fight with the

Pinkertons began on the 6th, burning all of its records for fear its members would be held responsible for any ensuing troubles. Still, the members of the Committee met clandestinely to discuss the progress of the fighting and the strike, and on Saturday, July 9th, the Committee surfaced again with its members as leaders of the Homestead commune. The Committee kept all of its old officers, but remained an informal body with Hugh O'Donnell as chairman in control of the conduct of the entire city.[3]

These leaders exercised their previous authority to preserve order, to prevent drunkenness, and to keep the more hot-headed of the locked-out men from committing any overt action without commands from the central leadership.

The new organization also followed the advice of the Amalgamated attorneys, so that now the workers were organized within the strictures of the law, instead of continuing their unauthorized occupation and "defense" of the mills as they had before the Committee was abandoned.

With Hugh O'Donnell still Chairman and W. L. Conegly as the new Secretary, the 100 new members of the Committee of Homestead citizens included businessmen as well as Amalgamated members. In place of the former 50 leaders of the Amalgamated lodges, the new Committee was, O'Donnell said, suppose to preserve order rather than, as before, to prosecute war with the Pinkertons. It was also to arrange for signaling and to keep the "riffraff," which allegedly followed all strikes, within bounds.[4]

The new Committee was not only a coalition of interests designed to preserve law and order in Homestead, then, but a force for control and repression within the city. It initiated a crackdown on all opposition from the local population, so that there could be neither open advocacy nor popular participation in making democratic decisions. In its singular attention to preserving order, the new Committee's weakness was that it would attend only to this, rather than arbitrating or settling the lockout.[5]

Thus in Homestead during the four-and-one-half months from July 6th to November 21st, it would quiet or keep out those with political ideas the dictatorial Advisory Committee thought would undermine the narrow demands they put forward. By stressing order, the Committee actually made easier the entrance of the State Militia with its goals of protecting nonunion workers brought into the mills.

**Socialists in Homestead**

There were those who opposed such "order," though.

Before and during the battle of July 6th, Homestead held a trove of Henry George socialists who stood for a basic change in ownership of the underlying means of life and labor. These socialists had a unique set of beliefs, viewing the government as a mediating institution that impoverished the working class for the benefit of the class of owners of the land and instruments of labor. In order to take these objects of labor from the owners and place them in the hands of the working population, these socialists believed a single-tax could be placed on wealth so as to drive it out of existence.

Some of them also believed in revolution, but most thought a peaceful reallocation of wealth could take place if they could gain control of the State. One of them, Rev. W. T. Galloway, Pastor of the First Baptist Church of Homestead, counseled his flock a few days after the armed struggle that: "As American citizens, it is time we begin to consider the truths of Henry George, the truths of socialism that are in harmony with the common principles of humanity, . . . that the laboring man has a righteous cause and he is simply pleading for practical application of Bible truths to society." Continuing, he turned to the root of the problem:

> He who studies the social conditions of our country sees capital becoming more dictatorial, tyrannical and oppressive. He also sees labor organizations multiplying all over the world. . . . The labor organizations are a force that no nation can overcome. If defeated in one quarter they will only rise again more powerful than ever in another.

But the laws did not provide equality for all:

> Manufacturers have protection. . . . [N]o one will dispute the fact that congress has passed laws to protect American manufacturers, neither will anyone dispute the fact that we have a large crop of millionaire manufacturers. The laws of the government have made these millionaires possible. The manufactured products of these millionaires are protected against the competition of other nations by a tariff on foreign goods. Who protects the laboring man against the millionaires when they attempt to oppress or defraud him out of his wages? Congress never passed any laws to protect the laboring man against the greed and insatiable rapacity of the money sharks. The capitalist has his money, the laboring man has his muscle – the one is capital as much as the other. If the government protects capital why not protect muscle. Capitalists all over the country combine and set the price of labor and no one seems to

object or raise an outcry against it. Why then should not labor combine and have something to say in fixing the scale of wages? Shortsighted conservatism says that if men are not satisfied at the wages offered at Homestead, let them go elsewhere. But suppose they do. They will find capitalists combined to lower their wages wherever they go.

The only alternative then, for the laboring man is to contend for his rights where he is – fight his battle on his own ground. He must also fight mental and moral battles as well as physical. He must strive for equal laws, which will compel the rich man as he grows richer to bear more of the public burdens. We can learn a lesson from ancient Athens, where the rich men by law were compelled to bear the chief burdens of the state.[6]

That lesson became the foundation of Henry George's single tax, which was suppose to alleviate all social ills and undermine accumulated wealth.

Far to the political left of the Henry George Socialists were the Anarchists, who believed that destruction of the State would enable the population of workers to directly control and run the instruments of mill and land production. A group of these radicals had arrived in Homestead on July 7th, and by 9 A.M. were distributing a handbill with a socialist message, making it sound as if Homestead workers had Leftists and Anarchists among them who were prepared to attack Carnegie's ownership of the instruments of production. Emphasizing the injustices placed upon working people, the leaflet said: "We strikers are obliged to see how tyranny is running rampant here in Homestead, how the biggest fraud and hypocrite in Pennsylvania, a philanthropist, has turned his iron plant which was created by our labor, but stolen from us by these thieves (Carnegie, Frick, and company), by means of law and privilege, into a military fortification, thus actually declaring war upon us workingmen from whom he has robbed all his ill-gotten wealth."

Recounting Frick's efforts to cut wages and criticizing his use of deputy sheriffs and Pinkertons, the leaflet polemicized that the "Carnegies, Fricks, and their class generally, are forcing the working people to revolutionary methods, actually dispelling all hopes of a peaceful settlement of the economic question." Thus, there was "no course left but to meet force with force," for peaceable measures would never undercut the capitalist system of production, exchange and communication.

Since the workers had created the tools of work, they should continue to occupy the mills that of right belonged to them by striking and defending themselves against any intruders, the leaflet insisted.

The same struggles would continue "until the tools belong to the toilers, the products to the producers and the workshops and mills to the workers." Therefore, the leaflet concluded: "When we resort to revolutionary methods, we should be consistent and change the infamous system of production, which is the cause of all our troubles. Brothers, become Anarchists."[7]

Not only was this a call for armed struggle to overthrow the wage system the Homestead workers accepted, but more than anything else, the last sent shivers down the spines of the members of the Advisory Committee. Many pointed to the fact that the Carnegies and the Sheriff could not possibly believe the demands in the leaflet were theirs, for the workers of Homestead had held the mills without incident for a full day in which they were in complete command.

Not only did many workers see the circular as designed to injure their cause,[8] but to stop the movement of the Anarchists, the Advisory Committee sent out squads and apprehended four men suspected of distributing the handbills. They put two of them in the lockup, and then, to avert any danger from workers who wanted them hung, placed the four suspects aboard a train and sent them out of town with a stern warning not to return.[9]

* * *

To this day it is not known precisely if any among the rank-and-file workers were socialists, responded to socialist goals, or understood that when they used armed force there was a critical distinction between seeking secure wages or jobs and gaining possession of Carnegie's instruments of production as their own. Yet, Anarchism and revolutionary socialism seem to have been outside the realm of Homestead politics, although elsewhere socialists cheered on the Homestead workers to make a new society.

Nor is there sufficient evidence to show that Homestead workers viewed Carnegie's ownership of these instruments Advisory Committee of labor as the source of their perennial oppression, though socialists in places like Milwaukee pointed out the weaknesses of the Homestead commune. The *Wisconsin Banner und Volksfreund* thus analyzed that Homestead led by Hugh O'Donnell had entered the contest with Carnegie and his allies in Washington and had emerged unscathed; but the delayed reorganization of the workers' battle plans had inadvertently involved the abandonment of the Carnegie works. To construct a great new society, the *Banner* argued, "commander" O'Donnell had a duty to prepare militarily for the coming life and death collision, and the commune had best hang on to its military posture, rather than play the fool in dealing with the rulers of Pittsburgh,

Harrisburg and Philadelphia.[10] Militarily Homestead workers agreed with this scenario, but it is nonetheless certain that the Advisory Committee kept all ideas beyond the realm of trade-unionism out of the struggle, limiting them to demands for a fair day's pay for a fair day's work, security of job places and the right of workers to belong to a union.

Hugh O'Donnell made the policy of the Advisory Committee very clear when he repeatedly denounced the Anarchists as having nothing to do with the struggle at Homestead. In fact, reading the letter of a New York Anarchist to the press, O'Donnell said that his suggestion that a hundred pounds of metal be tied to Manager Frick for casting into the Monongahela was "calculated to injure our cause," adding: "the world knows I have no anarchistic tendencies. A fair field and no favor is my idea."[11] By this O'Donnell meant that labor should be treated fairly by Carnegie and, without favor, be represented by the Amalgamated in bargaining for jobs and wages.

### The Contradictions of Authority: Inane Aficionados and Figurehead Leaders

Trade union policies for order and fair bargaining were the original purposes of the struggle. But they were now used by the trade union leaders to prepare the workers for the occupation of the mills by the Sheriff, his Deputies, the Militia, and, ultimately, the non-union workers.

On July 8th, when from 200 to 300 locked-out workers were recruited by the Advisory Committee to attend a mass meeting at the Opera House, union officials presented the full weight of Amalgamated theory. Attorney W. J. Brennan warned the workers that compromise was needed, calling for a drastic alteration of their strike tactics:

> The owners know the law is on their side and their hearts cannot be touched. They have civil protection. When the militia comes to Homestead the mill will be surrounded and new men put in, and any man interfering with the civil authorities will be shot down. If any one thinks for a moment that the militia will stand around like hitching posts he is badly mistaken. When they do come it will not be for dress parades, but to shoot.[12]

Nonetheless, at this time the rank-and-file were not convinced they had to alter their organization to legally keep the Militia out of Homestead. They were only willing to allow Sheriff McCleary and his deputies into the Homestead Works on the dual conditions that no armed authorities – neither county nor Pinkertons – would take possession; and that well-known businessmen of Homestead and Pittsburgh – who were of good reputation and known to the workers – be appointed to do guard duty at the Works without arms.

But these were impossible conditions, as the workers knew. One prominent citizen who qualified, Dr. Stodden, told Sheriff McCleary he would only agree to recruit a sufficient number of reputable citizens for occupation of the Works as a guarantee for protection against violence if Mr. Frick and the other interested parties would also go there. But the latter the workers would never permit and the Sheriff would give no promises. He returned to Pittsburgh with the news that, of the 20 reputable citizens of Homestead who were suppose to act as deputies, only six responded and they positively refused to serve.[13]

On the workers' side, meanwhile, the Advisory Committee was gradually transformed into a figurehead of a leaderless movement.

By July 10th, with the Advisory Committee officially no longer in existence as far as the strikers were concerned, proposals to reorganize the Committee had not been carried out. The old leadership simply continued the functions they originally performed in the Committee, still acting authoritatively, but now attempting to disavow legal responsibility for anything which could not be personally attributed to them. At the same time, they lost touch with the needs and hopes of the rank-and-file.

"They are not leaders of a force that is totally controllable," wrote one reporter, "but of an unorganized and large body which may at any time break away and act as a mob on the volition of individual members." Community control levers had shifted to the eight lodges, each with narrow craft union demands, so power was now being "exercised unofficially and through separate lodges in which secrecy obligations hold as well as through a committee."[14] Since no one was openly willing to assume responsibility, and the workers themselves had absolute control over the mills, the *Cleveland Leader* agreed that the body of strikers were tenuously "held in check by the uncertain authority of a committee of leaders."[15] Thus individuals gave free reign to their own ideas of what should be done, many feeling that Sheriff McCleary, the Pennsylvania Militia, and all outsiders should be kept from Homestead. And yet, there was no organized resistance to their presence.

One of the reasons for the disbandment of the Advisory Committee in the first place had been the developing alliance of Sheriff McCleary and the Committee leaders. The Sheriff and others had tried to use the

Committee by obtaining its endorsement of the presence of deputies in Homestead, and, through the Committee, the endorsement of the rank-and-file, who were on the whole against the idea.

Committee members now openly admitted the workers were acting together without their directives, though still in a proper and orderly fashion, so that an organization was no longer absolutely necessary. O'Donnell hedged on the possibility of any reorganization of the Committee, saying it was expected that "every man do his duty in this crisis."[16]

The underlying malaise, the basic disorganizing factor, rested on the continual lack of democracy – a defect that had existed from the commune's inception. Homestead may have been "as much in a state of siege as though a public enemy were encamped upon the heights around the town and their gunboats lying in the river," as one reporter stated. And the workers' martial law of the city certainly functioned smoothly.[17] But the autocratic and dominating leadership still had goals and interests quite different from those of the rank-and-file.

As one reporter surmised, the Committee alliance of Amalgamated craft workers and petty businessmen would in a future military confrontation impede and be opposed by popular demands, thus acting as a source for stop-gap pacification at best:

> This Advisory Committee is an informal body of gentlemen, of whom Mr. O'Donnell is chairman, who have undertaken the conduct of the town of Homestead for the time being. Its members are conservative and well-known businessmen and others, and as long as no imminent danger of riot or violence threatens, there can be little doubt of its power to govern the town.
>
> In the presence of danger, however, the great majority of people in Homestead seem to think that they would be more apt to be their own Advisory Committee. In other words it seems to be quite certain that during the present lull before the coming storm the Advisory Committee is as good a way of preserving order as can be found, but if there should be any fighting to be done the locked out workmen will probably go to other leaders.[18]

Rank-and-file distrust of the Committee had been building. O'Donnell's conduct in protecting the Pinkertons after their surrender had continually drawn and fed doubts about the authority of the Committee: "His efforts to secure the safety of the detectives has seriously shaken his influence with the unthinking part of his following, and this is one of the elements in the problem of the Advisory Committee," noted one reporter. "The people generally seem to be glad that . . . the Committee has taken up its work again, but they do not believe that it will be long before it again dissolves itself and burns its records."[19]

The split between the leaders and the rank-and-file was further widened after the Pinkerton surrender. By July 8th, Homestead deaths symbolized by crepe fluttering from the doorknobs of so many humble cottages had driven men to the saloons to drown their hatred – or increase it. Although the saloons had been closed the previous day, on the 8th they opened their back doors to the growing patronage.

"It is one step downward," deciphered a reporter in Homestead. "Once let 5,000 idle men get started in that direction and it will take more than Pinkerton detectives, deputy sheriffs or militia to stop them. Fortunately the strikers are not all Hungarians and Slavs. Fortunately the men in the lead are men who know their success depends largely on their bravery. But the talk of the men themselves will show which way the wind is blowing. They gave their honest opinions unbiased by any crowd or agitator, for they were away from their companions and at home," he explained his expedition in Homestead's mews and byways.[20]

The leaders could no longer moderate the anger of the rank-and-file, and by the 8th the reorganized Advisory Committee represented itself alone. The legal scenarios that were to accompany a future occupation were being played out by these leaders in autocratic fashion, with McLuckie and O'Donnell proselytizing to let the Militia enter Homestead unmolested. But the rank-and-file worried that scab workers might be brought in under the Militia's protection. Moreover, they feared that Pinkertons and arms would be introduced under the occupation by the Militia, thereby empowering Carnegie to defeat them even if the Militia withdrew.

By July 10th, there were only three possibilities facing the strikers: a limited military presence, armed encounter with the Militia, or military repression of labor. Since the State would direct the size of the Militia, the question resolved itself in terms of a fight or not. And by the evening of the 10th, the strikers were in an uneasy frame of mind, one reporter predicting that it was "possible that a feeling of desperation over the prospects may cause the hotter heads to counsel violence to the militia," adding that "such a suicidal cause would not be approved, and if endorsed by the more ignorant workmen will be carried out only in the face of determined opposition from the leaders."[21]

Once the Governor ordered them in, the purposes of the militia would act as a critical factor in the workers' response, declared the pro-union *Milwaukee Daily News*: "If they were to come simply to protect the property, there is no occasion for their presence. They would probably be unmolested, saving for the possibility of some unauthorized action causing trouble. Should they attempt to act as an escort and as a means whereby nonunion men were to be introduced, trouble would result, unless the militia made such a strong demonstration as to overawe the men who are as a class more intelligent than the average workingman of a determined nature and of English, Scotch and Welsh stock to a large extent."[22]

Many strikers believed that any troops brought in would speedily be withdrawn and that the Governor would refuse to assist the Company in bringing in nonunion workers, leaving the situation practically the same as before their arrival. Since the strikers believing this scenario fully expected that they themselves would continue to bar the entrance of nonunion workers,[23] they opened themselves to be called upon by the Advisory Committee to observe its recommendations.[24]

**Preparing to Defend the Commune**

The division between the Committee and the army of strikers was thus momentarily glossed over. The Advisory Committee, concentrating on job and wage preservation, sought to abide by the laws that authorized government officials to secure Carnegie's private ownership of factories and mills. The rank-and-file at first opposed cooperating with any government officials for fear that Militia ordered to Homestead would play the role of the Pinkertons in protecting scab workers. To win over the strikers, the Advisory Committee then concentrated on preparations to repel the Pinkertons, thereby hoping to persuade them to accept the Militia as an ally.

In the process, the leaders of the commune covered one eye and blinded the other, making two costly tactical errors. First, thinking the Pinkertons would make a major second attack, these leaders prepared the workers only against this. Second, viewing the established institutions protected under the law as supreme, they were also in favor of allowing the Militia to enter Homestead to enforce Carnegie's legal rights of ownership in the mills.

* * *

O'Donnell said the Pinkertons were surrounding Homestead and the city of Pittsburgh. Yet, he expected the attack to be delayed until there was a sufficient number of nonunion men collected to start the Homestead mills running again. A strong Pinkerton force would then open their way.

The workers again prepared to resist. Through the railmen who were sympathetic to their cause, the strikers now could secure advance information of any movements by the Carnegie-hired Pinkertons. The workers also planned to keep up the fight even if the Pinkertons took over the facilities. Though the sacrifices would be heavy, the workers thought they could hold out as long as a year.[25]

In the meantime, it was commonly believed that the Pinkertons would be collecting the names of the strike leaders and the Carnegie Company would be pressing legal action. "There seems to be no doubt that there are a few Pinkerton men in town who are employed in getting a list of the wounded and others who were active participants in the riot," reported the pro-labor paper, the *Homestead Daily Messenger*. "As a word of warning, it behooves our people to be quiet on the subject and give them no opportunity to find the object of their search. In talking of the matter, let no names be mentioned and their errand will be fruitless."[26]

Since nearly everyone in Homestead believed the Pinkertons were collecting information, that they had to be suspicious of the unusual and that the Advisory Committee's initial press regulations were designed to keep the Pinkerton agents out of Homestead, "Every presumed intimation of an unsympathetic feeling for the cause of the locked-out men is resented," an observer said – so that "an open attempt to turn them from their present course would be dangerous."[27]

Action began on July 8th, in the wee hours of the morning, when the Advisory Committee's Hugh O'Donnell was alerted by a telegram from Cincinnati stating that under orders from Carnegie and the Pinkerton Detective Agency, workmen had been employed to remodel a number of freight cars to be used at Homestead. The telegram gave details of cars being made bullet-and fire-proof, with rifle holes provided. Each car was being stocked with rations sufficient to feed a small army for several days, and Pinkerton officials were preparing to send the cars in with 300 to 400 men just as soon as everything was in readiness.

O'Donnell sounded the alarm, and committees were immediately appointed to scrutinize every train approaching the place. At the same time, several reliable men hurried to Pittsburgh to take up posts in the railway yards and wire information to Homestead as soon as the trains put in an appearance.[28]

Hugh O'Donnell himself went to Pittsburgh on the night of July 8th to learn the intentions of the Pinkerton men and to try to checkmate them with legal injunctions. The tactics were varied. "The men," wrote a *Sun* reporter, "are watching every point and have assistance from many quarters in which valuable aid can be rendered. The railroad men are in heavy sympathy with them in the strike, and for this reason it would be hard to surprise the strikers. They have an abundance of money, the treasury of the Amalgamated Association being well supplied with funds, and they are able to employ the best legal assistance."[29]

Not every warning about approaching Pinkertons proved accurate, though; false alarms came so frequently that the believers themselves began to doubt that the Pinkertons were as powerfully prepared as they were first thought to be. On the night of the 8th, for example, loud whistles warned that detectives were on their way, and people poured out of every street of Homestead, although no danger really threatened. Yet the incident showed the city organization had already been honed to the point where the leaders in each block hallooed, collected their forces, and tried to control the trigger-happy.

"On two corners," a reporter observed, "a couple of young men began blazing away with pistols, firing at the ground. This was not part of the program, and angered one lieutenant, who rushed down the street and yelled, 'Stop that shooting, you fools. If there was any need for shooting you would not be around.'"

A family support network emerged as well. "One wife hurried to her husband, who was doing patrol duty along the railroad tracks," the *Cleveland Leader* reported. "She put a rifle in his hand. 'I loaded it myself,' I was all she said, and with this turned away and joined a group of women who had collected to learn the trouble."[30]

On July 9th a new alert was telegraphed from East Buffalo to Committee Headquarters, warning that – "Three cars of Pinkertons and two loads of ammunition passed here at 9 o'clock today. Be on guard." Sent at 12 o'clock noon, the telegram alerted the leaders of the commune. Six hours later, five special passenger coaches were attached to the B. & 0. train from Chicago on its way to Pittsburgh, though the coaches were not run into the depot. Instead, following the secretive Pinkerton style, they were switched directly onto the Pittsburgh branch, with special precautions to keep the coaches from attracting attention.

Labor leaders believed they contained Pinkertons on their way to Homestead, and so late Saturday night, the 9th, they flashed the news to the locked-out men to watch for the approaching railroad cars.

The workers were ready for a new Pinkerton attempt to take over the Works. They expected the Pinkertons to disembark from their train on the unused Wheeling branch of the B. & 0. Railroad at a point Northwest of Homestead near the Carnegie property, and then to come over the hills under cover of the Company's houses to secretly enter the idle mill.

To prevent such a surprise, scouts went to the homes of 300 sleeping mill men, and quickly assembled them at Committee Headquarters. Then five squads armed with confiscated Winchester rifles were sent out to scout along the hilltops overlooking the tracks of the Wheeling. By 1 A.M., some 250 picked men were positioned in the hills overlooking the B. & 0. Railroad tracks, ready for a fight, while in Homestead word was passed from house to house and armed men hurried towards pre-selected vantage points. A bloody conflict was expected before daybreak,[31] but like so many tactical maneuvers on both sides, it never materialized.

Substantial grounds supported such readiness. By July 10th, a dozen telegrams from friendly sources at various points had warned that the Pinkertons were mobilizing for a massive move into Homestead. Some 800 Pinkertons were believed ready to face 3,000 to 5,000 able-bodied Homesteaders. Since Homestead leaders accepted most of this information as authentic, they concluded that a decisive struggle was not far distant. Alarm after alarm swept the town, keeping the people on a razor's edge in preparation for an impending battle.[32]

The threat was real. As one strike leader reported, the Pinkertons were assembling in back of the Carnegie Mills. "There is no question about the presence of armed men just South of the range of the hills," his report stated:

> Our scouts have located them in a dozen places. The are in groups and are moving about. They are gradually getting together, and will in all probability make a feint upon some point in our lines which they believe weak. I do not think their first attack will be a serious one. They are back two or three miles from the River, and are evidently waiting for reinforcements. They are arriving in small detachments, so that they will not attract notice. Not one of us doubts the presence of these men and that they intend to make an attempt to reach the mill.[33]

A new readiness began. Guards were posted with an especially strong body of men in the "hollow" South of the strike Headquarters. Men not on duty were warned to hold themselves ready to obey a quick call.

The workers' web of surveillance was spread out. Across the river the locked-out men picked up and questioned loiterers, and, if they could not satisfactorily explain their presence, took them to Braddock and put them on the train bound outward.[34] On the 9th, a detachment of Homestead guards passed through McKeesport, unmolested, armed with Winchesters and marching through Reynoldtown right down Fifth Avenue. Saying nothing, they moved up the river to watch for supposed or rumored arrivals of Pinkertons or nonunion men.[35]

The heightened vigilance continued, as picket duty continued for all men and, on the morning of the 9th, guards at the mill gates were doubled when it was reported that the firm was about to send in some laboring men and carpenters to make repairs.[36]

Meanwhile, a hospital was equipped in the Amalgamated Building, under the direction of a young physician who came to O'Donnell with a note of instructions from Sister Magdalene of the Sisters of Mercy – an order that had charge of the West Pennsylvania Hospital to which some of the injured of the battle of the 7th had been sent.[37]

Men were dispatched by their companies to go to committee Headquarters to learn the latest news. Armed, rifle muzzles pointed towards the ground as a precaution against accidents, the strikers, waited anxiously to hear the latest developments. In the streets and along the roads, young boys tracked all unknown persons, under the authority of leaders standing at street corners, sending squads in all directions. A single leader would dispatch 100 or more men and each would carefully follow orders. Guards were everywhere in Homestead, and were soon sent across the River as well to prevent surprise. All along the trestles connecting Homestead and the opposite river bank, dynamite was stored to blow up the tracks and all other communications should the need arise.[38]

The locked-out men now held total control of Homestead. "They are the autocrats of the town, the *Cleveland Leader* criticized, "and many of them beneath their slouch hats wear the dignity and knowledge of sovereignty." The traditional system of local government was overturned, the *Leader* continued:

> The town is under what is known as the borough system of government, and the representative of the state is a burgess. His name is McLuckie, but from him can be obtained no news official, nor does he give orders. To all inquires he refers to headquarters, a large room in the third story of the brick structure owned by the Amalgamated Association of iron and Steel workers, where around a table laid in shavings and sawdust orders are given by the advisory, or some other committee of the workmen, of what shall be done.[39]

What was done was carried out by the rank-and-file, the report explained:

> Many of the men . . . with no special leadership of the locked out workmen, undertake to assert their right to question the proceeding of the people in the town, to ask of them the object of their presence and to

> insist on an explanation. These explanations are usually accepted, but the fact remains that this is probably the only place tiday where an American citizen is not free to go and come as he pleases.[40]

Even this critical journalist conceded that order was strictly maintained, and the locked-out men themselves committed no "act of violence, plunder."[41]

Such was the regimen that on July 9th, the commune was described as follows:

> A discipline that might well excite the envy of a militia organization, is alleged by the strikers to exist in their ranks, and while their policy is to wait and watch, they declare they are well prepared for the enemy. So perfect is their organization and so numerous are their sentinels up and down the river that surprise seems next to impossible from any force. The men have apparently made up their minds that sooner or later the mill owners will make another effort to resume their dominion here, and every preparation is being made to repel the effort when it comes.[42]

Newcomers into Homestead were now followed quietly from arrival to departure, so they would never suspect the clever shadowing and "espionage" that was actually going on. Unless a visitor ventured too near the mill, he would never notice the martial law that had been declared by the Amalgamated Association." Approaching the mill, however, any visitor would instantly be stopped, not by a hand of steel, but by "the touch of velvet," as a sentry crossed his path and courteously said: "Thus far shall you go, and no further," a command that was rigorously enforced.[43]

* * *

On July 8th, Hugh O'Donnell accompanied by three businessmen from Homestead, formed a committee to visit Governor Pattison. They went with and at the request of Burgess McLuckie to inform the Governor of the conditions prevailing in their Borough. They carried with them the Pinkerton roll book captured in the surrender, containing the names of all the men sent to Homestead and the contracts with their employers.

During the meeting O'Donnell told Attorney General Hensel that there was no peril to any property in Homestead, no attempt had been made by the Sheriff to execute any legal process asking them to leave the mills, and he realized the Amalgamated men could not prevent nonunion men from taking their jobs because the strikers had no legal rights to keep non-unionists out.

The Governor had already twice stated that peace and order and the observance of the law had to be adhered to and would be maintained even if it required all the force of the State or an appeal to the Federal Government. He now added that, under the forms of the law and civil authority, this would be done to protect property and the rights of all parties in the struggle, without regard to the merits or demerits of the business difference between them. Agreeing to this, O'Donnell and the others from Homestead replied they hoped the Governor would arbitrate the troubles – especially the question of the employer's recognition of the Amalgamated as the bargaining agent for the workers.[44]

The Attorney General responded by asking, in the event the Company persisted in putting in nonunion men did the Amalgamated think they could keep them out. O'Donnell answered that, in that case, the Association would be compelled to yield and recognize the legal right of the Company to employ whomever it chose.

By proposing that the strike be arbitrated and that the Amalgamated recognize the legal right of the Company to employ scab laborers, O'Donnell acted in a way that was totally unresponsive to the rank-and-file. Yet, on his return to Homestead, O'Donnell tried to pacify the workers into believing the Militia would not be used to bring in nonunion workers, claiming:

> The policy of this administration [under Governor Pattison] is to use the militia only as a last resort and only for execution. They will not be used to take possession of the Works to permit others to work in the Carnegie mills.
>
> Now, understand me; I am not saying this on the authority of the Governor, but I have arrived at the conclusion from what I have seen. Heretofore, whenever a manufacturing corporation or firm has seen fit to reduce the wages of their men they have forced an issue with them, incited them to some trifling overt act of violence, and then called in the military and used them to operate their Works under protection.
>
> In other words, the manufacturers have been able to call out the militia to accomplish their personal ends regardless of the rights in the case. But that era is at an end, the militia, if called out at all, will perform their proper function – that of suppressing insurrection.[45]

While the workers must resist the Pinkertons, O'Donnell said, they should also respect the actions of the officers discharging their legal duties.[46] Should the Militia be ordered in, O'Donnell continued, he would be in the forefront of encouraging the people of Homestead to accept them.

In fact, the entire Advisory Council had adopted a similar stance by this time, called a meeting the same day at the Opera House; and Amalgamated Ex-President Weihe counseled permitting the Sheriff to take peaceable possession of the mill property: "The Sheriff and the men he will bring here will be in sympathy with the people of Homestead. He will not be foolhardy enough to bring a lot of men to shoot us down. We have their sympathy now and will have it when they come. The Sheriff must do his duty and will do it, but not in a spirit of animosity."

An unnamed speaker then added: "If the Amalgamated Association does not win at Homestead, it will not amount to a row of pins in the future. The question is, will you permit the Sheriff to take possession of the mill as our friend? We will not be compelled to go to our homes and remain there, and we will know we have some friends among the Sheriff's deputies. Time will show that the best thing for us to do is to let the Sheriff take possession of the mill. We all know the law is against us."[47]

Some of the leaders of the Amalgamated and the commune had now sided with the law against the strikers' stated goals. The rank-and-file recognizing this began to withdraw their support from the self-appointed leaders.

By July 9th, the split had widened, one reporter noting that: "Flushed with victory, the striking steel workers are becoming defiant. The leaders are losing their control over the men, who are becoming restless. On all sides can be heard groups of strikers denouncing the Sheriff as a political trickster. They are suspicious of the leaders, who at a meeting today advised them all to let the Sheriff take charge of the mills."[48]

Propaganda for compromise continued: Ex-President Wiehe again advised the men to allow the legal authorities full power to act, and to permit the Sheriff to take charge of the Works. He said there were but two courses for the strikers to follow: First, to permit the Sheriff and his deputies to enter the mills and take possession; second, to forbid the Sheriff to go into the mills and, if he attempted to use force, to keep him out – a step which would lead the Governor to call out the Militia.

"In this event," Wiehe concluded, "you will be shot down like dogs. You cannot withstand the law. It is too powerful. Attempt it and you will be defeated in the strike." The men disagreed; several expressed the conviction that the only thing they could now do was to hold the Works until some arrangements to resume work were made. As an experienced leader, Wiehe took their discontent into consideration, tactically planning step-by-step to pacify the rank-and-file to reconcile them with the Sheriff and his goals.

Thus the night before, on the 8th, Wiehe had called a meeting which proceeded to appoint the president, vice-president, and secretary of the Amalgamated Association to act as a committee to present the strikers' demands to Sheriff McCleary. The strikers insist that the Sheriff select his deputies from among the strikers or that he allow a committee of strikers to select the deputies. Only if Sheriff McCleary agreed to this, was he to be permitted to take charge of the Works; otherwise not.

Since McCleary was committed to protect the Carnegie Works by law, he did not confer with the workers of the Amalgamated at this point. He simply drew up his own narrow list of deputies responsible for the Carnegie property, excluding any mill workers. He did go to Homestead at the strikers' invitation, but failed to call on them.

"I did not see them," he told a reporter. "There was no use; I could do nothing, because I had nothing to do it with. What could I do if I had possession of the property? As soon as the Company tried to put in non-unionists the union men would resist and I could give the non-unionists no assistance, because I have no men, and I can't get any. The Governor will not give me aid, and there I am at the end of my rope."[49]

Thus the Sheriff alleged he could not secure the mills; the Governor and leaders of the Amalgamated asserted the workers must bend to the law in allowing the Militia entrance to protect the Steel Works; and the rank-and-file watched the noose encircling them tighten, while maintaining their own respect for the wage system – believing in the role of their union in future negotiations over wages and jobs, and adhering to their underlying commitment to Carnegie's ultimate ownership of their means of labor.

All advocates now relied on the same principles: The law protected the mills and required that the workers turn over the property to the rightful owner and his agents in an orderly fashion. The rank-and-file had no real choice: Their acceptance of the wage system and Carnegie's right of ownership meant they had to relinquish possession of the Works to the constituted legal authorities. Even if they opposed the commune's self-appointed leaders and denied Sheriff McCleary's entrance, they still could not keep the Governor from ordering in the Militia to guard the mills and allow the Company to run them with nonunion labor.

The only alternative for the rank-and-file was to oppose the wage system, seek to make the mills their own, and keep the Militia out of Homestead by armed force. To win, they would have had to stage a local revolution. But that was not really an option, given their goals and beliefs.

## Notes

[1] Rosa Luxemburg, *The Accumulation of Capital*, Agnes Schwarzschild, trans. (New York: Monthly Review Press, 1964).
[2] *The Milwaukee Daily News*, July 7, 1892.
[3] *The Cleveland Leader*, July, 11, 1892; see also *The New York Times*, July 10, 11, 12, 1892; *The Press*, (Pittsburgh, Pennsylvania), July 8, 9, 11, 14, 1892.
[4] *The Post*, July 10, 12, 1892; *The Sun* (Baltimore, Maryland), July 11, 1892.
[5] *The New York Times*, July 10, 11, 1892; *The Mail and Express*, July 10, 11, 1892; *The Sun* (Baltimore, Maryland), July 10, 11, 1892.
[6] *The Local News*, July 23, 1892.
[7] *The Cleveland Leader*, July 8, 1892.
[8] *The Globe* (Toronto, Canada), July 8, 1892.
[9] *The Cleveland Leader*, July 8, 1892.
[10] *Wisconsin Banner und Volksfreund* (Milwaukee, Wisconsin), July 12, 1892.
[11] *The Cleveland Leader*, July 11, 1892.
[12] *The Sun* (Baltimore, Maryland), July 9, 1892.
[13] *The Local News*, July 9, 15, 23, 1892; see also *The New York Times*, July 7, 8, 9, 10, 11, 1892.
[14] *The Cleveland Leader*, July 11, 1892.
[15] *Ibid.*
[16] *Ibid.*
[17] *Ibid.*
[18] *Ibid.*
[19] *Ibid.*
[20] *The Milwaukee Daily News*, July 8, 1892.
[21] *The Cleveland Leader*, July 11, 1892.
[22] *The Milwaukee Daily News*, July 8, 1892.
[23] *The Post*, July 11, 12, 1892; *The New York Times*, July 11, 12, 1892; *The Cleveland Leader*, July 11, 1892.
[24] *The Cleveland Leader*, July 11, 1892; see Burgoyne, *passim*.
[25] *The Cleveland Leader*, July 10, 1892.
[26] *The Homestead Daily Messenger*, July 9, 1892.
[27] *The Cleveland Leader*, July 11, 1892.
[28] *The Sun* (Baltimore, Maryland), July 9, 1892; *The Local News*, July 9, 1892; *The Chicago Tribune*, July 9, 10 ,11, 1892.
[29] *The Sun* (Baltimore, Marylandl, July 9, 1892.
[30] *Ibid.*
[31] *The Cleveland Leader*, July 10, 11, 1892; *The New York Times*, July 10, 11, 1892.
[32] *The Post*, July 10, 11, 1892; *The Sun*, (Baltimore, Maryland), July 11, 1892.
[33] *The Cleveland Leader*, July 11, 1892; *The New York Times*, July 10, 11, 1892.

[34] *The Sun,* July 11, 1892.
[35] *The Cleveland Leader*, July 10, 1892; Burgoyne, 94.
[36] *The Cleveland Leader*, July 10, 1892.
[37] *Ibid.*
[38] *The Sun*, July 9, 1892; *The New York Times*, July 9, 10, 1892.
[39] *The Cleveland Leader*, July 9, 1892.
[40] *Ibid.*
[41] *Ibid.*
[42] *Ibid.*, July, 10, 1892.
[43] *Ibid.*
[44] *The Philadelphia Press*, July 9, 1892; *The Post*, July 9, 1892; *The Daily Dispatch*, July 9, 1892, *The Daily Gazette*, July 9, 1892.
[45] *The Cleveland Leader*, July 10, 1892.
[46] *Ibid.*; *The Press* (Cleveland, Ohio), July 10, 1892; *The World* (Baltimore, Maryland), July 10, 11, 1892.
[47] *The Philadelphia Press*, July 9, 1892.
[48] *The Milwaukee Daily News*, July 9, 1892.
[49] *Ibid.*

Distributing the Dynamite

Governor Robert E. Pattison of Pennsylvania
(Circa 1892)

# Chapter 7

# Homesteaders, Carnegie Property & the State: Legal Scenarios Illuminating the Military Occupation

Yes, quiet at Homestead – no sound at the mill
No hum in the bee hive of labor,
But all through the town, and over the hills,
the gleam of the rifle and saber.
And down in the hearts of the men as they wait
How deep are the passions that slumber,
How dark the revenge and the demons of hate,
that fright, love, and reason encumber!

– L. Leinad[1]

In the progress from primitive equity to final injustice, the steps are silent, the shades are almost imperceptible, and the absolute monopoly is guarded by positive laws and artificial reason.

– Edward Gibbon[2]

On the surface, all remained quiet at Homestead after the great battle of July 6th. For those living far away, relying on daily newspaper reports, Homestead was one event among many that would soon be forgotten with the new arrivals at the mill town's graveyard. The Militia occupied the city on the 12th, presaging an end to the struggle. Yet many were seething against the injustices put upon them – determined to maintain themselves and the jobs of the mill hands.

They had won the battle, but the war went on – and observers of the day saw it as war, the *Sun* described:

> To the thousands of men to whom the loss of their customary employment in the Carnegie Mills meant loss of bread for themselves and their families, the "Pinkerton men" represented no principle or authority in which they felt either obliged or inclined to yield the slightest respect. It was simply a challenge to a duel with their fellows - to a struggle of brute force.
>
> On the one side were Carnegie's unemployed workmen, and on the other Carnegie's men newly employed to defend by force, if necessary, not only his property, but such laborers and workmen as he might be able to engage to take the places of the unemployed. It was a private war between two hostile interests and parties, with society in the person of its constituted representatives and authorities quietly looking on.[3]

In truth the authorities were hardly quiet or disinterested, and the workers' victory gave the nation a harsh lesson: Only military power could hold off these hired gunmen. Not a few saw the Carnegie mercenary troops as a danger to the nation as well as to the strikers. In Congress Senator Palmer argued for his resolution to investigate the Pinkertons, saying that, in fact:

> The Pinkerton force was not a new thing. It was as well known in this country as the army of the United States. It had existed for years. Its commander-in-chief, like the barons of the Middle Ages, had a force which he could increase at pleasure and which was at the service of those who paid him. That force had been employed in New York and Illinois as well as in Pennsylvania and it might have been employed in other states. The presence of the Pinkerton armed force at Homestead was a contempt of the authority of the state of Pennsylvania. Its manner was menacing and insulting. It was difficult for American citizens (whether they were in the right or in the wrong) to submit to being driven by an armed force. The citizens of Homestead had a right to be there – had a right to employment and of living there. Hereafter, large manufacturing establishments would be regarded as political establishments with the owners holding their property subject to the correlative rights of those without whose services their property would

be utterly valueless. The owners were entitled to only a reasonable profit on the capital invested in their enterprises. And those workingmen, having spent their lives in that service, had the right to insist on the permanency of their employment and on reasonable compensation for their services. At the time of the strike at Homestead, the workers thus had a right to be where they were and had a right to defend the ground they held. The owners had the right to operate their own mill, while the workers had the right to assist in that operation – with labor entitled to reasonable compensation and capital entitled to profit.[4]

## The Law of the Manufacturers

One famous lawyer and politician of the day, Benjamin F. Butler, agreed, declaring that Carnegie managers had violated the law and the Homestead defenders believed they had the right to "drive off the invaders who were acting without right and against law."

"Deserted by the state authorities the higher law of self-preservation might be invoked," Butler continued, "not in legal justification for what they [the workers] did, but certainly in palliation . . . . Pinkerton's body of cut-throats should be disbanded by law if the lesson they got at Homestead will not disband them. The existence of such an organization under any form of pretext should be made felonious in [sic] whosoever takes part in it."

Who should be charged, then? Butler pondered: "I. . . , as a lawyer, believe that those having charge of the Carnegie Company and organizing this riotous invasion should be indicted and punished with great severity under the present law for conspiracy to break the peace and commit murder, and I hope they may be if there is any law or justice in the State of Pennsylvania not overshadowed and controlled by miserable political considerations."[5]

Others in high places echoed the same sentiments. Congressman Jerry Simpson told 20,000 attending the American Federation of Labor protest rally in Philadelphia that the introduction of Pinkerton men was the "head of high treason," and that they and Mr. Carnegie should be tried and hung for murder. "Nevertheless," the Congressman philosophized, "the law of right and justice is the vindication of the people who shed their blood at Homestead; it is higher than the laws of the manufacturers."[6]

Philosophy aside, "the laws of the manufacturers" were followed by both the Republican and Democratic parties, the pro-labor newspapers insisted. The workers had helped put the Republicans in office, and this party had double-crossed labor. But the Democrats were no better, these sentinels argued, suspecting ulterior motives even in their efforts to investigate the Pinkertons. The *Milwaukee Daily News* hotly editorialized that:

> The millionaire is dear to the hearts of both old parties. The Democrats have regretted thousands of times that Carnegie was a Republican and made heavy contributions to that party. But they have their revenge now. They could not get Carnegie's contributions in money but they expect and pray that his actions in reference to his workmen will cause a big contribution in votes.

Its conclusions were painted broadly:

> The great celerity with which the Democratic House hastens to appoint a committee to inquire into the Homestead difficulty of yesterday suggests that the democracy is about to appoint a committee to drum up votes. The misery and suffering at Homestead does not trouble the democracy; it is the votes of the laboring men there that worries it.[7]

Beyond the question of look-alike parties currying labor's vote was the underlying one of who held the power to deploy violence. The workers had certainly defended the ground they held against the hired Pinkertons. But, concluded the House of Representatives Report: "Whether assumedly legal or not, the employment of armed bodies of men for private purposes, either by employers or employees, is to be deprecated and should not be resorted to. Such use of private armed men is an assumption of the State's authority by private citizens. If the State is incapable of protecting its citizens in their rights of person and property then anarchy is the result and the original law of force should neither be approved, encouraged or tolerated until all known legal processes have failed."[8]

**The Milieu of State Military Authority**

The mobilization of government military authority was a legal and political maneuver. Militia had been deployed often after the Civil War, and labor had increasingly been put on the defensive as factories and other means of production were centralized, while the possessors of capital moved to control labor by attacking their unions, bringing in scab laborers, and using state and federal troops.

In many states politicians, generals and employers alike had demanded a total reorganization of the military in response to the inadequacy and disorganization of local police, state militias and federal troops in the Great Railway Strikes of 1877. By 1892 the Pennsylvania Militia had thus been reorganized and centralized. Its more than 8,000 officers and soldiers were newly equipped with Springfield .45's, cannon, Gatling guns and sabers. There were companies of infantry, cavalry, specialized troops and Provost Patrols, all to be commanded by General Snowden for service at Homestead.[9]

With such arms and preparations afoot, workers nationwide knew that employers could and would use state troops against them, as was done in Pennsylvania's coke fields in 1891, and in Tennessee in early 1892, when striking coal workers and miners at Coal Creek and Oliver Springs fought mine owners bringing in convicts to replace them - also battling state troops protecting scabs. When the workers, weapons proved no match for the Gatling guns turned on them, the miners had fled. But their leaders clandestinely continued the battle, mapping out the entire State and the movement of troops in an effort to stop the influx of chain-gang labor replacing the strikers. Against Militia re-enforcements and rearmament, the leaders organized, calling on the rank-and-file at their February 12th celebration of Lincoln's birthday "to defend themselves against the troops and to wipe them off the face of the earth if any blood is shed." As best they could, crowds of men then spread out among the hills of Coal Creek, bravely making guerrilla attacks on troops manning howitzers, Gatling guns and smaller arms.[10]

Such battles were repeated elsewhere; Homestead was thus the rule, not an exception, with alternate military plans being laid far in advance by Carnegie, the State authorities, and generals.

* * *

In the sweltering heat of July, once General, now Republican President, Benjamin Harrison – still at loggerheads with Carnegie over U.S. intervention in Chile – watched helpless as events unfolded at Homestead.

The President was keeping track of the battle, the *San Francisco Chronicle* pointing to his serious concern "at the grave turn matters have taken." The Executive knew that if he was called upon to provide troops there or elsewhere, it would be a political drawback in the coming national election.

Some cause for relief came when General Schofield reported that no appeal had yet been made to the Federal military authorities for aid to suppress the "riots" at Homestead. Nor did he think such aid would be necessary, stating that the Federal Government could not interfere unless the Governor of Pennsylvania informed the President that the case was beyond the control of the state and municipal authorities.

Governor Pattison made no such request, trying rather to shift all responsibility to the local level.[11]

So far as the Militia of Pennsylvania was concerned, Colonel W. J. Volkmar of the National Guard reported to the U. S. War Department that the state guard was alert to the situation at Homestead. It was using the Associated Press reports to monitor the workers' preparations and to thereby warn the troops. Volkmar added that links between the Governor, the military and the railroad owners would also make possible rapid deployment of his armed forces for occupation. "The incredible rapidity with which this little army of between 8,000 and 9,000 troops was assembled and transported so far with all its paraphernalia may be attributed to Governor Pattison, through whose two administrations the guard has received his best thought, he having once been a soldier himself in it before he became commander-in-chief," this report congratulated. "With sagacity and forethought he has habitually drawn around him, among his military staff, gentlemen closely identified with the great railway lines of the Keystone State."[12]

**The Politics of Power**

Governor Pattison had successfully sold both sides of the street. On one side, he had assured the mill owners that the generals under him could mobilize the National Guardsmen and transport them by rail to the scene of any strike in Pennsylvania. On the other side, he had cleverly gained the support of a great many workers and other Pennsylvanians for investigating the affairs of the Pacific Railroad. He had also become popular with the voters in New York and elsewhere and even aspired to become a presidential candidate.[13] In February of 1892, Pennsylvania Democrats backed Pattison at their convention, and under the "unit rule" of the state (providing that all delegates had to cast block votes for candidates at the National Democratic Convention)[14] secured his Presidential bid.

Having built an extensive popular base, Pattison was also able to win new friends from among Pennsylvania's leading capitalist groups. He supported their interests, refusing, for example, to interfere with the takeover of Pennsylvania railways by the Reading line, which merged the Reading, the Lehigh Valley and the Jersey Central. Pattison insisted this merger was in the interest of the people, reasoning that it would lower the price of Pennsylvania coal at home and only raise it to people located in distant cities and states.[15] Without the cheap and uninterrupted transport of low-cost coking coal, iron ore and limestone on integrated rail-lines, cheap steel could not be produced, giving mill owners a compelling economic reason to support the merger scheme.

With these merger plans, the democratic politicians and newspapers also agreed. The *Philadelphia Press* wrote in favor of the merger, seeing "no danger of monopoly" and rate-setting.[16] National politicians tried to sidestep responsibility, Interstate Commerce Commissioner Morrison arguing the Reading merger would not unfairly raise the price of Pennsylvania coal or resources sent to other states, throwing the burden on Governor Pattison and state politicians to look into and justify any aspects of monopolization. "I believe," Morrison solemnly declared, "there is law in Pennsylvania against consolidating railway lines, but I do not understand it fully. It seems to me that a railroad has the same right to extend its lines, so long as it maintains equitable rates, as a private individual has to extend his business."[17]

Pattison's aid to the railroads and other large businesses inspired opposition by many Pennsylvania delegates to the Democratic Party Convention, however. They charged him with patronage and moved to break the "unit rule" by bringing it before the convention as the Tammany Hall delegates had in 1884.[18] This put Pattison out of the running, though both he and New York candidate Grover Cleveland were prepared to transfer their political support to one another, depending on the preponderance of bloc power at the Convention.[19] In other words, no matter which candidate the Convention chose, he would favor big business just as much as the Democrat who remained Governor of Pennsylvania or New York.

On the tariff question, by contrast, there could be less illusive footwork on the national level. The Republicans gave lip service to a tariff on iron and steel and then earned labor's criticism by lowering it for their own purposes. The Democrats insisted they opposed the tariff, but party members like Pattison said nothing about it. The radicals wished a pox on both parties, the pro-labor *Milwaukee Daily News* noting that: "If the iron workers of the country vote as they shot there will be some tired protectionists in these United States before the ides of November."[20]

For Pattison, however, the road to political success in his own party was to attack the Republicans and assuage both manufacturers and workers. As a Democrat he hoped to make the Republican tariff seem the cause of low wages for labor. As an image-maker, he sought to convince the employers that he would deploy the Pennsylvania Militia when needed, knowing that General Schofield in Washington would not use Federal Troops, the General having advised that the State Militia, better organized since the 1877 railroad strike (when Federal troops were used), could easily concentrate in a force of 8,000 at

Pittsburgh or elsewhere on short notice.[21] As a supposed friend of labor, Pattison would follow the legal steps incumbent on any Governor before he could call out the guard.

In Homestead, then, there was a scenario that had to play itself out. First, we have seen, Frick locked out his workers and tried to use Pinkertons to bring in nonunion men; then, according to the legal form, he applied for the Sheriff's aid to guard the Works; and, when that failed, the Sheriff requested Pattison to mobilize the State Militia, ultimately to protect the introduction of scab workers. It was Pattison's move, then, and, as expected, he called out the Pennsylvania Militia.

**Mobilizing the Militia**

General Snowden, Adjutant-General Greenland and Quarter-Master General McClellan immediately mobilized the Guard. The Governor telegraphed McCleary that the National Guard was being sent "to your support at once." The three military leaders then converged on the capitol to meet with the Governor and State officials.

"I do not think the National Guard will meet with resistance at Homestead," Adjutant-General Greenland told a reporter. "Preparations for the approaching annual encampment have put the troops in first-class shape and they can be concentrated in a very short time."

This was certainly so. Snowden quickly readied the First Brigade under Brigadier-General Dechert: "Concentrate your command in camp at Mount Gretna by Monday [July 12th] afternoon and there wait further orders. Battery horsed. Take with you three day's ration and all ammunition on hand. First Troop, mounted, will move on first train available on Pennsylvania Railroad bound West." Similarly, the Second Brigade was concentrated at an unknown place – presumably Homestead – and the Third at Lewistown, to move west towards Homestead. Instructions to all regimental commanders were telegraphed. And it was not until 11 P.M. of the 10th that the news of the mobilization broke in Homestead – 8,000 troops had been ordered there!

The news was received with incredulity. Some hot-headed residents were at first disposed to be a bit defiant, and there were occasional declarations that the Militia would be opposed. But when it was learned that the division ordered out numbered 8,000 men, these declarations promptly gave way to statements like: "Oh, well, we have beaten the Pinkertons, anyway. They did not dare to come." There was popular consolation in the reflection that the most hated enemy of organized labor, the Pinkertons, had been obliged to confess defeat and give up the battle.[22]

By midnight Hugh O'Donnell was nonetheless distraught. Informed that Governor Pattison had ordered out the Militia after his interview the day before, he could not believe that in 24 hours Homestead would be the bivouac of 8,000 militiamen, ending the reign of the millmen. But facts were facts, and shortly after midnight on the 10th, he assembled the Advisory Committee to decide what attitude the population should take toward the Militia.

Though this meeting was at a private residence behind closed doors, the recommendations of the Committee were to become the attitude displayed by the great army of strikers. Operating on the premise that their control had been perfectly legal, the Committee believed that the Governor and the Militia would respect their right to control the city. They reasoned that Burgess McLuckie was the highest officer of Homestead and expressed the interests of the people. Moreover, the regular police, composed of a chief and two patrolmen, were officially supplemented by twenty sworn officers. Many of these were millmen out of work due to the lockout. And the Town Council was also composed in part by some of the leaders of the millmen. Legal authority thus constituted, the Committee was certain that the liberality of the millmen in permitting a few clerks and other employees at the mill to pass in and out unmolested spoke well for the legality of their control of the city. By also refraining from placing guards within the Works, the strikers claimed there was no excuse for intervention by State troops.

But all this was self-serving, for the Works were in fact surrounded, and all movements of watchmen or others within the Carnegie grounds could be observed from the strikers' headquarters in their new three-story brick building overlooking the Works within pistol-shot of the celebrated fence Frick had erected around the mill.[23]

The Governor and the generals had been prepared for this set of circumstances. They had caught the strike leaders in the awkward position of having to welcome them in order to remain legally-endowed with authority. And, of course, the military procession of troops and their strength in weapons and ammunition could not have been successfully opposed.

* * *

If it had come to a shooting, any invading force would have had a hard time in Homestead. The population had the finest make of repeating Winchesters taken from the Pinkertons, and an efficient and effective military structure.

They also had plenty of time to prepare for a fight. The Militia could not enter until certain cumbersome steps were taken by the Sheriff, the deputies, the strikers, the Governor and a string of lesser-known politicians. Moreover, the Federal Government had to follow precedent as well. As General Schofield had said on July 6th in Washington, it could not interfere in the matter, unless the Governor of the State informed the President of the United States that the case was beyond the control of the State and the municipal authorities. Schofield added that indeed such a case had occurred during the Pennsylvania Railroad strike in 1877, when the State Militia was unable to cope with the strikers and the Governor asked for Federal troops. But, he continued, the troubles at Homestead would likely be less serious than 1877. Even if they grew worse, the State Militia was vastly improved compared with that earlier strike. The State would have to demonstrate over several days that the Militia could not suppress the outbreaks, so he did not anticipate any immediate appeal to the Federal Government. Indeed, the government would interfere in any State matter only in the greatest extremity and was unmoved by the allegation that the rioters were invading its rights in obstructing Carnegie bridges, merchant shipping and the navigation of the Monongahela River by the use of burning oil against the Pinkertons.[24]

District Attorney Lyon concurred: The Federal Government had nothing to do with slaughter on the Monongahela River and along its banks. "It is not interstate and Pennsylvania must deal with the difficulty," he said. "If the State was not able to cope with the rioters, then the Governor would have to appeal to Uncle Sam, but under no other circumstances."[25]

But Carnegie officials stretched the scope of federal authority: the assault on the *Little Bill* was itself an offense against the government of the United States, as the tug towing the barges carried the American flag and was engaged in the United States service in transporting various government barges. The *Little Bill*, they argued, was also on a navigable stream at the time of the of the fight, and so United States troops should be used to keep the waterways open.[26]

Officers in Washington had speculated about the possibilities of the Federal authorities intervening in the event of an emergency. But this individual speculation was overridden by General Schofield, who again explained that no Federal troops would be brought into use without a request from the Governor. In the latter case, the War Department would have to receive official information that the rioters were imperiling navigation, endangering merchant shipping, etc., and it would be up to the local army Engineer to indicate the need to open navigation before Federal troops of infantry and a battery of light artillery from New York harbor would be sent to Homestead.[27]

The possibilities for Federal intervention remote, on the 7th large amounts of ammunition had already been shipped from the State arsenal to Pittsburgh, to be on hand if needed by the troops. But two days later General Greenland was still saying that the Sheriff had not exhausted all means in his power to preserve the peace and no State troops would be taken in until this was done.[28]

As the Federal Government would not enter the dispute without the State using all its power to suppress the "rioters" at Homestead, it fell upon the Governor of Pennsylvania to take action. But even he was restrained by the necessity that local authorities first exhaust their powers. Thus, by July 8th Sheriff McCleary attempted to peacefully enter the mills and assume charge, following the opinion of his legal adviser, Attorney Perry, that the Carnegie firm was neither legally nor actually in possession of the Works.

"The men are intensely suspicious and are afraid the proposition of the Sheriff has behind it some sinister motive, and that the introduction of deputy sheriffs is but a cloak for more important actions," wrote the *Cleveland Leader*. "They fear the Pinkertons will follow the deputies and that nonunion men will follow the Pinkertons. Both the latter ones they will resist with all their force. The cooler men are advising the strikers to permit the deputies to come in peaceable and take possession of the Works, but if it be granted, the strikers insist as conditions precedent that the deputies shall be neither numerous nor well armed, and that they shall not make a move to protect men whom the firm may wish to introduce."[29]

Of the balancing act, wrote the *Leader*:

> It will thus be seen that the move is really but a continuation of the strike and of the forcible prevention of the, continuance of mill work. . . . The unwillingness of the men to turn the Works over to the Sheriff and his deputies is increased by their distrust, openly expressed of Sheriff McCleary. This course does not seem to have met the approval of the firm, and the strikers openly accuse him of being with the company and of acting a double part, but the specter of the appearance of a long line of soldiers with no purpose but to sternly obey any orders they may receive is behind the Sheriff's failure and some of the men realize this.[30]

The leadership of the Advisory Committee immediately moved to pacify the rank-and-file by presenting them a choice only of letting McCleary into the mills or enduring a military occupation and the introduction of scabs. Hugh O'Donnell and others had returned on July 8th from the Capitol at Harrisburg and presented McCleary's proposal that he and his deputies take the Works and hold them for the Carnegie Company. Amalgamated President Weihe, President-elect Garland, Secretary Kilgannon, W. J. Brennan, now legal counsel for the workers, and the Advisory Committee all strongly favored this proposal.[31]

At a meeting called at the Skating Rink by these officials, but attended by only 200 men, President Weihe argued that the State was compelled to enforce observances of the law, and could not tolerate a government in Homestead unrecognized by the law. Since the firm was legally entitled to possession of its property, possession would eventually be guaranteed, either by the Sheriff or troops sent in by the Governor.

Attorney Brennan wanted the Sheriff given immediate control, arguing that if the Sheriff was not allowed in, then the Militia would be a grimmer alternative:

> The owners know that whatever may be the moral position of the men, they have the law on their side and that the Sheriff must put them in possession of the property. The people are against the Pinkertons, but they will not be with you in resistance to the Sheriff. He and his posse are the legal representatives of the peace, and are not like the Pinkertons, a hired band who are to act as their employer directs.

Others echoed the legal approach to McCleary's takeover of the mills. One speaker insisted that the workers would "gain friends by showing that we are not Anarchists, but law abiding citizens, fighting for our rights as workingmen," adding: "We cannot afford to fight the civil authorities."

The small attendance at the meeting was in part due to the opposition of many men to the idea of yielding the mills, their distrust of McCleary, and their insistence that he would have to enter the mill unarmed, allow the workers to control the situation, and not use deputies to introduce strikebreakers – assurances McCleary could not give under the law. "The strikers claim," said the July 8th Homestead *Daily Messenger*, "that the Company's property is as safe in their hands as in the hands of the deputies, and that if the deputies are permitted inside the Works the Pinkertons will be rushed in and the men will be forced to fight for their rights." [32]

This distrust prevailed for the moment, as the few who attended gave little mandate to the conservative leaders in the Amalgamated, a reporter noting their feelings, that "a great number of the younger and hotter elements are opposed to any yielding."[33] Finding no solution, the leaders left for Pittsburgh to confer with the Adjutant-General and McCleary to learn what the Sheriff would do if control over the mills were given to him. But by the 10th, rank-and-file opposition to the Sheriff still unyielding, the Adjutant-General concluded that though McCleary could not protect the mills, the National Guard would not be resisted.[34]

* * *

Under orders to prepare the way for the Militia, McCleary was in no mood to negotiate, and the men defended by keeping him out of the mill.

The Homestead *Daily Messenger* pointed out the resulting alternatives - to let the troops come in with the duty to keep out non-unionists until the Company and the strikers settled their affair, or to let them in to protect both Carnegie's Works and his scab workers employed inside the mill. Though the workers did not want to fight the troops and hoped the Militia would be neutral in their battle with the Company and the Pinkertons, the *Messenger* believed the Governor might order the troops to Homestead at the strikers' expense.

O'Donnell and the other leaders were playing a two-faced game – telling the authorities that neither the Militia nor the scabs would be resisted, yet also telling the rank-and-file that their visit with the Governor was "entirely satisfactory" towards fulfilling the workers' goals. "From casual remarks, though," the *Messenger* saw through the veil, "it seems probable that notwithstanding what is said about the Governor's reception of them, no assurances were obtained, and that the troops may be ordered here."[35]

A moment of uncertainty reigned. Not only was McCleary refused power over the mills at the Amalgamated "mass" meeting of the 9th, but the *Subcommittee on the Press* lost its hold over rank-and-file who bounced reporters, while the leaders themselves were nowhere to be found. "The control of the leaders has been greatly loosened in their absence from the town on many missions of importance," the *Cleveland Leader* dismayed, adding that "other work has practically dissolved the Advisory Committee, which for a time kept the reins of authority well held."

By the 10th, however, most of the leaders were again in Homestead, "endeavoring to assert their influence to secure unity of [the population in] intelligent and well directed action."[36] Towards this end, the commune was again aligned under the dictatorial powers of the Advisory Committee.

First, said a reporter:

> Arrangements are being made to reorganize the Advisory Committee and again get things in order so that none but the proper representatives of the mill men will assume to act for the men. A number of the locked out men deprecate the actions of yesterday in holding up newspapermen and others, and say that means will be devised so that people may establish their identity and not be suspected of being disguised detectives.[37]

Though the power of the Advisory Committee had lapsed, now it was reorganized, imposed strict discipline on the men, and absolutely denied reports that the men intended to use stored dynamite to blow up Carnegie property on the approach of hostile forces – resenting the imputation that they were Anarchists. They also denied reports that two cannon were on the way to them, though on the 9th two twelve-pound Napoleon field pieces had been shipped from Brenham, Texas, one of them originally issued by the Government to the Brenham Field Artillery, with tags directing them to the "Strikers, Homestead, Pa." Too, they took issue with the statement that they were fortifying the mill and property to resist attack, though reporters were excluded from the mill and those with field glasses observed that the strikers had fortified weak points on the river bank – creating a position it would take a large force to conquer.[38]

Though the Advisory Committee also portrayed the commune as pacific and law-abiding, the fact is that the rank-and-file were armed, had stored dynamite, had stepped outside the law, and were preparing for a long siege against all outsiders – be they Pinkertons, newsmen, or State militiamen. Recognizing this, the Advisory Committee gave immediate attention to dissuading the workers that the Militia was part of the enemy.

## Opening the Way for the Militia

Sheriff McCleary meanwhile played out his role in taking the legal steps necessary to gain support from the State Guard, three events indicating the tactics involved.

On July 6th, Governor Pattison had still refused to order out the Militia: "the local authorities must exhaust their resources first," said he. "The Sheriff can appoint ten thousand deputies if he needs them."

In response, on the 7th the Sheriff took a total of 2,000 names from the Pittsburgh City Directory, sending summons without response. And, meanwhile, General Wylie planned to go to Pittsburgh to represent the Governor in resolving the labor troubles.[39]

These steps were, in retrospect, a calculated setup for ordering out the Militia. A day later, on the 8th, the Governor took note of their reconnaissance. General Wylie went to Homestead with his staff to go over the terrain and was fully advised of the lay of the land. Every phase of the situation from a military and legal standpoint was then thoroughly discussed. General Greenland, hoping to settle the lockout peaceably, advised exhausting every legal resource before resorting to a call for Militias.

Considering the suggestions of securing an Allegheny County court injunction denying possession of the Works to the strikers, the State's legal counsel insisted on proof that the men at the mill were trespassing – though such was not the case, for the men were not in actual possession of the mills, but only patrolling along the road nearby and picketing on the heights overlooking the Steel Works and the Monongahela.[40] Clearly, as an essential precondition for the Militia's occupation, McCleary had to show he could not control the strikers.

Thus on the morning of July 8th, he again summoned 500 men to act as deputy sheriffs at Homestead. He had called the entire list, one by one. But in all only 30 men appeared, the majority of the others filing excuses, and some of those present also tendering doctor's certificates testifying to their physical inability to serve. The stack of filed excuses indicated to the Sheriff that few would go to Homestead: "After calling the list of summons sent out and waiting a considerable time for responses," said the dour officer, "only twenty-three have signified their willingness to go to Homestead. It would be useless to go there with this small number of men. I thank you all for coming here, and ask you to hold yourselves in readiness to answer any call I may make on you for assistance. I am going myself. None of those who appeared are armed, and I have no arms to give you. I will, therefore, not ask you to go along, and excuse you for this morning." Thereby, the 23 prominent business and professional men recruited were judiciously relieved of service by Sheriff McCleary himself.

The Sheriff was playing for time so the Militia could legally be brought in. Governor Pattison had appointed McCleary and it was McCleary's task to preserve law and order. Only if he could not do so could the Governor order the troops out. The Sheriff had not yet shown he could not do his job. The Committee of Homestead Citizens, moreover, had indicated their willingness to cooperate with the legal authorities. Hugh O'Donnell of the Amalgamated had led their delegation to the capitol at Harrisburg to meet the Governor at 10:30 P.M. on the 8th.

The Committee now stated that the Sheriff could take charge of the property belonging to Mr. Carnegie and that there had never been any attempt to execute a "writ" or to serve a "legal process" on them and that the Sheriff would be permitted to do so at any time. One Committee member offered that the people of Homestead, without regard to political party or organization, commended the Governor's action in declining to send troops and requested that he adhere to this position.[41] No assurances were given that he would.

* * *

Ten hours earlier, at 1:00 A.M., Adjutant-General W. W. Greenland had arrived in Pittsburgh and agreed that the replacement of Carnegie Steel Company watchmen by Homestead workers in the mills "ought to settle the matter." The historical frame was clear, he continued, as

> The National Guard of Pennsylvania was not organized and does not exist simply to perform police duty. To compel it to do such work four or five weeks every year would demoralize it. There is no danger that it will be called out at present and the Governor is not coming to Pittsburgh or Homestead.
>
> These people at Homestead are not the kind that ran rampant in the coke region a little over a year ago when the troops were called out. Over sixty percent of the Homestead workers are intelligent Americans who own their own homes and are good citizens, except under great provocation.
>
> In the coke region seventy-odd percent of the workers were ignorant foreigners, and the troops were not sent among them until the Sheriff of the county had made a generous effort to preserve the peace and had been unsuccessful. He deputized the members of the Militia companies, to which I am opposed, but I presume a Sheriff might compel them to serve until they should be called upon by the Governor.
>
> It is a notable fact that not a judge of the county has sent a message, either for or against calling out the National Guard. This is a big county, with a population of over half a million people, and it can handle a disturbance like the present one. It has not become a riot like that of 1877.[42]

Sometime between 10:00 A.M. on the 8th and the llth, a big change took place in the thinking of generals and politicians. The Adjutant-General had spent the day of the 8th conferring with the Sheriff, the officials of the Amalgamated Association, and prominent citizens. No Militia was thought needed. "If it is demonstrated that the Civil Authorities have exhausted all their resources to put down the disorder and that the trouble cannot be amicably settled," wrote the Baltimore *Sun*, "it is probable the Governor will order out the National Guard. Up to the present time, however, the efforts of the Sheriff to secure a posse have been very feeble."[43]

Without a posse, the Sheriff could not even test his authority to establish law and order. "I think," stammered the Sheriff, "the best plan is to let matters alone until the excitement subsides. All the means at my disposal I have exhausted. I do not know whether I shall make another call on the Governor. It is impossible to agree to let the strikers guard the Works. The Company would never get possession of them in that case. I am not prepared to say if I will place my own deputies in the Mill."[44]

By Friday evening, July 8th, the workers no longer trusted Sheriff McCleary. Some said they believed McCleary was simply waiting for his opportunity to make more trouble in order to turn the trend of popular sympathy away from the strikers. Harsh feelings by Homestead people were intensified by the common suspicion that the Sheriff knew ahead of time about the Pinkertons, plans to take possession of the Carnegie Works.[45]

McCleary was more or less at the end of his legally prescribed options. He had asked for assistance from the Governor four times by July 9th, and had been denied help each time. He could not raise a posse, and even if he could raise 500 men, McCleary was sure they would be resisted, as the strikers, he insisted, "will give up possession of the mill only conditionally, and I, as Sheriff, could not take it that way. . . . I positively will not take men up there to be killed."[46]

On the 10th, the generals and the Governor again conferred and agreed McCleary had followed his legal script well. The Adjutant-General of Pennsylvania said that while the Sheriff had not done what he could at the onset of the trouble, it was beginning to be apparent that his power to maintain the rights of the Company was gone. Though the Sheriff would be permitted to take charge of the mills with 25 to 30 deputies, he questioned whether a larger force would be permitted to enter by the workers. He also conceded that the Sheriff's possession would be in theory only, and that the workers would maintain real control. And though a local war would begin if the Carnegie Company introduced new men into the mill, the General thought that the National Guard of Pennsylvania would not be resisted.

To this opinion listened the Governor, Major General Snowden, Quartermaster General McClelland, and Colonel C. T. O'Neil, the Superintendent of the State Arsenal. Snowden had come to obtain from General Greenland a clear statement of the situation and to talk over probable disposition of troops in the event that they were called out.[47] Given subsequent events, he apparently learned what was to be done.

Sheriff McCleary suddenly came up with the final appeal to bring in the Militia, pleading total inability to control the workers at Homestead.

"The situation at Homestead has not improved," he telegraphed Governor Pattison on the 10th. "While all is quiet there the strikers are in control and openly express to me and to the public their determination that the Works shall not be operated unless by themselves. After making all efforts in my power I have failed to secure a posse respectable enough in number to accomplish anything, and I am satisfied that no posse raised by civil authority can do anything to change the condition of affairs, and that any attempt by an inadequate force to restore the right of law will only result in further armed resistance, and consequent loss of life. Only a large military force will enable me to control matters. I believe if such a force is sent the disorderly element will be overawed and order will be restored. I, therefore, call upon you to furnish me assistance."[48]

The planned legal scenario had been played out. On July 10th, the Governor, as Commander-in-Chief of the Pennsylvania National Guard, ordered out 8,000 troops to Homestead for the purpose of supporting the Sheriff, his mandate reading:

> George R. Snowden, Major-General Commanding the National Guard of Pennsylvania: Put the division under arms and move at once with ammunition to the support of the Sheriff of Allegheny County, at Homestead. Maintain the peace – protect all citizens in their rights under the constitution and the laws of the State. Communicate with me.[49]

## Leader Preparations for the Occupation

On July 12th, the Militia was received by many millworkers and the people of Homestead with some misgivings. It was clearly not an exuberant welcome by everyone. Some could see the connection between Militia, Pinkertons and the bringing in of nonunion workers in the future. Yet, by and large, the people had been quieted by persuasion or threats.

"Pleasantries Between Troops and Strikers," the free-trade Baltimore *Sun* cynically announced the coming meeting of Homesteaders and Militia:

> The march of the Pennsylvania troops to Homestead will be a picnic if the strikers do all the amiable things for them that are predicted. It is said they will meet the troops at the railway station with a brass band and cheers. That is very well. Ice cream should be the next thing in order. The military should reciprocate with recitations from the McKinley bill, showing how it 'protects' the workman. It might be well

> also to quote this passage from Mr. McKinley's speech in presenting the [Republican] Conference Report on the bill in September, 1890: 'Nothing has been done in the metal schedule that can result in a loss of business and nothing that can call for a reduction of wages.' That would be pleasant to hear, now that the chief beneficiary of the metal schedule is cutting wages from 30 to 50 percent.[50]

Preparations for the Militia were extensive. A mass meeting was first called by the Advisory Committee for the afternoon of the llth. It was attended by about five thousand people, including just about everyone who had fought the Pinkertons on the 6th. The meeting was called to order by Hugh O'Donnell, who began appeasing the press for the indignities they had to put up with during the days of the commune, now inviting them to share the platform with the leadership. "They are the people," he announced to the onlookers, "who have created a public sympathy for us all over the United States. They have been our friends." Great applause followed, while Burgess McLuckie made his way to the podium as chairman of the meeting.

The crowd was not of a single mind, though. Many, shouldering rifles fresh from the threatening war with the Pinkertons, held grave doubts about bringing in the Pennsylvania Militia. The Advisory Committee had tried to convince the people that the State Militia would protect them and that the Pinkertons would not again try to occupy the Works. Since Homesteaders had themselves routed the Pinkertons and made detailed preparations for any sneak attacks, it did seem the crisis had passed and that Homestead would know hostilities no more. Yet, many wondered if the State Militia would attempt to smuggle Pinkertons into the Works. Still more had questions about the real interests of Governor Pattison. Did he favor mediating the workers' interests against Carnegie's? Would his ordering out of the Militia have that effect? Burgess McLuckie tried to allay their fears:

> The object of this meeting is to take suitable action for a reception of our friends, the military, who will arrive here in a day or two. The general impression is abroad that the military people are a dangerous people. This is a mistake, and in order that the dignity and honor that they so nobly and faithfully represent should not be insulted, we have come here to take whatever action may be deemed feasible to prevent that calamity. We want to arrange for a reception – one of the kind that the military has never thought of.

Peals of laughter rolled from the audience. Relaxed, McLuckie continued:

> We want our own people . . . – in fact, every man, woman and child in this town – to welcome the military people with open arms. Brothers, just as true as we stand here, I pledge you my judgment that this is one of the best steps that could possibly have been taken for our people - this calling out of the military.

Until now the Democrats had made hay out of the Pinkerton attack, by associating it with the Republican's tariff-favoritism for Carnegie against the interests of the workers. So McLuckie called attention to the willingness of Democratic Governor Pattison to listen to and negotiate with the strikers after the Pinkerton onslaught, portraying the Militia sent by the Governor as an ally.

> I want to say to you that I think this man, Governor Pattison, is acting wisely and judiciously. He understands our cause and he understands our position; so does the entire civilized world. He is a just man and a man who will not cater to monopoly; a man who will not permit the troops of this State, the servants of the people, the defenders of the dignity of the Commonwealth, which is ours and as near and dear to us as to any human being that ever God let breathe – I say Robert E. Pattison will never permit an outrage to be penetrated upon such a people as those of Homestead and this surrounding community.

The audience broke into great applause, never thinking that Governor Pattison had connections and interests quite different than those McLuckie portrayed. And McLuckie cited the Pinkertons as the danger, not mentioning the State or its agents:

> It is also evident that the unwashed horde of Pinkerton people are again mobilizing their forces. That is also evident to our Governor. He does not want any bloodshed here, and he knows by your notion of the past that if that unclean horde strikes our shores there must be bloodshed.

This brought still more applause, McLuckie further embellishing the Governor's character in ordering out the Militia:

> People said last night: 'What is the matter with the Governor?' They do not know as much about this as the Governor does. This is why they ask this question. The Governor knows that these unwashed hordes are mobilizing again. It is the most desperate blow ever struck to the Pinkertons since the organizing of that institution.

More applause followed, McLuckie having focused well the people's hatreds, bringing forth cries: "We will give them another!"

As for the Governor's vision, McLuckie insisted:

> He does not want us to be slaughtered, or to receive further injury at the hands of that illegal, unlawful, dirty, sneaking institution.

Applause and laughter joined, Homestead's *Burgermeister* taking the offensive to acclaim the Military:

> And so your friends are about to come here – the safest people that you could possibly be in with. You know you do not want H. C. Frick or his horde; neither do you want Bob Pinkerton or his horde. The military are here to defend that which is right to defend – the State and the Constitution, and the people and laws.

Cheers greeted this facile view embellishing the State and securing Homestead's interests, McLuckie grasping the opportunity to stress that the people of Homestead must receive the Militia well:

> I am here to ask the citizens of this town to receive our friends with open arms and to tender them one of the grandest receptions that have ever been tendered to any people. [Applause] Go down with the reputation that you have already established as the most generous, noble and manly of men. [Applause] I do not want the mobilized representatives that may put in an appearance here to be insulted by even a hoot or a howl or the least indication of displeasure. I for one stand here and say that if any man is indiscreet enough to do an act of that kind in this emergency he should be taken to the river and be ducked, and I will be the one to help do it.[51]

"Good! good!" the crowd echoed with applause, bringing McLuckie's smart reply: "You are not going to interfere with our best interests, so our ducking committee will scarcely be necessary to be appointed at this meeting." Laughter led McLuckie into his *finale* of approval for the military occupation, a serious resolution being carried to establish a "ducking committee," with special policemen who were later instructed to carry it out to the letter, it also being ordered that a committee should be appointed to inform all the mill workers not present that such action binding them had been taken.[52]

Trying to interject another view on the merits of opening Homestead to the Militia, an indiscreet striker suggested that a committee be appointed to make complaint if any member of the military should be guilty of unbecoming conduct at any time. But McLuckie insisted otherwise: "If it should be found at any time that there has been any breach of the peace by the military people we will look it up then, but we are certainly not going to criticize their conduct until they commit some act."

With the great applause of approval that followed, McLuckie announced he would issue a proclamation to keep out of Homestead any persons not connected with the mills and having no business in the city. These people would be kept away while the Militia was there, a measure that was passed by unanimous vote.[53]

The proclamation McLuckie promised was designed to compel all strangers with no business in Homestead to depart. The Burgess issued it immediately after the mass meeting, indicating he had already prepared it as part of a planned sequence of events:

> Whereas, many strangers are coming to Homestead with no other purpose in view than to gratify a curiosity aroused by the pending settlement of the *wage question*, and inasmuch as the presence of these idle strangers only tends to further complicate the adjustment of the difficulties and maintenance of order,
>
> Therefore, I, the Burgess of Homestead, do earnestly request and warn all persons not having important business in Homestead at this time to remain away from town; and further notice is hereby given that all strangers within the town will be held strictly accountable for their conduct and be dealt with in a manner warranted by the gravity of the situation.

The proclamation then turned its force against the people of the city, restraining them along lines of outrage, opposition, age and gender:

> I further request and demand that the citizens of the borough assist to maintain order and quiet by themselves refraining from violent talking and unnecessary assembling in public places and by restraining all others to obey the officers of the borough. I also enjoin the women and children to remain off the streets, and especially during the prevalence of unusual excitement.
>
> I hereby order and direct that all the saloons in the borough close and remain closed until further orders. I deem this of the first importance, owing to the large number of strangers and unemployed persons now in the town, and I request a full compliance with this order.[54]

To enforce this order for the original goal of settling the wage question, a special police force, one hundred strong had been summoned by McLuckie. Never a democratic thinker, McLuckie himself had hand-picked men from among the strikers who were the best representatives of brawn and muscle Homestead could offer. With commendable impartiality implementing McLuckie's proclamation, these police immediately set about performing their duty, McLuckie insisting in the afternoon on the 11th that the "special police" were appointed to see that his orders were obeyed and "any unseemly manifestations . . . suppressed."[55] By evening the special police had silenced several people they considered "boisterous," marching the disrespectful off to the lockup. Most of those arrested were mill workers, one of whom, a reporter said, was a lesser leader of the strike, "obstinately slow in recognizing that the war is over and that the era of peace has arrived."[56]

Another reporter witnessed strict surveillance on the streets: "All loud talkers, whether strikers or sympathizers, were promptly suppressed, and to give an insolent answer to these guards was to be unceremoniously hustled off to the lockup. Several of the fellow-workmen of these impromptu police presumed upon past familiarity to the extent of mildly guying the tin-starred officers, but in every instance they had occasion to regret their temerity."[57]

Now Homestead's "little Caesar," McLuckie's special police may have changed no one's inner thoughts about the merits of bringing in the Militia for the purpose of "settling" the wage question, but change their outward actions they did. What was really essential to counterbalance the leaders' point of view and repressive measures was another view from a group of rank-and-file workers or others, pointing out the limited nature of the economic demands the leaders sought, as well as their use of repressive measures against the workers to enforce their will and their misguided policy of viewing the Militia as a force in their favor.

Though such rank-and-file efforts did not emerge, few dissenters missed the meaning of the speech by Dan Harris, president of the New York Cigar Makers, Union, at the rally. "You," he told the throng, "are the pioneers of a movement that must eradicate from the civilization of the United States a blot upon its escutcheon – the Pinkerton detective system."

With him other speakers counseled order and sobriety, when the Militia entered the city.[58]

The links were obvious. Both the Cigarmakers' Union and the Amalgamated Association were part of the American Federation of Labor, following its policy of a "fair share for labor" in wage negotiations. They accepted the wage system and the employers' legal rights in the instruments of production. They would oppose an illegal force of Pinkertons, as they stigmatized these as a private army of capital. But they would not do battle with the State Militia, as the protector of private property. Rather, they would roll out the welcome carpet and put down all dissenters to the Militia's occupation. They would create their own special police force to enforce their will against the rank-and-file. And they would make it appear that, rather than a hostile force, the Militia would protect the workers' interests and should be welcomed as a friend, not a foe on the side of the employers.

This was the plan the leaders followed at Homestead. The special police had been organized to see that no indignity was offered the soldiers by any of the injudicious or hotheaded workmen. And the efficacy of the police had been demonstrated within a half hour after the adjournment of the mass meeting called by the Advisory Committee, as they summarily escorted and locked up an intemperate mill man who dared question the wisdom of receiving the Militia as friends and brothers – citing Biblical scripture that the leaders were killing the fatted calf in the enemy's honor![59]The workers' unanimity thus secured, the leaders had used the decision of the mass meeting to prepare for the Militia.

McLuckie had set the tone for this welcome at the mass meeting, declaring that the plan on foot was "to get out every band in Homestead." Five thousand had applauded, and a representative of Homestead's musicians had shouted: "I am authorized to say that the bands have already volunteered their services." Popular enthusiasm exuding from the crowd, the president of each Amalgamated lodge was then instructed to make the proper preparations for the ceremonies.

Burgess McLuckie announced he would ascertain exactly when the Militia was expected and give notice of their approach by town criers.[60] In contrast with the reception to the hated Pinkertons, the Militia was to be welcomed by the jubilant blasts of trumpets and the music of bands!

## Welcoming the Militia

"Brass Bands! An Enthusiastic Greeting to be Tendered the Militia at Homestead. The Mill Men Will All Turn out and There'll be Music in the Air," the *Cleveland Leader* headlined the coming event.

The population retired at an early hour in order to be up at sunrise to participate in the reception ceremonies. But the band stayed awake to rehearse that evening, determining upon the "airs of triumph" that should be played on the arrival of the soldiers the next morning.

Also awake was a committee representing Burgess McLuckie and the Advisory Committee. They established telegraphic communications with General Snowden to try to ascertain the exact hour at which the Militia was expected. They informed the General of the action of the mass meeting, asking for information about the Militia's plans in order to time the reception with the arrival of the troops.[61] Then, after giving

General Snowden a picture of the situation at Homestead, they went about making final preparations for the morning ceremonies.[62]

The city slept, the Militia prepared, and the Advisory Committee's power of bureaucratic organization reigned the whole night of the 11th of July. Yet Committee members pondered whether or not they could confine Snowden's troops strictly to the Carnegie Works and leave the preservation of order in Homestead to the town authorities, the Burgess, the Chief of Police, and the Fire Department. Since the hostilities of the 6th were at the Carnegie mills outside the Borough, McLuckie and the others hoped to preserve their local autonomy.[63]

Yet McLuckie's plan to bring in the Militia to oversee the population and keep out the radicals and other concerned people was a recipe both to undercut local rule and for disaster. Nonetheless, his thoughts ruled the day, and detailed preparations for welcoming the Militia turned into an expected festival.

Although the Advisory Committee found it impossible to disarm the population by collecting the $5,000 worth of firearms that they had promised to return on the Pinkerton surrender, many people, thinking the military struggle was over in Homestead, tuned to momentary support of their families by hawking ammunition and even weapons to visitors looking for mementos from the battle scene. Hundreds now whiled their time away standing on the railroad tracks awaiting the troops. At the platform of the little railway station several hundred other men crowded, some standing on the curbs or in the doorways. Women and children also violated McLuckie's stern order to stay indoors – calico gowns and knee breeches appeared everywhere, women seeming to show more interest in the expected celebration than the men.

Two hundred brave policemen were now solemnly sworn in by the Burgess to patrol the town and keep strangers out. But even they could not stop the massive flood of visitors from neighboring boroughs and the countryside, coming on horses, in vehicles of all sorts and by train. Some stayed overnight when the troops did not appear; others went to nearby Brinton to locate the soldiers in blue, only to return to Homestead disappointed, while Burgess McLuckie and the Advisory Committee tried to reach General Snowden so – the troops would arrive for the crowd's enjoyment between daybreak and sundown.[64]

"I wish," muttered the undisputed executive of Homestead, Hugh O'Donnell, "the fact could be carried to every friend of ours in the land that we are really glad that the troops are coming, and we propose to testify our satisfaction in an unmistakable manner. The Militia relieves us of a responsibility which has borne on us heavily, and we gladly give over to them the preservation of order and the protection of the Carnegie property."[65]

Had Homestead officials and the Advisory Committee a level head among them, it might have been a prime moment to herald the dangers to the strikers in bringing in the enemy. But the Militia was seen otherwise, and support for it had been nearly universalized. Reporters more skeptical, one engaged O'Donnell in an interview to which every thinking unionist might raise an eyebrow.

> Q: Will the next step be the bringing in of Pinkertons and nonunion workingmen?
>
> A: We have not come to that yet. It is a long way in the future. I do not think it will be a healthy climate for Pinkertons for some months to come. But the question is not to be decided now. The present duty is to attend to the reception of the troops. Everyone of our men will put on his Sunday clothes, get a clean shirt and come down to do honor to the military power of the State of Pennsylvania. It will be one of the most pleasant and popular receptions ever seen in Homestead. We have no quarrel with the troops, and as I said before, they are relieving us from a great responsibility.
>
> Q: But when the Pinkertons come; you know it is said that there are 400 massed in different places within two hours' ride of Homestead?
>
> A: Well, the Pinkertons' force coming here would be an illegal attack on our town. If they attempt to come here the Militia will disarm them and send them away, or if they do not very likely some of our people may.
>
> Q: And how about nonunion workmen in the mills?

Hugh O'Donnell laughed:

> A: Nonunion men are not got so easily as you would think. That is a question that is not for this Summer, but for next. We are not worrying. All we are thinking about now is the best way to welcome the Pennsylvania Militia to our town.[66]

And so the plans were laid and the Militia under Snowden marched into Homestead unmolested.

**Military Tactics**

The troops arrived in Homestead as a State Militia, not as a Federal

Guard. This distinction was important, so far as the process of legal domination over the people of Homestead was concerned. The authorities at the U. S. War Department had considered it prudent to avoid any connection between America's regulars and the State Militia so long as the Homestead fight continued.

The decision was largely political. It was well known that in an emergency like a general strike or lockout, effective troop strength in any particular state could only be mobilized if there were no other drains on its section of the National Guard. With the heavy industry of Pennsylvania at the eye of labor conflict, General Snowden had staunchly opposed U.S. Senate legislation that would have diluted the Pennsylvania Guard by allowing members to enlist in a volunteer regiment for foreign duty. "Such remedial legislation," Snowden fulminated, "is by no means necessary. Should a call for volunteers be made the Governor [of Pennsylvania] would easily discharge a percentage of the State militia for that purpose proportionate with the state' population towards the [Federal] Government. . . . Of course a number must be held in reserve for home defense. It is likely that, if a call were made, only one or two regiments of the National Guard would go from this State, and though re-enlisted as volunteers, they would be still be practically the Guard in organization."[67]

Accordingly, the War Department did not meddle, not even to detail officers to visit Homestead to witness the operations of the Pennsylvania Militia – not even in the capacity of inspectors and reporting observers for the War Department.

With President McKinley's tariff bill favoring Carnegie against his workers as backdrop, Major W. J. Volkmar, the Federal Adjutant-General, also took special precautions to ensure that no one could say the War Department under the President had planned the Homestead occupation. Volkmar originally was supposed to leave Washington, D.C. during the week starting July 11th, but conveniently delayed his departure for Pennsylvania until after Governor Pattison had ordered Homestead occupied by the Militia. Thereafter, Volkmar cautiously joined the Governor for an inspection trip of the various state regimental encampments. The implications were clear, though the newspapers went out of their way to assert that Volkmar's orders did not mean the Federal Government was detailing its officers to assist Governor Pattison.[68]

The Militia was ordered out for "riot." General orders for the mobilization at Homestead were issued: "The movement of troops will only be in compact bodies. No firing will be permitted except by order of the officer in charge. When all other means of quelling the riot and restoring order have been exhausted, the officers commanding the troops shall notify the rioters that they will be fired upon unless they promptly disperse. Should they not disperse the order to fire will be deliberately given, and every soldier will be expected to fire with effect and to continue firing until the mob disappears."[69]

Not understanding the intricacies of military logic, neither the Homestead leaders nor the community had expected anything like this.

Yet, the ordering of the entire Pennsylvania Guard to Homestead was based on the calculation that thereby the Militia could be kept from fraternizing with the local population. Soldiers living in other areas of the State would not likely know anyone living in Homestead and would, in the opinion of the military authorities, remain detached in carrying out their orders to disperse any mob with firepower.

Warlike preparations had been made, as if the workers at Homestead were in total opposition to the Militia. The inpour of troops was thus to be fast, massive – a *blitzkrieg*. Five large trains were to carry the Blairsville troops alone, with the First Brigade at Mount Gretna and the Third at Lewistown yet to come.

"Within twelve hours of the issuing of orders to mobilize the First Brigade at Mount Gretna," one account says, "1,900 men or within a couple of hundred of the brigade's full strength, had left this City. . . . Quickly followed the Third, the Crack First Regiment with 608 men out of 630. . . . All was bustle and activity. Long trains of cars stood upon the tracks and as the men entered they were rapidly gotten aboard. The two trains with these commands had hardly left before the Second Regiment marched into the Station. . . , 500 out of 632 men. Among the Second's baggage were ten thousand cartridges stored in boxes. The battalion of State Fencibles, 280 strong, were the next arrivals. They followed the Gray Invincibles, a colored company, who of their 60 men had 46 in line. The Second Regiment completely filled one train in twelve cars and the battalion and the Invincibles took up another train of cars."[70] With such a formidable force, General Snowden was confident he could assume a posture of arrogant and unopposed control over Homestead.

And that is what he did. The population was trapped. The rank-and-file that initially agreed that the Militia would be received without any hostility, then said that an armistice with Carnegie would not be declared until after the Militia departed. As for the Militia's purpose, conjectured one perceptive leader of a little group congregated near the Depot on July 11th, "they will have to go away some time, and when they do we would like to see them run the mill nonunion."[71]

Nonetheless, by midnight, the troops on their way, the workers began to fear that the Company might sneak in Pinkerton men in disguise of workmen while the Militia held possession. Such a maneuver would mean a practical triumph over the strikers and this harsh suspicion left most Homesteaders in an uneasy frame of mind.

Burgess McLuckie, always the City Father, tried to calm them, responding to the press as well:

> Q: How do you regard the latest news - the calling out of the Militia?
>
> A: That is right on our line. It suits us first rate.
>
> Q: How will the Militia be received?
>
> A: With open arms.
>
> Q: There will be no hostility?
>
> A: None whatever. We will receive them in a fraternal spirit and accord them the respect due to the representatives of the grandeur and dignity of the great State of Pennsylvania. We have been expecting them and we are glad they are coming. There will not be the slightest opposition to their presence, nor the least manifestation of unfriendliness.
>
> They will be kept here a few days, and when it is seen how peaceable everything is I predict they will be withdrawn. The Sheriff has been insisting that he could not keep the peace, and I suppose it has been finally determined to send the troops here to see how necessary they are. They will be received with respect. Our fight was against the Pinkertons -- against the invasion of our homes by an armed, illegal and disreputable private army.[72]

McLuckie, maybe honestly, maybe blindly, was attempting to pacify the population for the occupation, while General Snowden planned to take Homestead in *toto caelo* and by surprise – with the whole Militia to be brought in early Tuesday, the 12th, so early in fact that the workers' reception with formalities would be next to impossible.

* * *

Though the Governor mobilized the Militia on the 10th it was already the 11th. Homestead had waited almost 48 hours, and the people were apprehensive. The City's musical *corps* stood at the ready to play for the arriving Militia, and, while awaiting word from McLuckie, passed the time on the evening of July 11th playing such favorites as "Mc-Ginty" and "Annie Rooney," to raise the falling spirits of the community.

The next day came without the slightest information from the Militia. "We could have blown up the mills and sacked Allegheny before the troops got here if we had not been peaceable and law-respecting," a brawny striker blustered, drawing smiles of approbation all around.[73] It was true: if the workers had not been becalmed by the leaders of the Amalgamated, and by McLuckie and his special police, the Militia might have found Homestead a camp armed against them. For this purpose McLuckie's police force was highly successful in repressing the population. Saloons continued to be regulated and had to close their doors an McLuckie's orders, supposedly to be given one hour before the arrival of the Militia, if sufficient notice could be had.

What happened was quite different indeed. For the troops under General Snowden arrived by train, not on Monday night the 11th, but at 9:30 a.m. on Tuesday morning the 12th, and not in Homestead where the celebrations were arranged, but at Munhall station in the heart of the Works. General Snowden and six of his officers accompanying Sheriff McCleary stepped from the parlor car, just as the crowd surged from Homestead to meet them. But, under Snowden's orders, they were stopped short and shoved back about 100 yards by the advance of Company E of the 18th Regiment.

At this point, the leaders of the Amalgamated and the Advisory Committee again deluded themselves. Three times striker representatives tried to speak with Snowden, and three times they were waved away or verbally abused. Snowden ordered O'Donnell and other members of the Advisory Committee to disband, also refusing to listen to Burgess McLuckie's carefully prepared welcome speech and the musicians' planned triumphant tunes.

Once Snowden had established his headquarters at the schoolhouse on Carnegie Hill, another deputation headed by Hugh O'Donnell representing the Advisory Committee sought to negotiate. "General Snowden," O'Donnell respectfully stated, "on behalf of the citizens of Homestead and Advisory Board of the Amalgamated Association of Iron and Steel Workers, we welcome the National Guard of the State of Pennsylvania to this borough, and proffer our assistance in maintaining order."[74] O'Donnell continued: "after suffering an attack of illegal authority, we are glad to have the legal authority of the State here."[75]

Snowden replied coldly: "The State of Pennsylvania needs no help from the Amalgamated Association in preserving peace." Besides, Snowden said, he did not recognize the Amalgamated Association – only the citizens of Homestead, whom he expected to submit to "law and order" under Sheriff McCleary.[76]

O'Donnell changed his stance, replying that in the capacity of a citizen of the State he welcomed the Army. But Snowden was firm: "As citizens of the Commonwealth, you did not assist Sheriff McCleary in maintaining order, did you?"[77] O'Donnell, standing directly in front of the General, diplomatically offered that he did submit to the silver-mustachioed Sheriff and had never questioned the Sheriff's orders. "I'm glad to hear it," replied the smiling Snowden with sarcasm: "We are here to preserve the peace. We represent the executive arm of the State of Pennsylvania, and I have nothing to say to you further than that the Sheriff must be obeyed."

O'Donnell was aghast: "But we have obeyed the Sheriff. Haven't we, Mr. McCleary?" "No you have not," replied the now confident Sheriff. "You refused to let my deputies enter the Works."[78]

This ended the *entente cordial.* After a short innocuous exchange on the altitude of the mercury, the Amalgamated representatives withdrew, leaving the Major-General to his own reflections and the new duty of taking entire charge of Homestead.[79]

"The crushing blow to the feelings of the men who had instituted what was designed to be a whole-souled reception," a reporter wrote, "came with a force that well nigh disheartened even the most hopeful, and lent color to the suspicions of those who had all along doubted the object of the grand display of military power."[80]

The occupation proceeded quickly: "Rapidly descending from the trains the troops formed in a column at the switchyards just behind Munhall station," a witness said. Their mechanical lock step demanded servility, with the first detachment of stiffened soldiers drawing not a gun from the disenchanted workmen, as the military maintained perfect order and responded silently and promptly to every order of their superiors.

Another on-the-spot reporter documented their encompassing movements:

> One company was at once detached on picket duty, and a line was immediately thrown out along the mill yards. Then the main body of the troops marched down the streets, headed by the regiment band, and along what is known as "scab hill." This is an eminence overlooking the mills and the scene of . . . hostilities, and its bluff was covered with spectators, mostly women. There was not the slightest manifestation of hostility. . . .

The troops proceeded to completely replace the commune defenders, first surrounding the Works, then marching into the heart of Homestead:

> In twenty minutes the Works were entirely surrounded, and the Carnegie mills were once more in the hands of their owners. All the surrounding streets were patrolled by pickets, and a guard at every street intersection kept the crowd at a respectful distance and prevented the assembly of groups of strikers.[81]

One company halted near the strikers' Headquarters, another marched to the vicinity of the depot, and a third occupied a slight eminence overlooking and commanding the commercial center of the town. Carefully trained, the militiamen jumped to the commands of their officers, so that, as a journalist penned, "Homestead was in the hands of the military and martial law had succeeded the reign of the Advisory Committee."[82] Thus was the peaceful program of the strike leaders observed to the letter under the new military *régime*.

* * *

In traditional military style, Snowden had selected in advance those points he would use to subjugate the City, thereby hoping to prevent fraternization of the people with the soldiers, and, ultimately, to compel the workers to submit to the Sheriff's efforts to comply with Frick's plan in bringing in scab laborers for the Works.[83] By the time the population perceived the difference between what O'Donnell and McLuckie said would take place when the Militia occupied Homestead and the overriding control Snowden established, they were trapped:

> Men gathered in groups on the streets and indulged in bitter murmurings. "We have been deceived," they said, "deceived by the Governor, deceived unwittingly, but nonetheless effectively, by our own advisers. We have stood idly by and let the town be occupied by soldiers who come here, not as our protectors, but as the protectors of nonunion men, for whom Frick is now scouring the country and who will be brought into the mills and installed in our places with the aid of .the commonwealth. Then, if we undertake to resist the seizure of our jobs, we will be shot down like dogs. It is the story of the coke regions over again."[84]

Indeed, there was no distinction between steel and coke workers in the minds of mill and mine owners, the Governor, the generals and other agents of the government.

Led by McLuckie, the leaders at Homestead had failed to view the state in its proper perspective and had dissuaded the population from protecting itself against the Militia. McLuckie had no clear thought or theory of classes and their connection with the apparatus of the State or its functions. His attachment to narrow trade unionism also left him cold to the thoughts of radicals determined to transform the social order and transfer the instruments of production from private to popular ownership.

Both sides in the struggle, leaders and state authorities, made Homestead's subjugation easier, more effective, and devastatingly complete.

## Notes

[1] *The Local News*, July 23, 1892.
[2] See Edward Gibbon, *The Decline and Fall of the Roman Empire* (New York: Wise & Co., 1943), 421ff., for this historical progression,
[3] *The Sun* (Baltimore, Maryland), July 8, 1892; for details see Wolff, *passim*.
[4] *The Sun* (Baltimore, Maryland), July 8, 1892.
[5] *The Milwaukee Daily News*, July 9, 1892.
[6] *Ibid.*
[7] *Ibid.*, July 7, 1892.
[8] U.S. Congress, House of Representatives, *Report 2447*, XV.
[9] *The Daily Record* (Baltimore, Maryland), June 8, 9, 10, 1892; *The Herald* (Baltimore, Maryland), June 8, 9, 10, 1892; *The Sun* (Baltimore, Maryland), June 10, 1892.
[10] *The Philadelphia Press*, February 8, 13, 14. 1892.
[11] *The San Francisco Chronicle*, July 7, 1892.
[12] *The Sun* (Baltimore,, Maryland),, September 7, 1892.
[13] *The Herald* (New York, N.Y.) January 5, 6, 7, 8, 1892; *The Democrat* (Niagara, New York)., January 5, 7, 8, 9, 1892; *The Philadelphia Press*, January 6, 7, 1892.
[14] *The Daily Dispatch*, February 11, 12, 13, 1892; *The Daily Gazette*, February 12, 1892; *The Philadelphia Press*, February 12, 1892.
[15] *The Philadelphia Press*, February 15, 1892.
[16] *Ibid.*, February 16, 1892.
[17] *Ibid.*
[18] *The Sun* (Baltimore, Maryland), June 20, 1892. See also *The Herald* (New York, N.Y) for the year 1884; and Ostrogorski, Vol. II, 179-227.
[19] *The Sun* (Baltimore, Maryland), June 17, 1892.
[20] *The Milwaukee Daily News*, July 9, 1892.
[21] *The Sun* (Baltimore, Maryland), July 7, 1892.
[22] *The Daily Dispatch*, July 10, ll, 1892; *The Post*, July 11, 1892; *The Sun* (Baltimore, Maryland), July 11, 1892.
[23] *The Press* (Pittsburgh, Pennsylvania), July 10, 11, 18; *The Times* (Pittsburgh, Pennsylvania), July 10, 11, 1892.
[24] See *The Cleveland Leader*, July 11, 1892.

[25] See *The World* (New York, N.Y.). July 7, 8, 9. 1892; *The Post*, July 8, 1892; *The Cleveland Leader*, July 8, 1892.
[26] *The Cleveland Leader*, July 7, 1892.
[27] *Ibid.*
[28] *The Daily Dispatch*, July 9, 10, 1892; *The Post*, July 10, 1892; *The Cleveland Leader*, July 10, 1892.
[29] *The Cleveland Leader*, July 9, 1892.
[30] *Ibid.*
[31] *Ibid.*
[32] *The Daily Messenger*, July 8, 1892.
[33] *The Cleveland Leader*, July 9, 1892.
[34] *The Times* (Pittsburgh, Pennsylvania), July 10, 11, 1892; see also *The New York Times*, July, 10, 11, 1892; *The World* (New York, N.Y.), July 10, 11,1892; *The Cleveland Leader*, July 11, 1892.
[35] *The Daily Messenger*, July 9, 1892.
[36] *The Cleveland Leader*, July 10, 1892.
[37] *Ibid.*
[38] *Ibid.*; see also *The World* (New York, N.Y) July 9, 10, 1892; *The Herald* (New York, N.Y.), July 10, 1892.
[39] *The Milwaukee Daily News*, July 6, 1892.
[40] *Ibid.*, July 9, 1892.
[41] *The World* (New York, N.Y), July, 9, 1892; *The Sun* (New York, N.Y.), July 9, 1892; *The Mail and Express*, July 9, 1892; *The Sun* (Baltimore, Maryland), July 9, 1892.
[42] *The Sun* (Baltimore, Maryland), July 9, 1892.
[43] *Ibid.*
[44] *Ibid.*
[45] *Ibid.*; *The World* (New York, N.Y.), July 8, 9, 10, 1892. This was certainly the case, ex-Sheriff Gray recalled his conversation with the men at Homestead on July 5th, for the Congressional Committee: "We came back to the Sheriff's office and were informed that Pinkerton men would be taken up there that night. I was to go along to preserve the peace." Gray then recounted the trip up the river on the boats, though he said he did not see any of the men armed until they arrived at Homestead. Gray contended that Homestead executive Potter was the Pinkerton's employer, that he, Gray, did not swear the Pinkertons in as deputies so there was no exercise of his authority as Deputy Sheriff, but that he reported the details to Sheriff McCleary, who then held a conference with Amalgamated President Weihe. See Wolff, 104, 107, 113, 121; *The Sun* (Baltimore, Maryland), July 14, 1892.
[46] *The Cleveland Leader*, July 10, 1892.
[47] *Ibid.*, July 11, 1892; *The Daily Dispatch*, July 10, 11, 1892.
[48] *The Sun* (Baltimore, Maryland), July, 11, 1892.
[49] *Ibid.*
[50] *Ibid.*, July 12, 1892
[51] *Ibid.*

[52] *Ibid.*
[53] *Ibid.*
[54] *The Cleveland Leader*, July 12, 1892, emphasis added.
[55] *Ibid.*
[56] *Ibid.*
[57] *The Sun* (Baltimore, Maryland), July 13, 1892.
[58] *The Cleveland Leader*, July 12, 1892.
[59] *The Sun* (Baltimore, Maryland), July 12, 1892.
[60] *The Cleveland Leader*, July 12, 1892.
[61] *Ibid.*
[62] Ibid.
[63] *Ibid.*
[64] *The Sun* (Baltimore, Maryland), July 12, 1892.
[65] *Ibid.*
[66] *Ibid.*
[67] *The Philadelphia Press*, January 25, 1892.
[68] *The Sun* (Baltimore, Maryland), July 12, 1892; *The Chronicle Telegraph*, July 11 and 12, 1892; *The Daily Gazette*, July 12, 1892; *The Post*, July 12, 1892; *The Times* (Pittsburgh, Pennsylvania), July 12, 1892.
[69] *The Sun* (Baltimore, Maryland), July 12, 1892.
[70] *Ibid.*
[71] *Ibid.*; *The Post*, July 11 and 12, 1892.
[72] *The Sun* (Baltimore, Maryland), July 12, 1892.
[73] *Ibid.*, July 13, 1892.
[74] Stowell, 129.
[75] Wolff, 149-150.
[76] *Ibid.*; Stowell, 130.
[77] Stowell, 130.
[78] Wolff, 149-150.
[79] Stowell, 130.
[80] *Ibid.*, 131.
[81] *Ibid.*
[82] *Ibid.*
[83] Burgoyne, 114-123.
[84] *Ibid.*, 123-24.

**Carnegie Mill, 1892**
**(Credit: *Pittsburgh Commercial Gazette*)**

**General George R. Snowden**
**(1892)**

# Chapter 8

# The Militia and the People

Imagine the horror and dismay that spread through this bower of peace when strangers from the outside world brought news of a fight at a place called Homestead between mill workers and Pinkerton detectives in which the latter were whipped. And the mill owners had now called on the other branch of their service, the National Guard of Pennsylvania, to protect them in their rights as guaranteed by the McKinley bill.

– Homestead resident Murray Scott, July 1892[1]

The troops soon proved who they came to protect. It was the Carnegie mills, not the Homestead workers.

– Anarchist Emma Goldman, 1892[2]

I am of the opinion that there would not have been a drop of blood shed if the proposition had been accepted to let the locked-out men guard the premises.

– Pennsylvania Governor Pattison[3]

On Tuesday morning, July 12th, when the National Guard of Pennsylvania marched into Homestead and occupied Carnegie Hill overlooking the Steel Works and all areas below, many thousands of people flocked to watch them file up the slope and take their positions along the face and top of the hill. By the next morning, many hundreds of tents dotted the hillside and regular camp life began. Arms were stacked, the soldiers settled themselves on the grass to rest, and Major General George R. Snowden and his staff established their headquarters in the little Carnegie School House, overlooking the town and the Steel Works, and affording a panoramic view of the Monongahela Valley.

Military preparations continued everywhere. Cannon and Gatling guns were brought in, a company of cavalry galloped through the streets, and a long freight train appeared at Sixth Avenue where, all night long and the next day, fifteen railroad cars were unloaded with the help of every teamster in Homestead hauling baggage and camp supplies, tents and other equipment.

For the moment, despite the armaments, the friendliness of the people enabled both soldiers and strikers to deal with each other more or less cordially. The town retained its holiday appearance; many displayed likable good humor. And contrary to the original plans of the military leaders, citizens were admitted to the camp and began to fraternize with the soldiers.[4]

Snowden had studiously ignored the accounts in both local and national newspapers that the strikers were prepared to receive the soldiers "with open arms," preparing instead to crush their opposition. He also carefully disregarded the strikers' enthusiastic welcome, festive plans and (what one eyewitness sarcastically called) "other terrible methods," noting instead that, following his general orders, "Under instructions from the provost marshal, Major Kay marched resolutely through the streets, but, contrary to his expectations, met with no opposition or uncivil behavior."[5]

**Methods of Occupation**

A pyramid of order givers and takers was immediately set in place to dominate Homestead. Under orders from Governor Pattison, General Snowden directed the movements of the National Guard in profound secrecy. Using his close connections with the Pennsylvania Railroad, Pattison placed Snowden and the Militia in position like chessmen on a board.

Early Tuesday morning, July 12th, trainloads of National Guardsmen were moved to Braddock, unloaded, and then marched to the other side of the river, arriving in Munhall at about nine o'clock.

"They were about the most thoroughly fatigued lot of men ever seen," eyewitness Stowell traced the first leg of their journey, "and as they marched up Shanty Hill, every visage showed the signs of hardship. This eminence reached, they rested; and, improvising tents of ponchos stretched on bayoneted rifles, they lay down to sleep in the boiling sun."[6]

Though the various Pittsburgh regiments knew why they had been called upon, none of the officers of rank less than Brigade Commander knew where they would go before they got to Homestead, for those from Pittsburgh had been sent East to Radebaugh, where, with the entire army of the Monongahela, they were mobilized preparatory to a short march on Homestead, stopping at the opposite side of the river, a few miles to the South and West.[7]

Just across the river, the Fourth, Tenth and Fourteenth Regiments were bivouacked on Swissvale hill. While the *impedimenta* of the army was arriving and half the well-disciplined regiments were at work pitching tents, the other half were placed on guard in the boroughs of Munhall and Homestead.[8]

The military occupation was both powerful and complete – forcing all Homesteaders into immediate acquiescence. Ninety-five train coaches of troops followed Snowden's initial entrance into the city; and when the militiamen detrained, they took up assigned positions in the surrounding hills to prevent the strikers from stopping Battery B's cannon from being unloaded from its open freight car.

The soldiers may have peacefully pitched their tents and stacked their bayonets. But by noon, Tuesday, the artillery was wheeled into position – trained on the Carnegie mill grounds and the strikers' homes. Munhall Hill, the highest point in Homestead, was meanwhile reserved for Snowden's pride, the 18th Regiment, and other units were stationed to the West, with but one brigade making camp on Shanty Hill across the river near Braddock.[9]

Both sides of the river thereby occupied, the Militia commanded the whole of Homestead. From the preeminence of his headquarters Snowden now trained his field glasses over the entire terrain of the occupied territory. His plan of action firmly fixed, he rejected outright the offer of the leaders of the Amalgamated to help preserve peace, planned to squeeze the enemy towards the river in the event of a military encounter, and indicated that the troops would be used to protect nonunion men.[10]

With topographical maps of Homestead and its vicinity in hand some thirteen days before the occupation, Snowden had thoroughly familiarized himself with the country surrounding the mills, as would any general conducting a difficult military campaign. "I and my officers are now familiar with the streets of the borough," he self-assuredly proclaimed to a reporter. "You see how the six streets lead from the apex of the hill on which the troops are encamped, diverging down to the river? If any conflict should arise (and I hope for the workmen's sake there will not), it can be seen how easily they could be forced down into the river. That was the secret of Sheridan's success – in knowing the country."[11]

Snowden had also thought out how the troops would be deployed to open the mills.

"General, is it intended to use your troops for the protection of nonunion men?" a reporter asked him.

"The gates are open," was the General's significant response, "and you may enter if the company permits it."[12]

"The reply," wrote a contemporary, "would admit of no misinterpretation. It means that if the steel company desired to start the mills with the assistance of nonunion labor, the guns of the militia would be trained upon any body of men who offered resistance, and it was the beginning of the end of the great strike. To some it sounded like a victory for capital, and many of the toilers felt they must sooner or later submit or witness the shedding of more blood."

Massively armed, the soldiers were meanwhile directed to control the population as well as protect the mills. The army was seemingly invulnerable, fifteen regiments of infantry, three troops of cavalry and three batteries equipped with the finest repeating Springfield rifles available, six Gatling guns and several smoothbore cannons. Generals Snowden and Greenland mapped out the campaign, carefully mobilizing the troops at strategic points. Pittsburgh's Eighteenth regiment had been mobilized for duty burnishing arms and gathering the paraphernalia for an active campaign; and the Mt. Gretna soldier boys were equally expeditious.

The National Guard hoped for a rapid solution at Homestead, a reporter concluding that: "Inasmuch as eighty-five percent of this army were said to be crack marksmen, and all were known to be well up in the tactics of warfare, no apprehensions were felt for them in the event of a conflict."[13]

Snowdon was taking no chances that the workers would search for new arms, either. And Pittsburgh Superintendent of Police O'Mara meanwhile took the extra precaution to judiciously guard all gun stores.

* * *

By nightfall, Homestead's holiday atmosphere had completely

disintegrated. The day had been filled with bands playing popular airs and tunes of patriotic sentiments. The Stars and Stripes had floated gracefully in the gentle breezes that now and then relieved the stifling atmosphere. Hundreds of men had marched to and fro, keeping pace with the inspiring refrains. Women and children had dutifully waived handkerchiefs, and almost everyone, including the leaders of the commune, had participated in the general rejoicing. Those who at first had not welcomed the Militia during the day were now quieted, and only the town's most exploited, the Hungarians and their families, regarded the soldiers as invaders entitled to their utter contempt.[14] But the first day of occupation under Snowden had also changed the popular mood and, by sundown, the people of Homestead, fathoming his plan, despised him.[15]

"Insomnia reigned supreme in Homestead," witness Stowell analyzed the town's pathology in terms of future labor, wages and survival, "and its ghost stalked through the town, invading every household." The workers' dream of victory had almost vanished, for that one fatal sentence – the gates are open and you may enter if the company permits it – "uttered by the general whose pen and sword could blast and butcher, revealed the secrets of the future, and Homestead's men and women for the first time realized their helplessness. In this pitiful condition they looked into each others' eyes," Stowell empathized, "yet found no consolation there."[16]

The inpour of militiamen only worsened the situation for Homesteaders. They could not avoid infringing on the rights of the people, given the fact that Homestead normally accommodated 12,000 people, while on July 12th alone, an additional 6,000 soldiers had marched into the city.[17] The great mass of troops had had nothing to eat the night of their march and train ride to Homestead, and some had not tasted food for thirty-six hours – their first military hardship. The majority, moreover, had fasted for a day and a night.[18]

Mindlessly caught up in his important mission, Snowden had neglected to order rations for his troops, so as quickly as possible squads of soldiers were given leave to visit the town and get something to eat and drink.[19] Hundreds now invaded the city, buying up all eatables and searching for water with canteens.[20] By late evening, at least five hundred of the lads in blue were in a state of alcoholic hilarity, visiting the saloons in large numbers and providing a thriving trade to the dealers in "wet goods." The brewery in back of town became a special center of attraction, closing for a time only when its liquid inventory was depleted.[21]

At first, local retailers and businessmen took cheer in the

commercial successes brought on by the Militia. Six thousand National Guardsmen made Homestead stores, bake shops, saloons and barber shops hum. Some stores had never sold as many goods since they set up in Homestead.[22] So, for a day and a half, the troops lived off the town, fraternizing with the people. This ended abruptly when a perturbed Snowden instructed Provost Marshal Greene to clear the soldiers out of Homestead, giving them leave only in groups escorted by officers who would stop their intermingling with the strikers.[23] Thus, Homestead's hospitality was checked, and after the day of the National Guard's arrival, there was little or no drunkenness.[24]

The occupation was also different from any that had taken place in the past. Old-timers recalled that, in 1877, the Tenth Regiment had encamped at Homestead for several days after suppressing laborers striking in Pittsburgh.[25] But now the "riots" were at Homestead, the city was making history, and Hugh O'Donnell and Burgess McLuckie had become the most talked about men in the United States.

The *Little Bill*, the tug used by Frick and Potter to transport the Pinkertons to and from Homestead, was also being deployed by the State of Pennsylvania to transport messengers between military camps.[26] And under Snowden's *régime*: "Even when the services of the steamer *Little Bill* were re-employed to ply between Pittsburgh and the army for the purposes of carrying the necessities of the hour, the strikers had to look upon the boat and bottle their wrath," a reporter painted the psychology of the subdued. "They said it was adding insult to injury, which proved something of a relief to them."[27]

Early on as well, Homestead ladies took an interest in the soldiers, and *vice versa*. But the interest quickly turned into disgust at the invaders' bold entreaties and fear they would act on them. In defense of virtue, the City Council met at the Amity Hotel parlors on the 12th, and passed a resolution requesting the military officers in command to keep the soldiers from frequenting the saloons, the streets and the suddenly enlarged red light district of the Borough. Prostitutes were now immigrating to accommodate the soldier boys; and the Homestead *Local News* was outraged, instructing the generals that "a drunken soldier on the streets is not calculated to increase admiration for the representatives of the majesty and dignity of the commonwealth." General Snowden then promised to take care of his soldiers, and the soldiers congratulated themselves on not having to fire on anyone.[28]

Meanwhile sympathy for the strikers was running high nationwide. Hundreds of telegrams and letters were received each day by Homestead residents from friends and relatives, inquiring after their safety and comfort.[29] Homestead residents also looked after themselves, one in a conversation with General Snowden relating to him that, truthfully, there had been no trouble or disturbance in the Borough of Homestead. Ignorant of the delicate distinction between the locale of Carnegie's property and the tax district of Homestead proper where most workers lived, Snowden haughtily asked – then "What are we here for?"

"Oh," said the citizen, "The Steel Works is in Mifflin Township and not in the Borough."[30]

But this subtlety of geography swayed Snowden not at all, and he proceeded to follow a legal script that put Homestead and its population under effective, but not formal, martial law.

He now used the city's civil authorities as the first line of offense to keep the population under control in preparation for the Militia's ultimate domination. The Advisory Committee had meanwhile taken it on itself to recognize the "necessity" of using its own police force "to preserve order and eject dangerous men as well as disreputable women." And those residents who insisted that the Militia should not receive civil treatment "were promptly whipped into line," supposedly "with such success that when the soldiers appeared, the fraternal feelings developed into proportions of conviviality." Not only were a mass of "adventurers," "notorious women, and many light-fingered gentlemen" arrested and kept out of town, but newspapermen, those wearing a "strange" countenance, and many commercial travelers as well as merchants who had business in Homestead were also forced to give an account of themselves and, in most cases, compelled to leave.[31]

"So far as the government of Homestead is concerned," editorialized the *Local News* on July 16th, "everything is going on as it always has. The Burgess and Justices of the Peace, constables and policemen continue to exercise the functions of their offices without interference. If anything, there is less constraint upon everyone than before the appearance of the militia."[32] Yet the precise meaning of this increased freedom, we will see, was that local officials had to loosen their grip on ultimate, and sometimes immediate, authority.

* * *

The Militia had come to Homestead with the evident expectation of meeting with some resistance and finding something to do. "They have," wrote the *Local News*, "been agreeably disappointed. Their visitation is as serene and peaceful as if they were enjoying their annual encampment and inspection – which indeed it may turn out to be."[33]

Unfortunately for the people of Homestead, the peace did not last.

## Martial Law

Though hitherto the impression had gained ground that the Militia had been called to prevent the landing of more Pinkertons, upon Snowden's arrival he had dispelled the delusion.[34]

"As soon as we came into the town we were in possession, and law and order is restored," Snowden had told O'Donnell and the Citizens' Committee, now adding another dimension to state military power: "The mill property is already in the hands of the Carnegie Company. I understand Mr. Childs [one of the managers] is now in the Works. But, with all that we have nothing to do. We are here to support the Sheriff in the execution of his duties, and the presence of the troops was enough to settle the whole question."

With the Militia now the sole authority, the strikers' guards, who had previously surrounded the mill, were replaced completely by the military's Provost Patrol. This patrol guarded all the streets leading to the mill and about it. With their protection, contrary to what General Snowden had said about having no role in securing Carnegie property, manager Otis Childs quietly took possession of the mills at 10 o'clock sharp on July 13th.

Two hours later, the workers awaited a new landing of Pinkertons, some swearing to resist their arrival at all costs. The workers now believed that the Carnegie Company had over a thousand nonunion men ready to come into the Works as soon as their safety was assured, in addition to Pinkertons coming from scatterings around Pittsburgh to renew their defense of the Works.[35]

For the Company had sent their agents to all parts of the nation, workers polemicized. "Soldiers protecting nonunion labor! men to take the places of those who rebelled. "Scouring the country for scabs!" the workers polemicized. "Soldiers protecting nonunion labor! Taking away our bread to feed the unworthy!" Without the military to protect them, the fate of nonunion workers brought into the mill would undoubtedly have been quickly settled.[36]

* * *

Whether or not martial law was formally established at Homestead, the effect on the population was the same.

"Martial law really consists of rules ordained for the government of an army or military force," an observer explained of General Snowden's mandates making him a "dictator of that helpless borough."

"It is a law of war, as adopted by belligerents for control of their respective armies and of those violating, in respect of such armies, the laws of war," this analyst went on. "It is distinguished from military law, martial law being used for the conduct of war itself, while military law is imposed by a *de facto* civil government for the control of a country which is the seat of war."[37]

By definition, Snowden established martial law, whereas earlier the leaders of Homestead had established a system of military law regulating the commune.

The difference was significant in terms of the military center's purpose and its effect on the people. The commune at first had been built on the fears of the striking workers about what could possibly happen to undermine the strike and the security of their jobs, wages and homes. Now that martial law was effectively in place, they could see that it was a form of domination designed to protect, not them, but the Carnegie interests. "While martial law has not been officially declared in Homestead," a reporter confirmed on July 15th, "it has occurred."

General Snowden immediately demonstrated its meaning by an investigation of an alleged assault on a member of the press. Though a special committee representing practically all of the news correspondents in Homestead thought that the event did not merit such an investigation, McLuckie's special policemen had made arrests and General Snowden took charge and planned to "take rigorous measures, even to the length of declaring martial law."[38] And though Snowden relented, temporarily deferring his edict, the process itself revealed *de facto* military rule had been established.

* * *

From the first day, on the 12th, Snowden had ordered the Fifteenth Regiment down the hill towards Homestead to take up positions near the Provost Marshal's tent, fronting on Eighth Avenue. Two companies of about ninety men were then sent into the city under the command of Major Crawford, and squads were placed in front of the Advisory Committee's headquarters, at the three railroad stations, and at eight other places in town where, until then, men had been in the habit of congregating.

Orders for military personnel were explicit. In case any arrests were made by McLuckie's special police for a manifest breach of the peace, drunkenness or any like offense, the Provost Marshal or his patrol would not interfere. But in an arrest for any other cause, the patrol would at once take charge of the arrested person and the man arresting him, bringing both before the Provost Marshal or his representative. Furthermore, if the arrest were proved an illegal one, the commune policeman would be promptly punished.

The citizen policemen were thus discouraged from arresting anyone, especially since they lacked arms to enforce their will – in contrast with the Militia patrol members, who were equipped with bayonets on rifles and twenty rounds of ball cartridges.

Snowden, maintaining his orders, enlarged his forces, then softened his rhetoric to keep the commune off center: "there has been no change whatever in the position save that the provost patrols have been somewhat increased in strength and the preservation of peace can now be somewhat more certainly assured," he announced. "Yet, the patrols will prevent any further illegal arrests. They will cooperate with the civil authorities to that end, but further than that nothing can be done."

Any margin of power left the civil authorities could now only aid the Militia's orders. "The town," Provost Marshal Colonel Green nonetheless contended, "is most distinctly not under martial law, but there has been so much demoralization that it has become necessary for us to cooperate with the civil authorities in preserving order."[39] No one doubted that the *régime* of General Snowden and his troops had been firmly implanted.

"Martial law is built on no settled principles," an observer summed up the do-what-it-wished occupation, "but is entirely arbitrary in its decisions. As applied to civilians by an army of soldiers, it is in reality no law at all, but something indulged simply because the man who promulgates it has an invincible force at his command, while those who suffer its consequences are weak and helpless, and hence unable to protect themselves."[40]

Snowden applied his invincible force as soon as the workers congregated to discuss the sympathy strikes supporting their cause taking place in other cities. "Crowds gathered in various portions of the town to talk about the matter," said an eyewitness, "and naturally the largest gathering appeared near headquarters on Eighth Avenue." But the General was prepared: "The faultless organization of the Militia was never more apparent than upon this occasion," said an observer. "Without the slightest excitement, troops suddenly appeared at points suited for controlling outbreaks, and so admirably were they handled that the town, the mills and the surrounding country were placed under surveillance with remarkable rapidity. If trouble had been brewing, there would have been no chance for it to assume tangible form. These movements indicated perfection so far as the army was concerned."[41]

Unable to combat Snowden's forces marshaled lock-step by General Wiley, many were deeply embittered, turning to sarcasm for relief; one new Homestead resident, Murray Scott, lampooning the child-like behavior, facile camaraderie and power illusions of the soldier boys in blue:

> . . . as there exists the National Guard of Pennsylvania, it has been the custom of many years for the more adventurous of us of the male population to join that organization, the chief reason being that they could leave home in a body; take gun along and in case they got lost could be identified, and shipped back. . . .
>
> Hitherto the duties of this band of bearded and bronzed warriors had been to spend eleven months and two weeks of each year learning to count up to four. The other two weeks they let their whiskers grow to astonish the girls when they came home from camp. Now to go away from home among men who had actually cared, with a few cheap pistols and shotguns to capture three hundred Winchester rifles with Pinkerton attachments. No wonder they rushed to [General] Wiley with their woes.[42]

Despite making fools of them, perhaps the troops under strict orders understood fairly little of the merits of the struggle at Homestead; and, in any case, their awareness was not their assigned purpose. With the Pinkerton defeat, wrote citizen Scott more seriously, "the mill owners had now called on the other branch of their service, the National Guard of Pennsylvania, to protect them in their rights as guaranteed by the McKinley bill."

**From Friend to Foe**

Homestead's civil authorities had cooperated in preserving order, both to keep the city's rank-and-file in line and to place some limits on the playfulness of the soldier boys away from home. Many soldiers of the Militia had not volunteered and resented the strict military regulations that interrupted the normal course of their relatively unrestricted lives.

True, some of the entering Militia were themselves former Homesteaders or residents from nearby towns, and felt strong ties with the people of the city. But most of them were more concerned about their personal needs and schedules - taken, as they were, from weddings, even honeymoons, home life, jobs and businesses - and were closer to everyday affairs than military ones. Anxious mothers also disrupted Homestead city life, arriving the day following the entrance of the troops, to see that no harm came to their tender, inexperienced boys.[43] Little Billy or Franky, still wet behind the ears, might run into an ambush by the maddened army of workers or by one of the city's light-fingered shylocks or enticing hussies.

Nonetheless the soldiers were in a holiday spirit. Some without passes led the Militia patrols on wild goose chases about Homestead, jumping fences, even begging to be sheltered by friendly Homesteaders until the patrol had passed. It had become a common sight to see a couple of soldiers rushing through a vacant lot and scaling a fence or two. Getting together in out-of-the-way places, playing cards, and amusing themselves more at night than in the day, was the purpose of many a militiaman.[44] Yet, despite a sense of uneasiness among most Homesteaders about invaders making the city their playground, they tried to be hospitable, assuming complacent countenances to put the rank-and-file militiamen at ease.

Though residents now knew the cards were stacked against them, outsiders still concerned themselves with the legality of bringing in the Militia in the name of the people, when clearly Homesteaders were the people of their own city. Regular constitutional lawyers were on hand to offer astutely reasoned opinions on the most intricate and complex questions involved in the months-old difficulties.

But legalities aside, peace had been imposed, and, as one local humorist noted, by July 13th the soldiers' only shocking experience had been the instant death of a grazing cow struck down by a powerful thunderbolt near the tents - God's wrath misdirected, he thought.

Local residents also poked fun at national reporters who wrote sensational stories. Some said they had taken up the pen and made it mightier than the Militia's sword. The editors of the *Local News* also mused that the newsmen had skillfully spun a yarn so far removed from the facts that the adage, "politicians were husbanding the truth with their wonted frugality and dispensing fiction with their usual lavishness," could justly be applied to correspondents.[45]

Yet people were kept on edge by a multitude of published rumors. Some claimed that hundreds of nonunion men would appear any minute; others that disaffection was spreading among the strikers – a recurring rumor as the lock-out wore on. Another sensational claim was that a tunnel under one of the principal thoroughfares was designed to connect the town and the Works, so that men and dynamite could be stored and placed within the big enclosure. Still another fable circulated by a Cincinnati paper claimed that a fearful explosive had just been discovered by a foreign chemist who was on his way to Homestead to wreck the mills.

Blacks were erroneously reported to be on their way from Alabama to perform Homestead mill work. And a piece of gossip with some plausibility held that the Pinkertons had joined an organized band of guards under a new private detective concern.

Another oft-repeated yarn was that hundreds of cots had been smuggled into the mills by mysterious means. "Governor Pattison and staff either already were in camp, or soon would be there, had been circulated at stated intervals," a contemporary wrote, "and scores of other alarming conditions were rehearsed until Homestead had about concluded that a shower of meteors, accompanied by general havoc and destruction, would be a necessary to complete the list of thrilling episodes and harrowing calamities yet unmentioned."[46]

Yet, the seriousness of the occupation was unmistakable. Four branches of the military were now operating in Homestead: the guards at the Carnegie Mills, the Sheriff, the Militia, and, under them, the police committee of the Borough. The latter maintained tranquillity[47] and soon added to their special police force to keep control over the multitude of sightseers coming to view the troops amassed on Carnegie Hill, while the Militia kept control over the Homestead population.

Even Allegheny trade unions, like the W. C. T. U., were in favor of maintaining order, passing austere resolutions endorsing Burgess McLuckie's proclamations for keeping tranquillity in the Borough. And the day after the Militia arrived, Local Union No. 288 of the United Brotherhood of Carpenters and Joiners in Homestead also unanimously resolved "that the Amalgamated Association of Iron and Steel Workers be tendered the moral and financial support of the Union in the struggle now going on and that we enter our protest against the importation of Pinkerton guards and other armed bodies of men foreign to our state."[48] But as the Militia were not foreign, but indigenous to Pennsylvania, Local 288 did not voice opposition to the legally vested militiamen.

Control by the authorities was meanwhile rigid and thorough. Though for 17 days after the Carnegie plant had been closed down, not a stroke of work had been done in the mills and not a pound of steel had been produced under the régime of the few watchmen who manned the gates.[49] These guardians stayed at their posts, functioning under Snowden's authority which encompassed the idle mill as well as the town. The General ruled the mill watch and the Borough police with such a stress on order that the officers quieted everyone,[50] especially dissenters who warned that nonunion workers would soon be brought into the mills.

Even the police of the Borough had become so thoroughly paranoid that they equated the unusual with danger, and variously imposed inequities on innocent suspects. Outward appearances seemed in command in one "infraction," as they locked up a helpless tramp, Joseph Guttro, with a long beard, disheveled hair, and who allegedly had a "wild haunted anarchist look in his eye," on a charge of vagrancy, having him sent to the county workhouse for 30 days.[51]

Burgess McLuckie now insisted he alone could maintain "law and order" in Homestead. But while the soldiers did as they pleased under Snowden's regime and mixed crude playfulness with petty larceny and the abuse of Homesteaders, McLuckie at best could only jail local residents and outsiders for drunkenness, disorders, or threatening language. Thus, in one peccadillo, the ex-chief of Police and High Constable of nearby McDonald, Pennsylvania, had come to Homestead, having had one too many drinks on his mission. When he audaciously displayed his supply of military hardware and said for what it might be used, he was quickly arrested for gun-toting and his opinions about what should be done at Homestead after the arrival of the troops. Visibly unimpressed, Burgess McLuckie chastised and summarily sentenced the High Constable to jail, then banishment on the next train out.[52]

Such were the limits of McLuckie's power under Snowden's martial law, while the soldiers' abuses led most Homesteaders to want nothing further to do with lads in blue.

## Oppressors and Repressors

H. C. Frick, cast by the public as an oppressor of the innocent, had been reviled in the popular press, both trade unions and political circles on the Left viewing him as a dangerous assailant of the rights of labor. Among Anarchist groups, moreover, it was reasoned that the state, Frick and other symbols of injustice must be destroyed.

To remove the personification of injustice at Homestead, Anarchists like Emma Goldman and Alexander Berkman planned carefully, believing that the impending struggle at Homestead had "sounded the awakening of the American worker, the long-awaited day of his resurrection." Goldman exalted: "The native toiler had arisen, he was beginning to feel his mighty strength, he was determined to break the chains that held him in bondage so long. . . . Our hearts were fired with admiration for the men at Homestead."

Berkman went to Pittsburgh, inspected the *Chronicle-Telegraph* building holding Frick's offices, and mapped out an attack for assassinating Chairman Frick. If successful, Goldman planned to justify: "I could explain that he had no personal grievance against Frick, that as a human being Frick was no less to him than anyone else. Frick was the symbol of wealth and power, of the injustice and wrong of the capitalist class, as well as personally responsible for the shedding of the workers' blood. Sasha's act would be directed against Frick, not as a man, but as the enemy of labour."

But their plans soon went to shreds. After he shot Frick three times and stabbed him, Berkman was arrested. Though turning out to be only minor wounds that failed to kill Frick, the loss of blood prostrated him and won him public and press sympathy. The Pittsburgh press screeched for the police to act, to round up the leading Anarchists, "the instigators, Johann Most, Emma Goldman, and their ilk." Goldman and their accomplices, Carl Nold, Henry Bauer and one Mollac were soon picked up in Homestead and Long Island New York. [53]

The newsprint media near-unanimously expressed sympathy for Mr. Frick, echoed in Homestead and its unions. Many Homestead workers felt that, although Berkman clearly did not represent them or their cause, his act was certain to prejudice their interests. Meanwhile Frick won admiration for his supposed courage in quarters where previously he had been condemned, taking the edge off the animosity generated by his harshness towards organized labor.[54]

The event also brought new pressure for subservience into the ranks of labor in Homestead. "Besides the money cost and the loss of precious lives," the *Local News* elaborated the implications of the strike, "the trouble stirred up those pests of society, the Anarchists, and led to the shooting of Chairman Frick."[55]

Though a few days earlier their own military acts were condemned by others as unlawful, and their hatred for Frick was unequaled, the Advisory Committee now publicly declared itself against violence:

> Whereas, We deplore the act of the individual not connected with our trouble or Association, therefore, be it
>
> Resolved, That we condemn the unlawful act of the wounding of Henry Clay Frick, and tender our sympathy hereby, and further
>
> Resolved, That we spread a copy of this resolution on our minutes, and give a copy to the press.[56]

Militant union action had come to a sudden, ironic end. Berkman's bullet, an Amalgamated officer exclaimed, "went straight through the heart of the Homestead Strike."[57]

The Homestead press went along, the *Local News* echoing: "None regret and condemn, more than the people of Homestead, the attempted assassination of H. C. Frick on Saturday. . . . It has been amply shown that Alex Berkman and his associates are unknown here and in no manner whatever connected with labor either organized or unorganized. Anarchists have purposes of their own. Homestead has never harbored any nor does it wish to."[58]

With Berkman and the Anarchists depicted as dangerous outsiders likely to cast a cloud over the cause of organized labor, silence fell on Homestead. No one dared to voice approval for the shooting. So not from a citizen of Homestead, but of a militiaman came a lone voice approving Berkman's act.

The soldier, Private W. L. Iams, was a farmer by trade, 24 years old and belonging to "one of the best families" in Greene County, where he was a resident of Waynesburg, Pennsylvania. Unwilling to grant military protocol control over his innermost thoughts as he lay at camp with the Tenth Regiment across the river from the Carnegie Works, Iams had the temerity to shout out approval for Berkman's act in shooting manager Frick. "Three cheers for the man who shot Frick!" he had exclaimed, and, when found out, attributed his remarks to his deep-seated dislike of the chairman. Ordered to apologize before the Tenth Regiment, Iams refused, was sent to the guardhouse and, without the benefit of a trial by court martial, strung up by the thumbs with notice that he would be cut down as soon as he apologized.[59]

The orders had come down from Major General Snowden's headquarters, and Colonel Hawkins, Colonel Streator and Surgeon Grim of the Tenth Regiment were present when they were executed.

Colonel Streator executed these directions to the letter! Iams recalled:

> When I was first tied I asked for a chew of tobacco; soon I began to grow tired; my legs cramped; my head ached; my eyes felt as if they were being pushed from their sockets; the muscles in the back of my neck seemed as if they were being torn from the base of my skull; I suffered terrible agony; it seemed to me that I had been suspended in air a week before being cut down.

For thirty minutes he hung there:

> I saw Colonel Streator there; when I was cut down I was in a dazed condition; everything seemed blurred; the first thing I remember was when someone gave me some ammonia; I was lying on the ground; had a terrible headache and someone gave me some whiskey. I remained in the guard tent all night. . . .

The next morning, ordered by General Snowden to be disgraced and banished from camp by a string of lesser officers, Iams said:

> . . . my hair was shaved off, my uniform taken away from me and I was given a pair of overalls, a jacket and a straw hat; the overalls were much too short; the sleeves of the jacket were also very short; I was paraded in this dress before two regiments, then drummed out of camp.[60]

After thus having half his head and mustache shaved, Iams was dishonorably discharged, stripped of his uniform, clothed in a suit of ill-fitting overalls, and drummed out of camp to the tune of the "Rogue's March," in the presence of the entire regiment.[61]

lams, later attempting to sue them in court, claiming that the good Surgeon helped to tie the rope around his thumbs. Captain H. C. Cuthbert also of the Tenth Regiment admitted seeing lams tied up by his thumbs; but, in defense of the officers present, claimed the balls of Iam's feet were on the ground and that his heels were not raised more than an inch. To ease the stress, he claimed he had given Iams the chew of tobacco as he was hanging there, while Surgeon Neff was ordered to see that Iams received no permanent injury and that he be cut down if the prisoner said he was suffering greatly.[62]

Iams allegedly replied that if they expected him to retract, they would have to cut him down dead. But, in defense of the officers, Dr. Ullman stated he told the prisoner to feign sickness so that the regiment would have the impression that he had been severely punished. In any case, Colonel Hawkins argued, the articles of war provided for such treatment under conditions of insurrection.[63]

Though the case became a national *cause célebre*, the judge and jury agreed: They were satisfied that military law permitted the punishment inflicted, whether or not it fitted the crime lams committed![64]

The public shocked, the press censured Colonel Streator and General Snowden for the, extreme punishment.[65] Yet, as justification, they had both military practice and legal precedent on their side. Though lams described what had happened to him, no court ever convicted the military officers.

## The Rigidities of the Militia

The Iams case symbolized the situation in Homestead: The militiamen and the local police were strictly accountable to their superiors and the subjugated were helpless. "Between the Borough police, the soldiers and the deputies," the *Local News* had written, "it is hard for a guilty man to escape."[66]

Yet guilt depended on who enforced the law for whom. For the Militia, guilt was being a Homesteader, and they gave no consideration to the just demands of the workers. After July 28th, three regiments of infantry, a troop of cavalry and a battery were booked to remain in service until the Carnegie mills were in undisturbed operation - a condition which was not to be finally established until more than two months later.[67]

Meanwhile, for doing their duty at Homestead, the soldiers were paid for each day on a military scale where a Major General received $25, a Brigadier General and a Colonel $15, a Lieutenant Colonel $10, a Major $7, a Captain $5, a lst Lieutenant $4.17, a 2nd Lieutenant $3.75, but a Sergeant only $2, a Corporal and re-enlisted man $1.75 and a private $1.50. "There seems to be as much difference between a private and an officer as between a laborer and a high priced steelworker's wages," the Homestead *Local News*.[68] These differences expressed gradations on the job ladder in both institutions which, under other circumstances, might have linked the privates of the Militia with some workers in the city.

Yet, the soldiers who came to Homestead were often outrageous. Military rations were poor; many stole goods. One group of soldiers misrepresented themselves as a patrol looking for deserters in order to dip into farmer George Brenn's larder. "Some of the soldiers crowded around him, while the others went out to the cabbage and potato patch and helped themselves," criticized the *Local News*. "They pried an opening into the chicken coop, and Mr. Brenn hearing the noise charged them with stealing when they all broke and run. Another neighbor, George Wilson, saw them pass his house about one o'clock laden down with chickens and vegetables. Other town residents are obliged to stay up all night watching with shotguns to keep the soldiers away."[69] the *News* reported.

So the solidarity that might have existed between the ordinary militiamen and the people aborted or quickly vanished. One member of the Fifth Regiment even deigned ahead of time to deliver to Homestead his arrogant soldier's farewell:

How we long to leave thee, Homestead,
    Not because the prey we fear -
But we hate your stinking alleys
    And the acts that brought us here.
Leaving boardwalks far behind us,
    With Huns and Poles and sour beer;
Keep good order, do your darndest, or
    Again you'll find us here.
Hark! we hear the bugle sounding,
    'Tis the signal for to leave;
We never felt sorry and
    For you we will not grieve.
By your sinks and filthy mudholes,
    Where the boys patrol all day,
We will not forget you, Homestead,
    When we are far away.[70]

Under orders, the soldiers had also imposed their will on the citizenry. Incidents became so common that many kept as much distance as possible between themselves and the blue-coated troops running Carnegie Hill. Outsiders were also in jeopardy. In one case a gentleman with his family, carrying a present in a basket, on a visit to Homestead to see, not a resident, but a military man, was confronted by guards ordered to arrest all peddlers. Assuming this gentleman to be such as he attempted to cross the guard line, they detained him half an hour in the scorching August sun, derailing his trifling errand 'till explained to the Corporal of the guard.[71] "There is no doubt," recounted the *Local News*, "about the fact that some of the soldiers, while moving about through the Borough, have done many improper and annoying things, even to the extent of insulting ladies."[72]

Until the Militia arrived, Councilman Lynch had been the head of McLuckie's special police. But now he too, with his wife and child, was insulted by the soldiers. Others reported similar annoyances, and Burgess McLuckie stated emphatically at the Council meeting that many citizens of the Borough, especially in the Third Ward near the soldiers' encampment, complained of the militiamen's frequent insults to women and other improper conduct. The Sheriff's deputies, McLuckie added, were also becoming a source of annoyance by congregating on the streets. Both the Militia and the deputies were taking entirely too many liberties, McLuckie concluded, and should be stopped. A serious motion was then passed, referring the matter to the solicitor[73] – as if that officer of the law could do anything at all to control the army in occupation.

## From Passive Resistance to Armed Watchfulness

McLuckie and O'Donnell had been frantic, but firm, in encouraging the people of Homestead to receive the Militia graciously. When the occupying army revealed its true purpose, however, they had to admit their misjudgment, albeit in a backhanded fashion. By July 13th, O'Donnell realized that effective martial law had been established, saying it was expected under the circumstances. And Amalgamated official Cummings opined that: "The next step will be the bringing in of nonunion men to the mills, which may be done today, but will certainly be effected before the end of the week. I do not believe they have anything like the four hundred [Frick claimed], however. I shall be greatly surprised if anything like that number turns up."

Unconvinced, the press asked Cummings: "Will there be any effort made to prevent the men from going to the mills?" Cummings reflected carefully:

A: We will certainly ask them not to go in, and we will use every argument we can to persuade them not to take part in the war which is being made upon us. I think a great many of them will turn back when they understand the merits of the case. Those who do come are being deceived, I am sure.

Q: And if pacific measures fail?

A: They will not fail; even if they did, they would be all I would be in favor of using, if there was not a bayonet within a hundred miles of Homestead.[74]

This expression of confidence and pacifism was bolstered by the refusal of foremen, superintendents of departments and boss mechanics – all men not on strike – to operate the Works under an armed guard. Their decision to stop work on July 14th led one leader of the Advisory Committee to judge it would be utterly impossible for the work to be resumed without the cooperation of the foremen, as no men could be found to take their places, even if nonunion steel workers were found. He added there was not a boarding house or hotel in Homestead willing to feed any scab worker. Nor could skilled mechanics be induced to live in tents or shanties inside the mill. Hence Carnegie would have to settle with his workers.[75]

The attitude of the workers had also changed. To a man they now viewed the Militia as replacing Frick's illegal mobilization of Pinkertons. Since General Snowden did not recognize the Advisory Committee, there was no reason for the Committee to repeat that the Militia represented legal authority as distinguished from the "illegal authority" of the Pinkertons. Snowden's Militia was there to restore "law and order" in the name of the citizens of Homestead. This was what the Amalgamated and the Advisory Committee had thought they had done in defeating the Pinkertons. But, for Snowden, the meaning was different. For he viewed the role of "law and order" as restoring Carnegie's ownership of means of production, his right to hire the workers he chose, and otherwise to control the forces for initiating production anew. Thus, if Carnegie chose to hire nonunion men, the Militia would protect the undertaking.[76]

The locked out men were not mistaken in their belief that the Carnegie Company would take advantage of the presence of the military to bring nonunion men into the mill.[77] "Troops were early detailed to garrison the yard," said eyewitness Burgoyne, "and a line of sentries was posted outside the fence, so there would be no interference by the workmen either from the railroad or the river."[78]

Expecting the scabs to arrive by river, the Advisory Committee now strengthened its patrols on the Monongahela to meet and pacifically argue with the newcomers. The Committee leaders urged that resort to violence would be practically treason, and no man coming into Carnegie's Works was to be threatened or intimidated. Recognizing this was the main potential collision point with the troops, nonetheless, the Committee believed Snowden would not prevent them from having access to the newcomers outside of the Carnegie Works.[79]

Yet, the Militia was not so disposed. Conferring with Governor Pattison at the Mount Gretna camp, Major Wright of the Third Brigade gave an ominous analysis: "The locked-out men are armed, and there will likely be bloodshed before the final settlement of affairs. The present peace is enforced by the cloud of awe which has hung over the striking workmen since the arrival of the troops."[80]

At Homestead, two Amalgamated members agreed: "Mr. Frick's statement that there are a large number of men who are ready to return and accept the reduced scale is not true. There is not a man who will go back, but there will be no collision with the Militia until we see nonunion men employed. In that case Homestead may witness greater slaughter than it has yet seen."[81]

The official Amalgamated posture was very different. Though strict military discipline was preserved by the Militia, they said the locked-out men would offer no violence to nonunion men, whom they were constantly expecting. However, the millmen seemed determined to urge any nonunion workers not to accept employment.[82] Hugh O'Donnell, as a member of the Amalgamated Advisory Board, repeated on July 15th that he was perfectly willing for the Sheriff to send deputies into the mill to assume direction of the Carnegie watchmen and that, in turn, the locked out men would not attempt to pass the gates. But the rank-and-file remained as watchful as ever.[83]

Meanwhile, the military watched the watchers. Thus, on the 15th at 1:30 A.M., one reporter noted "there was a hurried beating to arms throughout the second brigade and a scurrying of officers which was very warlike. Information had been received that the camp was to be attacked from the rear by the locked-out men before daylight. . . . [I]t was sufficient to arouse the entire brigade, to have the guards doubled everywhere and to keep the Sixteenth Regiment under arms all night. . . . At an early hour this morning the guards around the mill fence were tripled, and later around the entire mill, so that now the guardsmen are only a few feet apart around the entire plant. . . . It is believed that the intention is to gradually place the borough under thorough military control."

The military move was interpreted differently by the people of Homestead, who prepared for what appeared to them to be an invasion. "The town was thrown into the wildest excitement last night by the sudden appearance of a skirmish line from the camp which on a double-quick surrounded the mill," another reporter said. "The men were certain then that persons to supplant them were about to be brought in under protection of the soldiers, and notwithstanding pacific assurances of the leaders a clash was narrowly averted. The feeling is not friendly towards the troops since the strict military discipline had been enforced," he concluded.[84]

Many Homesteaders now asked how they could defend themselves against the onslaught of nonunion workers protected by Pinkertons under the Militia's patronage. Their answer was the workers' quiet possession of weapons – 2,000 assorted guns, including captured Pinkerton equipment. "An amazing fact," the same reporter noted, "is that much of the ammunition is manifestly the property of the United States government. The boxes of cartridges bear the label of the Frankford Arsenal, and nobody seems to be able to explain whether these are part of the Pinkerton equipment, or whether they were differently acquired. The Frankford cartridges are all 50 caliber."[85] Federal support of the Pinkertons, clandestine sympathy from the occupying troops, or guerrilla appropriations from the government storehouse were all possibilities.

Opinions also differed as to the possibility of a final struggle with the Carnegies and their hired and politically compliant gunmen. Striker James O'Boyce thought workers would fight: "Most of the men expecting continuous employment here have put their savings into homes which will be lost if we are to be driven away from this town. The Carnegie mills were built up by us, the great profits of the concern were made by us. Our labor was expended for Scotch castles and library advertising. We do not say that Carnegie, Phipps & Co. do not own the mill property, but we do say that we have some rights in it ourselves. We have been constantly led [by the company] to believe that the Amalgamated Association would be permitted to control and direct our interests [concerning the pace of the introduction of new machinery, job security and the work process]. We have bought Mr. Carnegie's land on that theory, and we are now in a position that we must lose all that we have or win this fight. I think we will win it."[86]

Clearly, Carnegie and his Militia could not disarm the workers by a house-to-house search, O'Boyce continued: "Until Mr. Carnegie secures two or three amendments to the Constitution our right to keep and bear arms will not be infringed. We are not children in Homestead;

we know our rights and we will maintain them. There are men here prepared to lay down their lives in defense of what is or they think is right, and surely we have the right to drive away an illegal force of hired assassins brought here for the sole purpose of intimidating us. . . . We can get along with the troops or without them. We are Americans, who are not so fond of life that we will not stack it up and call anybody, we care not who, that attempts oppression."[87]

The fact is, however, that the Militia ruled Homestead most completely. Though people responded to abuse by withdrawal, passive resistance had only been effective up to a point. Hostility to the Militia had mounted, though it largely remained unstated or was pacifically acted out. Thus it was said among the people that the firing of Snowden's cannon in periodic gun salutes on Carnegie Hill had knocked all the plaster off the ceiling of the Carnegie School House where Snowden had his headquarters.[88] No other way to actively resist, many a Homesteader hoped reverently that the good General was standing beneath.

## Notes

[1] *The Local News*, August 20, 1892.
[2] Emma Goldman, *Living My Life* (Salt Lake City, Utah: Peregrine Smith, 1982), 122-123.
[3] John Steuben, *Strike Strategy* (New York: Gaer Associates, Inc., 1950), 194.
[4] *The Local News*, July 16, 1892.
[5] Stowell, 172.
[6] *Ibid.*, 127.
[7] *Ibid.*, 126-127.
[8] *Ibid.*, 127-128.
[9] Wolff, 149-151.
[10] *Ibid.*, 151.
[11] Stowell, 170.
[12] *Ibid.*, 131.
[13] *Ibid.*, 116-117, 131-132.
[14] *Ibid.*, 130-131.
[15] *The Local News*, July,16, 1892.
[16] Stowell, 132.
[17] *The Local News*, July,16, 1892.
[18] Stowell, 127.
[19] *Ibid.*, 128.

---

[20] *The Local News*, July 16, 1892.
[21] Stowell, 128.
[22] *The Local News*, July 16, 1892.
[23] Wolff, 152-53.
[24] Stowell, 128.
[25] *The Local News*, July 16, 1892.
[26] *Ibid.*
[27] Stowell, 143.
[28] *The Local News*, July, 16, 1892.
[29] *Ibid.*
[30] *Ibid.*
[31] Stowell, 123-124.
[32] *The Local News*, July 16, 1892.
[33] *Ibid.*
[34] Wolff, 149-150; Stowell,126.
[35] *The Sun* (Baltimore, Maryland), July 13, 1892.
[36] Stowell, 138.
[37] *Ibid.*, 142.
[38] *The Sun* (Baltimore, Maryland), July 15, 1892.
[39] *Ibid.*
[40] Stowell, 143.
[41] *Ibid.*, 180.
[42] *The Local News*, August 20, 1892.
[43] *Ibid.*, July 16, 1892.
[44] *Ibid.*
[45] *Ibid.*, July 9, 1892.
[46] Stowell, 180-181.
[47] *The Local News*, July 16, 1892.
[48] *Ibid.*
[49] *Ibid.*
[50] *Ibid.*
[51] Burgoyne, 147-148; *The Local News*, July 30, 1892.
[52] *The Local News*, July 30, 1892.
[53] *Ibid.*; Burgoyne, 150-151; see also Goldman, *passim*.
[54] Burgoyne, 150-151; Berkman, Chapters 2, 3.
[55] *The Local News*, July 30, 1892.
[56] *Ibid.*
[57] *Ibid.*
[58] *Ibid.*.
[59] Burgoyne, 152.
[60] *Ibid.*, 152-153.
[61] *The Sun* (Baltimore, Maryland), October 29, 1892; see also Burgoyne, 147-154.
[62] *The Sun* (Baltimore, Maryland), October 29, 1892; see also Burgoyne, 147-154.

[63] *The Sun* (Baltimore, Maryland), November 8, 1892
[64] Burgoyne, 154.
[65] *The Local News*, July 30, 1892
[66] *Ibid.*, August 6, 1892, 3.
[67] Burgoyne, 154.
[68] *The Local News*, July 30, 1892.
[69] *Ibid.*
[70] *Ibid.*
[71] *Ibid.*, August 6, 1892.
[72] *Ibid.*
[73] *Ibid.*
[74] *The Sun* (Baltimore, Maryland), July 15, 1892.
[75] *Ibid.*; *The Post*, July 14, 15, 1892; *The World* (New York, New York), July 14, 15, 16, 1892.
[76] Burgoyne, 121-124, 133-134; Wolff, 150-151; *The Sun* (Baltimore, Maryland), September 7, 1892; *The Philadelphia Press*, February 8, 13, 1892.
[77] Wolff, 151; Burgoyne, 102-145; U.S. Congress, House of Representatives, *Report 2447*, X-XIV,
[78] Burgoyne, 133.
[79] *The Sun* (Baltimore, Maryland), July 15, 1892; *The Tribune* (Chicago, Illinois) July 14, 15, 1892.
[80] *The Sun* (Baltimore, Maryland), July 15, 1892.
[81] *Ibid.*
[82] *The World* (New York, N.Y.), July 13, 15, 16, 17, 1892; *The Sun* (Baltimore, Maryland), July 15, 16, 1892.
[83] *The Daily Dispatch*, July 16, 1892; *The National Labor Tribune*, July 15, 16, 1892; *The Sun* (Baltimore, Maryland), July 16, 1892.
[84] *The Sun* (Baltimore, Maryland), July 16, 1892.
[85] *Ibid.*
[86] *Ibid.*
[87] *Ibid.*
[88] *The Local News*, August 13, 1892; see also Stowell, 185-187.

H.C. Frick

Sheriff McCleary

# Chapter 9

# Count Down for Production

1. No one will be permitted to interfere with the Civil, Religious or Political opinions of the workmen, and no political notices or posters will be allowed to be circulated or posted on the property of this Association.

2. All employees wishing to absent themselves for a turn, or longer, must first apply to and receive permission from their foreman; and all persons working on night turn must make their application before 4 o'clock p.m. All persons violating this rule will be subject to discharge.

3. Employees are requested to exercise economy in the use of all machinery, and to keep the machinery and Works neat and clean.

4. Any employee, who on account of violation of the criminal laws of the country is arrested, and reason of his arrest leaves his position vacant, will be discharged and his position filled.

5. Any employee, who through gross carelessness or malice, destroys the property of this Association, or is found stealing or carrying away the property of this Association, will be discharged.

6. Any employee who habitually neglects or refuses to pay his debts will be subject to discharge.

7. The use of intoxicating liquor by any employee while on duty, is absolutely forbidden, under penalty of immediate discharge.

8. All superintendents and foremen must pay strict attention to the rights and privileges of employees. When a position is vacant, the employee of the longest service at the Homestead Steel Works, and in the line of promotion, must fill the vacancy, without regard to his political or

religious opinions; provided, however, he is fully competent to fill the higher position.

9. Excepting only where a special contract has been made, as for advice, counsel, etc., every salaried employee of this association is expected to devote his entire service to the interests of his employer; and while no restriction is sought to be placed upon investments made by any employee, the taking of any active part in the conduct of the business in which such investment is made will not be permitted.

10. Department superintendents and foremen shall give hearing and prompt attention to any reasonable complaint or claim of redress, and if unable to amicably adjust the matter, shall refer the same to the General Superintendent.

11. All department superintendents and foremen must see that the above rules and regulations are strictly complied with and rigidly enforced.

By order of the Board of Managers.
– H. C. Frick, Chairman[1]

Sometime around July 16th, the Carnegie Steel Company had begun to post large cards about its offices and Works, listing rules and regulations that were to govern employees. Dated retroactively, July lst, and signed by Chairman Frick, the rigidity of the rules and their anti-union stance were unmistakable to every Homestead worker.

Frick's order was primarily designed to control the labor force in the immediate future. It thus called for the company's unimpeded regulation of the production process, as well as for worker efficiency and the workers' protection of mill machinery. It sought to make the workers believe they were secure in their jobs and advancement opportunities based on the quality of their work and seniority. It also endeavored to make them believe they could redress their grievances – a point no worker in Homestead could possibly believe. Workers were now stopped from posting notices of their grievances in the mills, or leaving work suddenly for a demonstration or strike. Nor could they take retribution by damaging Carnegie machinery or by protest that resulted in their arrest and thus absence from work. Promises of liberality were thus part of Frick's scheme to alter the relations with the labor force on his own terms: a hierarchical structure of unbending authority over the entire labor force.

Workers not complying and continuing on strike without wages would not only lose their job places, but also could not pay their bills, rent or mortgages, thereby forfeiting the right to work for Frick, live in

Carnegie housing or in Homestead itself. Nor did Frick's new rules allow them to prepare for a transition by investing and working in a business venture while employed by Carnegie. What was still lacking was a way to enforce these terms.

For this, Frick did not have long to wait. Governor Pattison soon arrived in Homestead and declared that the troops would remain as long as might be necessary. This, clarified the Homestead *Local News* on July 23rd, "is taken as being favorable to the steel firm bringing in new men to operate the Works. Up to this time, only a comparatively small number have entered the mills, and only a few of these are understood to be skilled workmen. The firm claims to have several thousand applications for work, which they do not wish to accept if their old employees can be induced to return."

The small number of skilled, non-Amalgamated members at Homestead aided the firm in starting up production again. The foremen, assistant foremen and all heads of department were not members of the Amalgamated Association. They had been mere onlookers in the strike, and, despite their initial refusal to work under armed guards on July 14th, were induced or threatened by Frick's order to report for work two days later when production began.

Yet their presence did not square with the statement from an informed Amalgamated leader explaining the smoke and steam issuing from the Carnegie mills on the morning of July 16th. "The cupolas are filled with dolomite, and are lighted to save the burned stone in them," he explained. "A few furnaces are fired also; this is being done to save them from dampness. There are no men in the Works but the regular watchmen and a few firemen. They have never been denied admittance, as our people thoroughly understand their business there."[2]

The firm meanwhile sought to induce its unskilled workers to return, posting that day, July 16th, a circular at all entrances to the mills, calling for individual applications by 6 P.M., July 21st: "It is our desire to retain in our service all employees whose past record is satisfactory, and who did not take part in the attempts which have been made to interfere with our rights to manage our business." If they did not apply, their old positions would be given to other men, and those first applying would "have the choice of unfilled positions for which they are suitable." The inducements to the most vulnerable laborers to return to work were obvious, and the company alleged some two thousand strikers were prepared to report to work on Monday, July 18th to resume their old positions.[3] The Secretary of the Company, Lovejoy, confidently stated that the mills would be running by July 23rd, with enough men "to operate them satisfactorily.[4]

The workers resolved to offer no harm to nonunion men, but to resort to "moral suasion" to prevent the mills from being started.[5] They were buoyed by a resolution passed unanimously at a meeting of machinists and laborers:

> We, the employees of the mechanical department and day laborers of the Carnegie Steel Works of Homestead, in meeting assembled, do offer . . . : That we are in sympathy with the Amalgamated Association and are willing to remain with them to the end. . . . That we consider it an injustice to the mechanical department and day laborers and an insult to their manhood to ask them to work under guard, believing that in this land of the free all things should be free.[6]

## Foreign Workers and Sympathetic Strikers

The real challenge was convincing the nonunion workers from outside Homestead who were being brought in by the Company.

The women of Homestead especially demanded that a peaceful solution be reached. "Their finer instincts," wrote a reporter, "foresaw the only possible end of an uprising, and they strove to avert violence and bloodshed."[7] With the blue-coated soldiers at hand, they had had enough of unequal fighting.

Even their men-folk's fear that European laborers would be brought in on Monday, July 18th, was momentarily offset by the belief that skilled American workers in sympathy with their cause would support them on principle: that the strike was in the interests of all labor. Only the foreigners might not appreciate the principle involved, the Amalgamated believed.[8]

It was commonly known, moreover, that the Homestead workers could not win without solidarity from other steel workers in Carnegie's empire, and by July 11th a company-wide strike was planned if nonunion men were brought into the Homestead mills under the Militia's protection. Such a strike would be less than industry-wide, however. "When it is known that the Homestead mills are less than one-eighth of the vast Carnegie interests," advised the *Cleveland Leader*, "the importance of a general strike may be appreciated. There is very little doubt that pressure will be brought upon the workers of the other mills to urge them to come out, but there is not much reason to hope that such pressure will be successful. In the first place one or more of the other Carnegie establishments are already operated by nonunion labor, and, of course, it would be folly to suppose that these men would come out because of sympathy with organized labor. Then, too, in the other Carnegie mills there is no dispute whatever upon wages or anything else, but the scale has been signed by both sides for the ensuing year. For either side to break the agreement now, whatever the pretext might be, would be to lay that side open to the charge of broken faith and injure their cause more than it could help it."[9]

The *Local News* disagreed, noting the Homestead workers had allies everywhere: "the locked out men continue firm with no signs of weakening. They have received assurances of support and backing from labor organizations all over the country as well as financial aid from many quarters. Their friends everywhere are bringing pressure to bear upon the firm by instituting boycotts against those using the manufactured products of the Homestead mills. The workingmen in the different Carnegie plants outside of Homestead express a strong sympathy and in some instances have refused to work until a settlement is effected."[10]

The jelling of sympathy strikes at other Carnegie mills, with plans for a boycott of Carnegie steel and appeals to nonunion workers, gave the strikers some glimmer of hope. Sympathetic demonstrations at other Carnegie Works had begun in Pittsburgh and Beaver Falls, where company employees requested a reopening of negotiations with the Homestead strikers. Chairman Frick remained silent, but Secretary Lovejoy curtly dismissed the question of negotiations under the threat of a potential company-wide strike: "It will have no effect whatever on the policy of the Carnegie Steel Company, nor will their request be granted. Our course has been set, and it will not be changed, no matter what the result. We will hold no more conferences with the Homestead mill workers, no matter what may occur. . . . If every man in our employ – and there are 16,000 of them – if every man in every department of every mill should refuse to work, it could not alter the case, nor would our determination to hold no more conferences be changed."[11]

This ultimatum helped precipitate the sympathy strike in Pittsburgh, at noon on July 14th, when the employees of the Upper and Lower Union Mills quit work and turned off the gas in the furnaces. Three hours later the men in all departments were out and the mills were shut down, though the firm had already signed the Amalgamated scale for the ensuing year. This was preceded by the workers at the Beaver Falls mills notifying the company of their intended sympathy strike because the company refused to bargain with the Homestead workers.[12]

The neighboring Carnegie plants began to go out on supportive strikes one after another. The workers at Beaver Falls struck on the 14th, those in the Upper and Lower Union Mills on the 15th, followed by those at Braddock, Lawrenceville and Duquesne.

These were rear-guard actions that helped to keep up the spirits of the Homestead workers. A strike in the Lawrenceville Mills, the refusal of carpenters to work on buildings where Carnegie steel was used, and extension of the boycott were all the Homestead workers asked.[13] At an open air rally of about 600 men at Duquesne Grove, stirring speeches were made and sympathy resolutions passed. Many workers present also applied to become members of the Amalgamated.[14] Others supported the strike as well, promising to hold out jointly against the Carnegie empire.[15]

Each group of workers in turn had warned Mr. Frick in writing to negotiate with the Amalgamated at Homestead, but Frick adamantly refused, and the workers struck immediately.[16] Employees at the Edgar Thompson as well as the Upper and Lower Union mills planned to remain out until the Homestead trouble was settled. It was rumored that all railroad employees were ready to strike against the handling of Carnegie products and that the employees H. C. Frick Coke Company were also ready to walk out.[17] The Carnegie mills on Twenty-ninth and Thirty-third Street in Pittsburgh also stuck in sympathy. And Duquesne, which had been nonunion from the beginning, was organized in July and most of the men had come out for a few weeks. Other mills in Pittsburgh that had no connection with the Carnegie Steel Company also ceased work. By and large, it was a summer of strikes in the Pittsburgh steel mills.[18]

The Advisory Committee at Homestead meanwhile tried to create a support system, issuing a circular to the nation's trade unions pointing out the dangers of the Homestead precedent to their own organizations. The wives and children of unionists killed by Pinkertons were left totally unprovided for, the circular lamented, adding that Homestead also had unskilled laborers who barely earned an existence when employed, and, without work, were in want. "Help us therefore," the circular concluded, "to gain a victory that will rebound to the name of organized labor, and wipe out that blot upon civilization, namely, the Pinkerton murderers. Therefore, again we ask you to assist us financially in our hour of need. . . ."[19]

**How Victory Might Be Won**

The question of how the struggle could be won by the strikers was more complex.

Everyone seemed to have their own solution. The Right Reverend Bishop Fallows, of the Episcopal diocese of Chicago, addressed a mass meeting in Homestead, counseling moderation on both sides and calling for arbitration, now that Homestead had "dealt the Pinkerton system a death blow!"[20] At this meeting a relief committee was also formed to care for the needy, and word was received from sympathy strikers that they would never surrender.

Homestead leaders received encouraging support from abroad, as well. And the secondary boycott of Carnegie iron and steel was taking hold. Carpenters, stonemasons and bricklayers employed on some buildings throughout the country had quit work because the owners of these buildings persisted in using iron and steel manufactured by Carnegie. "This kind of assistance we need," an orator at the meeting exclaimed. "We propose to fight our battles on legitimate lines. The great weapon for us from this time is the boycott."[21]

The boycott was begun in earnest. From Homestead a circular, addressed to all workingmen, was mailed to all parts of the nation, requesting them not to work on materials shipped from the Carnegie Steel Company, and by their refusal, to help the struggle for organization. Thereafter many unions all over the country refused to handle Carnegie steel.[22]

Nationwide labor prepared to support the Homestead workers. In New York City, the Knights of Labor and the American Federation of Labor planned to meet with other labor groups to map out a strategy. In Pittsburgh, the Knights offered the support of all their lodges to aid the Homestead strikers in any way. Labor organizations took measures to keep track of the Pinkertons nationwide. Throughout Pennsylvania the Federation of Trades worked closely with the locked-out men; and in Illinois the Federation detailed men to monitor all Pinkerton movements with the purpose of having these mercenaries arrested and forced to put up money for bail – money that would be lost if they failed to "keep the peace," illegally crossing state lines. As well, the Chicago Trades and Labor Assembly took legal steps to prevent the Pinkerton headquarters there from sending agents to Homestead, relying on an Illinois statute regulating the formation of military bodies.[23]

The Knights of Labor also worked nationally to aid the Amalgamated, and made an effort to form a complete alliance of all organized labor.[24] In Philadelphia on the 8th, 8,000 workmen protested Carnegie's treatment of the workers at Homestead. Kansas Congressman Simpson telegraphed the crowd that the issue at Homestead was a "struggle whether organized labor would go down or be recognized," adding: "The laws are made for the capitalists, not for laborers."[25]

Milwaukee, hotbed of socialism, saw workers gathering in little knots to discuss the Homestead invasion, some heatedly demanding – "down with the Pinkertons!"[26] The theme at the meeting of the Milwaukee Federated Trades Council on July 7th had been general approval of the initial refusal of Governor Pattison to call out the Militia; the Council asserted that if Wisconsin ex-Governor Rusk had done likewise during the Milwaukee May Day troubles in 1886, no one would have been killed. "I seen my duty and I done it!" Rusk had projected as he called out the troops to shoot the strikers down.[27]

The Pinkertons were also condemned, Congress being asked to pass laws against them, the Council resolving: "That the Pinkerton thugs must go, and that we approve of the action of the workmen at Homestead in repulsing the invasion of their . . . city by these lawless thugs. . . ."[28]

Labor in Milwaukee was almost unanimous in supporting the workers at Homestead; the *Daily News* noted that: "H. C. Frick, the manager of the Carnegie Works at Homestead is the same party who imported a lot of Hungarians a number of years ago to take the place of strikers in the Pennsylvania mines and was instrumental in having the men shot down like dogs by Pinkerton thugs. He evidently had a plan for a repetition of those deeds, and among local workmen regret is expressed that he is not among the number who were killed."[29] The editors added that: "No band of hired men, without authority save that of a detective bureau, should be permitted by law to arm themselves and march into any city or stronghold to shoot down citizens who may chance to differ with the employers of such hired assassins on questions of wages or otherwise."[30]

H. E. Taubeneck, the chairman of the National Committee of the People's Party, also noted that while he deplored strikes and opposed them at all times, "in the case of the Homestead affair, the pressure has been so great that the props have fallen away until the condition of affairs similar to those that existed here prior to the war of the revolution has been forced upon the people."[31]

A unionist, H. D. Lloyd of Little Comptons, Rhode Island, wrote Samuel Gompers to ask: "Are the relations between the Amalgamated Association of Iron and Steel Workers and the Federation of Labor such that the latter could lend a hand in the recent struggles?" He put the struggle in its social context, saying that:

> Carnegie is universally disliked, even among his own class; he is here and abroad perhaps the most conspicuous representative of the wage system in its most fanatical form – that which holds the laborer to be a

> mere commodity to be cheapened to the last cent – his recent attack on the union is the most dangerous threat labor has yet had to encounter of the persistent attempts to make it appear that the tariff is the fundamental question involved. . . .

This success could be blocked, the writer believed, since no other comparable "opportunity to arouse and unite the workingmen of the whole country" was likely: "This can be made the most important conflict in the history of organized labor *and it can be won* . . . . Carnegie can be used to teach the Captains of Industry that men who treat their 'brother' laborers like sponges and rats to be shot cannot continue doing business in that style in this country."[32]

Under such pressures Gompers responded. As President of the American Federation of Labor, to which the Amalgamated was affiliated, he addressed a mass meeting in Homestead,[33] solemnly describing the dimensions of the situation:

> I ascribe the whole trouble to the invasion of the Pinkerton men. Every workingman feels it as an insult when these armed mercenaries are hired by the employers during strikes to be arrayed against citizens in defiance of the Constitution of the United States.
>
> This is a conflict which has been provoked without just cause by the Carnegie Company. The men, when the new labor saving machinery came into use, recognized its importance. They knew that it involved a radical change in the methods of producing iron and steel and signified their willingness to submit to a reduction of wages during the experimental stage. When such a wholesale reduction as a 40 percent reduction was made, however, they refused to submit to it. The lock-out and all the deplorable results which followed it were the consequence.
>
> The people conducted themselves peaceably until Wednesday, but the sweeping reduction of wages made them paupers. . . . The company wanted the people to shoulder everything, to bear all the loss, if it meant starvation for the families of the men who had worked so long and so faithfully.

Believing that the 300 Pinkertons were the advance guard of a contingent of Hungarians and Poles who were to be brought in to displace American citizens, Gompers added: "I am opposed to strikes, but in such a case as this I am tempted to say that if men are going to be starved slowly, it is better for them to die heroically."

Yet Gompers did not think any such measures as a boycott would be necessary, given the sheer strength of the union. The Amalgamated was one of the strongest organizations in the nation, with a membership of 48,000, some 19,000 residing in Allegheny County alone. The

Amalgamated also had a full treasury. And Gompers thought the capitalists who had combined to extinguish labor organizations had undertaken too powerful an opponent.[34]

Various labor alliances added to that power. On July 9th, a secret meeting had been held in Chicago between T. V. Powderly, Grand Master Workman of the Knights of Labor, J. W. Hays, Grand Secretary and Treasurer of the Knights, A. W. Wright, Executive Council of the same organization, M. Wilkinson, President of the Brotherhood of Railway Trainmen, and E. Clark, Chief Conductor of the Brotherhood of Railway Conductors. Thereafter they conferred privately with workers from Chicago and a delegation left for Homestead.[35] The implication was clear that these organizations were lending their aid to the workers of Homestead.

Material support had also come on foot from all the mills near Homestead. During the afternoon of the 6th, some 5,000 men had walked into Homestead from the South Side mills of Pittsburgh, many of them armed. In one delegation alone there were about 2,000 mill workers, carrying American flags and proclaiming they would stand by Homestead labor in defending the mill at any cost. Late that afternoon another 1,000 brawny workmen from the Jones & Laughlin mill marched down the Pemicky trestle structure with an American flag at their head, coming to tender their sympathy and encouragement to the locked-out men.[36]

Others outside Homestead echoed their solidarity, calling on the farmers of America to prevent the Homestead workers being crushed. "The fight," orated Grand Master Workman Powderly of the Knights of Labor, "is but the rumbling of the coming revolution that is to say whether Wall Street shall or shall not control the country. It is the fight of labor, and labor organizations of the country should stand firmly at the back of these men who are fighting at Homestead. If it must be a protracted struggle, the farmers of the country alone will furnish provision enough to keep those men for ten years, if necessary, for they are acting with the Knights of Labor now, and through the Knights with organized labor in general."[37] These were firm words, but a tall order which the farmers never delivered on and the Knights of Labor quickly forgot.

### The State Digs In

Meanwhile, every calculating, Henry Frick planned to balance his new work force between some scab laborers and some former employees from Homestead, but no union men. Looking at the number required, he stated succinctly that it would "be necessary to obtain a certain amount of outside labor to fill the places of the men who, as members of the Amalgamated Association, will refuse to go to work under the new wage system, and whom, in fact, we could not consistently take again into our employ."[38]

Labor might organize nationally to initiate a boycott and offer material and financial aid from the outside, but spirits dropped as Governor Pattison and General Snowden repeatedly warned that the Militia would remain in Homestead until everything was settled.[39] Local zeal also began to flag as Homestead increasingly became a vaudeville for visitors coming on special holiday trains and artists who, like Frank Norcross of New York City, portrayed the stirring events of the strike, but could change nothing in Homestead itself.[40]

Pattison meanwhile did everything possible to ensure that his plan of domination was carried out. Appropriating public tax revenues, he cleaned up the facilities the Militia were using in Homestead and ordered that the gutters of the City Farm Lane be cleansed and disinfected[41]- state amenities that Homestead residents had never received in the past. Snowden also became a public relations specialist, ever-available to correspondents, talking, writing for publications, and openly admitting the troops were intended to protect the nonunion entrants to Carnegie's mill.[42]

When Pattison finally came to Snowden's encampment on Carnegie Hill, he solemnized the occupation with pomp. "Seventeen guns were fired in a short time thereafter in honor of Governor Pattison and staff, whose visit had been anticipated for several days. The Governor reviewed the troops, listened respectfully to the Advisory Committee.

Everyone had been pacified: Visitors had crammed into the camp to see Pattison with the soldiers in tents or on parade.[43] They had not realized or articulated that Pattison's main purpose was not to change anything, but to inspect the troops in occupation and, under the aura of an official visit, to make congenial small talk with visiting citizens at camp, expressing sympathy and dismay at the events besetting Homestead.[44]

Organizing was also undercut at Homestead, as the leaders of the Amalgamated were subjected to legal harassment and jail terms. The Carnegie Company swore out murder charges against these officials for both Pinkertons and strikers killed in the battle of the 6th. Two companies, each soldier with twenty rounds of ammunition, then marched after three constables, who trailed, as if they were "desperate revolutionists," the half dozen indicted men, who in turn "had not enough weapons in their possession to start a shooting gallery," Stowell wrote. These leaders were jailed, with bail set at $10,000 on each. The charges were outrageous. The workers collected bail money, and some of the leaders were "sprung."

Burgess McLuckie, having grandly invited the Militia into Homestead, was one of the first indicted, surrendering himself to the authorities in Pittsburgh, saving the brave constables the trouble of reading the warrant of arrest to place him behind bars. "The magnificent display of military force, which was again thrust upon notice of Homesteaders when the constables at the head of the flower of the army, were seeking for the indicted men, in Burgess McLuckie's case, was a grand fizzle," Stowell reported from the scene. "How essential two companies of soldiers were to complete so ordinary an event, has never been fully appreciated by those who understood the situation. It was in line with almost every move made after martial law had been illegally inflicted. . . . One plucky, level-headed Pittsburgh policeman would have made the arrests without an atom of difficulty!"[45]

Despite the bombast, the trio of officers had at first departed for Pittsburgh empty-handed because the indicted did not happen to be at home. Yet the military authorities feared trouble and at once began to make preparations to prevent an outbreak, one laudatory report delineating elaborate preparations to take on what was imaged as a bellicose force: "An order was sent to the cavalry battalion, which resulted in that admirable organization being on the streets of Homestead within a few minutes, and at the same time there was a good deal of hustling around the headquarters of Battery B. The provost guard was also strengthened. From division headquarters General Snowden watched through a field glass the movements of the posse, and had there been the slightest disturbance two regiments would have been thrown into the town almost instantly."[46]

To survive, the strikers and the Advisory Committee had had to change their tactics. They no longer demanded military retribution, but sought to have the Pinkertons and the Carnegie managers brought into court. "It was the revenge that modern law could and would afford," wrote a contemporary, "and the strikers demanded the fulfillment of the old Mosaic law, 'An eye for an eye and a tooth for a tooth.'"[47]

Meanwhile, the indicted strikers turned themselves in; the prisoners acquired a large following in the courts and made bail.[48] Judge Magee had fixed Burgess McLuckie's bail at $10,000, and had stated that the prisoner could not be held on a higher charge than murder in the second degree.

Bond was then raised, the news that McLuckie was returning to

Homestead spread, the people arranged for a demonstration which Snowden reluctantly "permitted," and 2,500 Homesteaders met McLuckie at the depot, marched triumphantly with him to the Rink, and extended an enthusiastic ovation.[49] The parade sent local zeal to a zenith. But the workers of Homestead were now clearly on the defensive.[50]

## Fight to the Finish

Meanwhile, to outflank the workers economically, the Carnegie empire shifted Homestead production to the Duquesne Works.

Although this mill had hitherto specialized in producing steel billets, Carnegie managers now planned to turn it into a producer of steel rails previously made at Homestead. They began by altering the Duquesne's Bessemer facilities, making necessary changes in the furnaces and machinery, so that 13 continuous furnaces would be run in a line, with four new ones erected on filled-up ground near transport facilities by river and rail. All orders for Duquesne steel billets were then canceled, and no more were to be made.[51] "The immense new foundry and machine shops, which will employ over 1,200 men," the *Local News* warned Homesteaders, "will be the largest in the world, will be near the river bank, and the new furnaces will join them."[52] Now that the high quality output of Homestead structural steel was to be shifted to Duquesne, pressure to maintain high production standards by the unskilled workers was removed.

The next step was to start up the Homestead mills with imported unskilled nonunion workers. Whereas the representatives of the mill confidently claimed that between twelve and fifteen hundred men were employed and turning out satisfactory work by August 6th, the locked out men maintained this work was not up to the previous standard of the firm. Only a small number of the striking workers had returned by then, and the strikers were morally bolstered in thinking their legal suits against the firm and the Pinkertons gave them an advantage – though in truth they had no chance the courts would ever convict Frick and his assistants. Money was also flowing into the treasury of the Amalgamated Association from all sides, and numerous boycotts of Carnegie products were going on at many points. "It is evident that the staying qualities of the locked-out men are far from being exhausted," the *Local News* concluded.[53]

Yet, Carnegie's millions meant the firm could hold its production facilities idle much longer than the financial resources of all the unions in the nation could sustain the idle workers. The firm might temporarily have idle plants at Homestead and relinquish some markets for its structural steel to competitors. It might also be temporarily unable to move its inventory of steel from the mills without battling with the locked-out men. But Carnegie was worth a half a billion dollars, while the burdens now placed on the workers would impoverish even the thriftiest. Perhaps as much as $50,000 to $100,000 was in the Homestead Amalgamated treasury, but this treasury would have to sustain a great many strikers. For two months, each striker was required under the rules of the Amalgamated to rely upon his own resources; then only $5 a week was to be paid out. By comparison with good times when there was work and the firm paid out some $90,000 each two weeks, now the Amalgamated pay-out would amount to $30,000 each two weeks if the entire 3,000 workers received Amalgamated funds. That would give the local Amalgamated treasury six weeks' staying power. Though there were contributions promised from other organizations to keep the population afloat, they would not amount to the wages many workers usually received. Comfort and even homes acquired by long years of toil could thus be lost relatively quickly.

Calculations differed among families, for many workers with homes also had savings in the bank, and the treasury of the Amalgamated was to be made available to those without savings. Yet the local organization's funds had definite limits.[54] Well-informed people in Homestead estimated it would take as little as four months to compel the locked out men to give in, emphasizing that a longer period would entail great sacrifice.[55] At best there would be six months' reliance on private resources, and then total dependence on Amalgamated moneys and contributions.[56]

During the count-down, the Company was also faced with three fundamental problems of its own: hiring a nonunion work force, achieving quality output, and overcoming possible boycotts of its production. By August 6th, there was still no material break in the ranks of the locked out men, so the company had to continue to recruit scabs, now getting as many as it required. Though the new men were certainly making steel, the quality was wanting and so were profits. The locked-out men thought the mistakes and accidents by the "green hands" were sure to embarrass the managers. And the Amalgamated set in motion a powerful boycott of Carnegie steel, though little output had yet been made at Homestead. There was no middle ground in the fight by this time – the recognition or non-recognition of the union was at stake, and as the *Local News* said: "It is a fight to the finish."[57]

**Bringing in the Scabs**

Passive resistance had proved an excellent way to raise the consciousness of the labor movement outside Homestead that the people of that city were refusing to cooperate with the Militia and its oppressive regime. But it failed to bar importation of scabs.

With the Militia guarding over Homestead and Carnegie's Works, the scabs could be brought directly into the mill without ever coming in contact with the population. Even when passive resistance was combined with actively trying to influence the scab laborers, manager Frick was able to outflank the unionists and their supporters.

The steamboat *Tide* was again put in service, this time as special transport for the strikebreakers. Under Colonel Hawkins' guns, the *Tide* crossed the river with the first squads of scabs who came from Swissvale on the Pennsylvania Railroad. "A posse of nonunion men arrived Wednesday," an eyewitness reported. "It was merely a handful of men – six in number – but was the advance guard. They were from the salt works, located on the B & 0 road, and knew as much about making steel billets as they did about the equity in law or the justice of a strike. They were there to make application for the position of one of the old men – no person supposed for an instance that six raw hands could take the place of more than one experienced worker."

They didn't last long either, once the strikers bent their ears. "Argument, that weapon of civilization," a reporter explained, "was used effectively upon the pilgrims from the salt works, and they turned their faces towards the jobs they had but recently abandoned. This victory was hailed as a good omen by the strikers, and they waited enthusiastically for other applicants to appear."[58]

The contest between the company and the strikers now assumed new proportions. The *Tide* continued its transport of cots and scabs, secretly ordered by Captain Rogers, infamous in the Pinkerton invasion, to steam up and embark for Homestead. The strikers meanwhile laid out a network to gain information from all the points in the U.S. and Europe about the departure of laboring men for Homestead. O'Donnell and his friends gained the support and cooperation of organizations in New York, Chicago, San Francisco, New Orleans and other cities. A system was soon perfected for sending and receiving critical information, and propagandists spread out their forces to intercede when the scabs were sent. "Men under the control of others with cast-iron instructions to suffer no interference were started for Homestead," Stowell wrote, "but many of them never arrived. Agents of the Amalgamated Association invaded every depot and gained access to almost every train containing the new men, and persuaded large numbers of them to turn back. The appeals of the strikers were listened to and heeded, to a large extent, but eventually capital claimed the victory."[59]

The union refused to budge for months, though. Chairman O'Donnell and his lieutenants received hundreds of letters and scores of telegrams every day about the feelings of their brethren in other cities and the movement of nonunion men destined for the mills.[60] They were needed, too. "Agents of the firm were set to work in every large city procuring recruits," said the roving witness Burgoyne, "and while it was found rather difficult to tempt men to risk the desperate chances of a trip to Homestead to take the places of the conquerors of the Pinkertons, a small number were secured each day, until, within a week after the first squad was brought in, nearly one hundred men were quartered within the mill enclosure."[61]

The first decisive steps were now taken to resume steel production. The furnaces previously fired required no particular skill or intelligence and had been accomplished by about 25 or 30 nonunion men and probably 40 or 50 clerks, protected by as many soldiers.[62]

The latter, with the constables, easily imposed their will. In a typical show of force, when striker Frank Tracy tried to interfere with the nonunion men, a constable arrested him on the spot, charging him with "larceny," and taking with him as witnesses to Alderman Oeffner's court four of the new men who had been at work in the mills several days. But this time, as 2,500 strikers proceeded after the witnesses, blockading the street in front of the courtroom, Major Crawford ordered out a small band of troops to disperse them. Insufficient, reinforcements of other troops and deputy sheriffs were needed before the order to disperse could be enforced. Even then, it was only after Homestead's own local police appeared that the crowd would consent to move on, another indication that the local government acted as an effective enforcement body to control the population under Snowden's occupation. "This demonstration indicated the hatred that existed for the nonunion men," a reporter said, "and doubtless had an influence in the decision of the powers that were, to retain the troops longer than had been intended."[63]

The scabs had been brought by the *Tide* from Pittsburgh for landing at the river bank within the Works, their new home. "A gang of men were waiting for her, and her freight, consisting of cots, blankets, canvas and camp utensils, were unloaded and placed inside the mill property," wrote the *Sun* on July 16th. "The poor farm property, which is inside the big fence and is now the property of the Carnegie Company, contains a big level field which will make a good camping ground for workmen, and the large building which has been used for the country poor will hold a number."[64] Residences, cots, food and supplies for hundreds thus provided, the Company also planned a new sewer system to dump refuse from the City Farm Lane into the Monongahela[65] – another measure to avoid deadly diseases Homesteaders had requested but not received before the strike.

With the Company's new City Farm to become the workers' housing center, the firm announced plans to discontinue sales of any of its lots[66] made ready to build one hundred new company homes,[67] and advertised for bids to furnish materials and build them. It clearly showed an intention to establish permanent, protected homes for nonunion men[68] to take the places of the old employees.[69]

Surprises were also in store for the Amalgamated once the trainloads and boatloads of "black sheep" had entered the mill. On Friday, July 15th, for the first time since the lockout began, a curl of smoke issued from one of the furnaces – the cupolas of the giant steel plant were being fired! The furnaces grew hotter, causing the news to circulate more rapidly in Homestead that the company had smuggled in sufficient non-unionists to renew operations. "An excited mob rushed pell-mell towards the mill yard," an observer reported, "but was stopped short at the picket lines, where the guards, with leveled bayonets, barred progress."[70]

Face to face with bayonets, held even by order-driven, sympathetic soldiers, the men had little choice: "the abrupt manner in which they were turned back . . . was taken by many as an indication that the troops are ready to fight, in fact, which may have hithertofore been disbelieved on account of the sympathy for the men expressed by the Militia," one reporter gave his offhand opinion.[71]

As the occupation of the troops dragged on, many came to a deeper understanding of the desperate plight of the population. But, however extensive it may have been, their sympathy did not square with military duty. The officers in command were carrying out the Governor's orders, and headquarters of each brigade issued suborders which were obeyed with alacrity. "Privates came forward with the determination of serving the Constitution, no matter how marked might be their sympathies for one side or the other," a reporter observed. "In fact, a large percentage of them did not attempt to disguise their feelings in the premises, but waived their personality and responded as soldiers should respond."[72]

The firm, protecting and housing its scab workers, then intensified the psychological tension placed on the strikers, trying to force them back to work. The rituals of the workday were ostentatiously resumed, black columns of smoke pouring from a score of the big chimneys and the workers once more trying to explain the ploy in terms of Carnegie protecting raw materials and the employment of unskilled workers. "At noon hour steam whistles sent their shrill notes into every corner of the town – and into the heart of every man, woman and child there was a little pain that grew and expanded as the shrill notes lengthened out and continued for some moments," wrote Stowell. "The whistles were being blown, doubtless, for this very purpose, as there was no practical occasion for the noise. The small force on the inside could have been notified with a tin whistle that it was time to satisfy the inner man."[73]

Could the Works be successfully operated by these unskilled, nonunion workers, then? Many Homesteaders thought not, especially after Allen Hubbard, the foreman of the mill's Armor-Plate Shop responded to the Company's order to superintendents and foremen to report for duty on the 17th by quitting. Hubbard refused to work under the Militia's "policing" in order to operate the mills. With his refusal, the entire force of the civil engineering department of the mills also quit work in sympathy with the locked-out men. Thirty-five Carnegie bricklayers who were not Amalgamated men also lay down their trowels, refusing to work until the strike was settled. Similar pledges were offered the Advisory Committee by the machinists, the carpenters, and the machine and furnace repair mechanics. Jubilant, the locked out men declared it impossible to start the mills without their assistance.[74]

The fact is the unskilled scab workers did not know how to use the sophisticated machines and to continue employing them had become expensive. By September, a serious breakdown had already taken place in 119-inch plate mill, involving the destruction of the big shears and a loss of $15,000, forcing operations to be suspended in this department.[75]

Yet Carnegie was prepared to spend great sums to recruit, train and protect his nonunion work force. As the scabs and their guards were enclosed inside the City Farm plant, a rear-guard-defense and refusal to cooperate were now all Homesteaders could hope for. Thus the cooks and women servants at the Frick Hotel – the best restaurant in Homestead – refused to serve Carnegie managers and the generals, requiring the officers to survive on army rations, so that now Militia headquarters imposed on the camp commissary for their grub. The "girlcott," as it was arrogantly disdained, was broken by replacing these women with a new force to serve the mighty.[76] Others who refused to cater to nonunion needs in town were similarly bypassed once the new work force was settled at Fort Frick.

Having turned its back on the town, the Company continued its nationwide recruitment of strikebreakers. Outside laborers poured in, though Homestead's own unskilled, locked-out workers remained on strike in solidarity with the Amalgamated. Despite the fact that the skilled had haughtily looked down on the unskilled as a lower caste before the strike, the latter proved faithful allies, hated Carnegie and Frick for their social class, and left Homestead rather than surrender to the Company. "The Hungarians began to turn their backs upon Homestead, many of them packing up their little chattels and deserting their humble abodes," said a witness. "They were not afraid of strife, – they were sick of it." The unequal contest was obvious, and most headed for Ohio and the other great manufacturing and mining centers of the country.[77]

Meanwhile, with Chairman Frick still laid-up with wounds from Berkman's assault, Secretary Lovejoy redoubled his efforts to recruit nonunion labor, announcing that "300 men were now in the Works and more were coming." To house them, the Company gave notice to the strikers that they had to vacate Carnegie cottages in ten days, and by the end of the week many old workers had begun to move to other quarters. Their places were quickly filled, as the steamer *Tide* made three trips each day, carrying scabs, cots and provisions to the mills.[78]

Nor was there any sign of relief for the old work force: The firm experienced little trouble in finding men nationwide from the then-present army of unemployed, as well as from the pool of underpaid workers in various lines – clerks, struggling young professional men and others – who were tempted by the high wages said to be paid at Homestead.

Though the commune leaders continued to contend that it would be impossible to operate the mills without a force of expert steel workers, "Mr. Frick seemed to care little about skill so long as he could secure men able and willing to learn the work entrusted to them," a witness concluded. "The mills, it appeared, were to be a school in which an absolutely new set of artisans were to be instructed, regardless of the losses that must be sustained in the process."[79]

## Notes

[1] *The Local News*, July 16, 1892.
[2] Stowell, 185.
[3] *Ibid.*, 185-186.

[4] *Ibid.*, 187.
[5] *Ibid.*
[6] *Ibid.*, 188.
[7] *Ibid.*, 189.
[8] *Ibid.*, 190-191.
[9] *Cleveland Leader*, July 12, 1892.
[10] *The Local News*, July 23, 1892.
[11] Stowell, 179.
[12] *Ibid.*, 179-80.
[13] *Ibid.*, 187.
[14] *Ibid.*, 188-189.
[15] *The Local News*, July 23, 1892.
[16] Wolff, 152.
[17] Fitch, 121-122.
[18] *Ibid.*, 132.
[19] Stowell, 150-151.
[20] *Ibid.*, 181-182.
[21] *Ibid.*, 182-183.
[22] *Ibid.*, 239.
[23] *The Milwaukee Daily News*, July 8, 1892.
[24] *The Cleveland Leader*, July 10, 1892.
[25] *The Philadephia Press*, July 9, 1892.
[26] *The Milwaukee Daily News*, July 7, 1892.
[27] *Wisconsin History Calendar*, (Madison, Wisconsin, 1974), graphic, May 1886.
[28] *The Milwaukee Daily News*, July 8, 1892.
[29] *Ibid.*
[30] *Ibid.*
[31] *Ibid.*
[32] Gompers File, "General Correspondence," " Copy of July 15, 1892, letter to Gompers from H.D. Lloyd,US, MSS, 117A, 11A, Box 1 (Madison, Wisconsin: Wisconsin State Historical Society), Original Emphasis.
[33] Stowell, 239.
[34] *The Milwaukee Daily News*, July 8, 1892.
[35] *The Globe* (Toronto, Canada), July 9, 1892,
[36] *The San Francisco Chronicle*, July 7, 1892,
[37] *The Sun* (Baltimore, Maryland), June 16, 1892.
[38] *Ibid.*
[39] *The Local News*,, July 23, 1892.
[40] *Ibid.*
[41] *Ibid.*
[42] Wolff, 151.
[43] *The Local News*, July 23, 1892.
[44] *Ibid.*
[45] Stowell, p. 195

[46] *Ibid.*, 195-196.
[47] *Ibid.*, 196-197.
[48] *Ibid.*, 197-198.
[49] *Ibid.*, 205.
[50] *The Local News*, July 23, 1892.
[51] *Ibid.*; see Stowell, 195-205.
[52] *The Local News*, August 13, 1892; see Stowell, 204-205.
[53] *The Local News*, August 6, 1892.
[54] *The Cleveland Leader*, July 10, 1892.
[55] *Ibid.*
[56] *Ibid.*
[57] *The Local News*, July 13, 1892.
[58] Stowell, 137-138.
[59] *Ibid.*, 141.
[60] *Ibid.*, 144
[61] Burgoyne, 133.
[62] Stowell, 206-207.
[63] *Ibid.*, 207.
[64] *The Sun* (Baltimore, Maryland), July 16, 1892.
[65] *The Local News*, August 13, 1892; see Burgoyne, 133.
[66] *The Local News*, August 13, 1892; see Burgoyne, 133.
[67] *The Local News*, July 23, 1892.
[68] Stowell, 186.
[69] Burgoyne, 133; see Stowell, 186-187.
[70] Burgoyne, 134.
[71] *The Sun* (Baltimore, Maryland), June 16, 1892.
[72] Stowell, 115-116.
[73] *Ibid.*, 202.
[74] *The Post*, July 15, 16, 1892; *The National Labor Tribune*, July 16, 17, 1892; *The Sun* (Baltimore, Maryland), July 16, 1892.
[75] Stowell, 252.
[76] *The Sun* (Baltimore, Maryland), July 16, 1892.
[77] Stowell, 126.
[78] *Ibid.*, 227-228.
[79] Burgoyne, 135-136.

Cutting Down Frick's Effigy

# Chapter 10

# Approaching the End

> All anxiety gone since you stand firm. Never employ one of these rioters. Let grass grow over the Works. . . . Use your discretion about terms and starting. . . .
>
> – Carnegie cable to Frick on the day following the clash[1]

> Capital has won. The methods employed in the name of labor have been the main cause of the inglorious defeat suffered by the Amalgamated Association.
>
> – Homestead *Local News*, November 22, 1892

By mid-July the commune at Homestead had been completely displaced by the Militia, and under General Snowden's *régime* the mills had been more or less started up by using non-union workers. Frick had developed tactics to win. He knew the strikers could be kept peaceful by the Militia, deputies, sheriffs and Pinkertons; and fear and tension would also help defeat the economically pressed workers. He understood as well that charity to keep labor on tap would be forthcoming locally from those with wealth and power. They wanted social stability and would give aid to get it.

Frick reasoned this would bring the Homestead work force, one-by-one, as individuals back into the mills. Then Carnegie and his managers could weed out the organizers of the strike and commune and keep those needed to reach full-scale production.

But could Frick succeed in breaking the Amalgamated, putting each worker to bargain on his own? And could the old skilled work force replace the scabs and be reintegrated into the new schedule of uninterrupted mill production?

## Scabs in the Mill

Even when most of the State troops were ordered home, scabs still arrived at Homestead from the East.[2] The recruits enjoyed no Utopia, though, most being as ignorant of Homestead as the lower-ranking Pinkertons had been during the invasion. "Many of these men claimed that they were induced to come to Homestead by false representations, and that, once inside the stockade, they were, to all intents and purposes, prisoners," one observer wrote.[3] Another added that: "The life of the men at work in the mill was by no means a pleasant one. They were closely confined, not caring to go out, for some time after the mill was first manned, unless on Saturdays, when those who cared to go to Pittsburgh were taken down on the *Tide* and allowed to remain away over Sunday."

Still, those lodged in Company houses near the mill under military protection had only reached the tiny number of 200 by mid-August; and military protection was clearly inadequate to successfully help even the first scab attempting to bring in his household goods. "A crowd of strikers met the teamster who was handling the goods at the ferryboat and threatened him with severe handling if he again ventured to haul the property of a black sheep through the streets of Homestead," one reporter noted the shameless methods of persuasion.[4]

To make matters worse for reinitiating smooth production, the abnormally high temperatures in the summer of '92 pushed the heat up to 150 degrees in Homestead's converting departments – and discontent to a boiling point in the shanties housing the strikebreakers. Day laborers began to desert in numbers; the lower-paid mechanics and repairmen participated in the daily exodus, threatening, a local paper noted, to "depopulate the mill in a week."[5]

Attorney for the Amalgamated, G. W. Argo, also reported that by September there were 1,500 non-union men at work in the mills, but they were mostly incompetents. Some of their number had died in the mills; others were about ready to succumb; and several had been maimed for life by the machinery of which they knew practically nothing. Together with bad water, heat ranging from one hundred to one hundred and fifty degrees dramatically increased the mortality of the non-unionists: the dead and wounded were carried nightly by the steamer *Tide* from the Works to Pittsburgh.[6]

Other scabs took sick due to the unsavory conditions of the mill kitchens. The environment in the mill in the months of August and September were themselves dangerous to human health. It was not only extremely hot weather, but the sewage system in and about the Works was poor, and many of the new steel workers were unaccustomed to the type of labor they were learning. There was also a great deal of sickness among the new men, fed in enormous mess halls erected on the river bank by the Company for the exigency of the lockout.

* * *

To explain the deaths and sickness, Frick cleverly took advantage of the situation, a reporter surmising that "the Carnegie attorneys, aided by the Pinkerton detectives, had furnished the country at large a nine day's sensation when they caused several arrests, on account of the charge that the sick men had been poisoned."[7]

The charges placed Hugh F. Dempsey, Master Workman of the Pittsburgh District Assembly No. 3 of the Knights of Labor, at the center of a conspiratorial network, with Dempsey supposedly issuing money, orders and poisons to others: poisons going directly to one Beattty; from Beatty to Davidson and Ford; and from Ford to Gallager.

It was alleged that Dempsey had hired one Robert F. Beatty and, through Beatty, a J. M. Davidson to poison the non-union mill workers. In September, the allegations continued, Beatty had also directly hired one of the cooks at the mills, Patrick Gallager, to put poison in the coffee and tea of Carnegie workers. Gallagher had already made a confession implicating Beatty and Dempsey, though Dempsey had no apparent motive, not being interested in the Homestead strike as such, the Knights having no direct connection with the Amalgamated Association.

But at the criminal trial put in motion, the chief witness for the prosecution, Gallagher testified that he and Davidson were paid to gain employment in the kitchen at the mills, to dose the coffee and tea from time to time with a powder to be furnished them by the defendants Dempsey and Beatty.

Gallager claimed he acted out his role, that Davidson replaced him in his part, receiving from him the stock of poison on hand, and that Dempsey had paid him various small amounts of money, $25 acknowledged in writing. He testified that Beatty went to Cincinnati to recruit more cooks for these ends and that the extensive sickness in the mills was caused by the mysterious powders provided by Dempsey. With Davidson making a similar exposition, the prosecution relied on Pinkerton J. H. Ford's statement that he overheard Beatty tell Gallagher the plan for poisoning the scabs.[8]

It was a profound press sensation; newspapers spun fabulous yarns of cynical conspirators lacing the mill with death. But the attempted frame-up was exposed at the trial. Gallagher and Davidson, who claimed to have been hired to lace the workers' larder, also claimed they had been promised immunity from punishment if they would make their "confessions," then found themselves double-crossed at the trial when they too were included in the circle accused of the alleged crime.

Experienced doctors, meanwhile, could not agree on the causes of death of the millmen. Some physicians solemnly testified the sick at the mills suffered from typhoid fever, others asserted they had dysentery, while still others authoritatively claimed they were poisoned – the conflicting testimony undermining the prosecution's proof.

Gallagher's insistence that the mill men had been poisoned was also brought into question when Baltimore coal merchant, W. C. Kaiss, claimed that prosecution expert witness Dr. Cooper had received $500 from the Carnegie people for his testimony. District Attorney Burleigh denied, but did not prove, that any money could have possibly been given Dr. Cooper, claiming instead that he had charge of the prosecution of Dempsey and would not allow "tampering with the witnesses."[9]

The defense was more thorough, subpoenaed some 57 witnesses who gravely testified that diarrhea was a common complaint among mill men and that they commonly used medicine to prevent it.[10] The case now turned on the anal regularities of the Homestead population. Several doctors, some mill men and a number of members of the National Guard seriously testified that diarrhea was prevalent in and outside the mill prior to the date when it was alleged the poisoning began.[11]

The prosecution, moreover, failed to prove that Dempsey or Beatty had given the alleged poisoned powders to either Gallagher or Davidson, or to show where these alleged powders had been obtained or what had become of the unused portions.

The defense then produced witnesses to prove the good characters of the accused and the unenviable character of the prosecution's witnesses. The defendant Dempsey also testified that the money claimed to have been paid to Gallagher was a loan to help out when the latter's household effects were in danger of being seized for debt.

Though the attorneys for Dempsey and Beatty admitted that Davidson and Gallagher were hired to go to Homestead, they claimed it was for the purpose of learning the number of men employed. Their trip was inspired by the New York headquarters of the Knights of Labor to determine the extent to which Carnegie's rose-colored reports of successful operations at Homestead with unskilled labor were true. Dempsey testified that others were also employed for this purpose and that Beatty was sent to Cincinnati, not to hire more cooks as Gallagher alleged, but to hire men to enter the mills as laborers to spy on the true conditions inside the mill.[12]

It was a grand passion play to tittle the nation while ignoring the central issue of the fate of the Amalgamated and the strike. Despite the cogency of the defense arguments, Dempsey and Beatty were convicted, each being sentenced to seven years in the penitentiary. Gallager and Davidson received sentences of 7 and 5 years respectively – though Gallagher admitted six months later his testimony was perjured and the entire case against Dempsey had been trumped by the Pinkertons, who had promised himself and Davidson immunity from prosecution if they would testify to convict the two men. Davidson admitted Gallagher's confession was true, but then reversed his stand on learning the punishment for perjury.

Though 90 percent of the people of Allegheny County believed Dempsey and Beatty were innocent and thousands signed petitions praying for executive clemency, the Governor did nothing and they unceremoniously went to jail.[13]

The point of the legal exercise in "guilt" attribution was simple enough: The Carnegies, the Pinkertons and the government apparatus would be used against the strikers in every way possible. The workers who were locked out would be placed in the most unfavorable light. And some workers, as well as others not connected with the Advisory Committee and the struggle at Homestead, would be used as scapegoats for the dangerous conditions in the mills. Though we may never know if some of the non-union men were poisoned by a conspiracy of workers and cooks, we do know there was an alliance of mill owners, State officials and Pinkertons who tried to make it appear the workers had turned to such measures.

Their scheme to divide those who supported Homestead workers and exonerate Carnegie, the Sheriff, the Governor and the Militia had its victory at home, as the *Local News* editorialized against those who had turned to unsavory methods: "The enemy of labor is not capital nor the state, but those misguided and vicious persons, who imagine that foul means can be successfully employed in accomplishing what they conceive to be the rights of the workingmen. The labor organization should join with the law in hunting down the worse enemy of both."[14]

**Workers on the Defensive**

Workers losing allies and angrily on the defensive throughout Allegheny County, they were also unemployed, lacking wages to buy amenities – and facing an intractable body of employers.

The coal miners in the vicinity of the Munhall mines had, for example, been on strike and gone wanting as long as the millmen at Homestead. In the nearby towns of Six Mile Ferry and Bellwood, many hundreds of the miners were also suffering greatly without publicity, benefits or offers of support.[15] Owners had mobilized local constables and sheriffs against labor throughout the valley. And despite sympathy strikes outside Homestead that had offered them initial support, Homestead workers were doubtful that they could long carry on the struggle.

The testimony of innumerable witnesses supporting Carnegie against Homestead workers was hardly relieved when one of them, Constable Isaac Jury, was found heavily under the influence on Eighth Avenue, was arrested, put in a wheel-barrow, dumped several times, allegedly kicked, took sick on the way to the lock-up, and "unexpectantly" died.[16] For it was clear that the legal tide was flowing against them: court trials of thirteen Duquesne workers who had supported the Homestead strike and engaged in a sympathy walkout resulted in one acquittal, two verdicts of guilty of unlawful assembly, and ten verdicts of rioting - the heavy penalty for the latter being a $500 fine and two years imprisonment. One of the rioters lived in Homestead; and the verdicts aroused a great deal of excitement in the Borough, as many thought they would have a heavy bearing on the Homestead cases.[17]

Buffeted by constant fear of their permanent replacement,[18] the unionists were also prey of imaginative scabs who, after receiving a few square meals and pocket money without too much labor within the confines of Frick's wooden fence, threw themselves into the arms of the strikers, saying they had been abused by the firm's agents, and asking for pecuniary aid from the Amalgamated Association.

Welcoming desertions, the strikers did not always look closely at the circumstances prompting the sudden change of heart by the deserters, thus leaving themselves open to tall tales by the wayward scabs that almost the entire corps of non-union men were preparing to quit work.[19]

Though this was untrue, conditions and the food in the mill were decidedly poor, the "mill sickness" continued, and a great number of mill hands were stricken dead or laid up for months.

For a change of diet, it took deputy sheriffs to occasionally escort the strikebreakers into Homestead where reluctant restaurant owners were compelled either to "stand and deliver" or have their licenses forfeited under Pennsylvania's Brooks law guaranteeing equal service to the public.[20] "During the week," the *Local News* outlined on September 10th, "large numbers of non-union workmen have been coming from the Works into the Borough for their meals, and with a few trifling excepts were in nowise molested. . . ."[21]

As one or more scabs usually slipped away after each meal, there was a continual need to replace them,[22] along with those uncooperative workers summarily discharged. As unhealthful mill conditions and incarceration produced discontent, escapes, complaints and ejection, a repetitious, vicious cycle was set in motion. Even the stories became familiar: One worker ejected from the Works by his superiors said he had been recruited by Carnegie agents to become a bricklayer in Pittsburgh, but with others "were made prisoners in the Works and guarded like convicts," adding that imprisonment had a dual character: "The more ignorant were told by the foremen that if they ventured outside the union men would shoot them down like dogs." But inside the mill enclosure, at least half of them were sick from heat, bad water and poor food.[23]

Some of those who were lied to or shanghaied chanced an escape.[24] Recruit Joseph Manning, misled and sent on the Baltimore & Ohio for Homestead, explained his last minute exit:

> There were fifty-six of us hired in Cincinnati by F. W. Nye, in answer to an advertisement promising easy work and good pay. We were told the work was in Pittsburg [sic], but had no idea we were intended for the Homestead mills. In the morning after boarding the cars we found the doors and windows locked and armed guards in charge. This roused the suspicions which were confirmed by the information, in reply to our .questions, that Homestead was our destination. This created great indignation, and we demanded the right to leave the train, which was refused. The doors were forced open and men began dropping off. When we got to the depot only twenty-one men were left. I was the

twenty-first man, and got away by mixing with the crowd.[25]

Others escaped from the mills by climbing the wall, one escapee, Hugh Sweeney, telling the Amalgamated there were not more than thirty or forty men actually working, although there were then nearly 700 men within the stockade. These, he continued, were mostly Hungarians, who knew nothing of the strike and were provided by the Company with their needs for food, tobacco and work clothes.[26]

Despite the heroics involved in harrowing escapes and the dismal accounts of the production potential of the unskilled, the force of mill employees, including office workers, increased rapidly. As wages offered were considerably higher than the average earnings of clerks and young men just beginning a professional career, many such were attracted to the Carnegie establishment. Skilled workers and experts of some renown were also employed for their education and background, including four prominent brokers of the Pittsburgh oil exchange! Though their real knowledge of steel making was slight at best, one was set to work running an engine; another became a foreman of the cold saw department; a third was placed in the armor-plate department, and the fourth, John McLaughlin, considered the "highest roller" on the Oil City and Pittsburgh floors, settled for subordinate duties.[27]

All Frick's recruits were clearly not ignorant or unskilled, then, though the quality of their work was clearly wanting. Bricklayer Henry Stocher, hired for the mills and ordered out after reading an account of the strike to other non-unionists, thus combined elements of fact and fiction in reporting that: "The men in the mills are perfectly ignorant of the Works, and are mostly foreigners, and absolutely no headway has been made in operating the machinery," adding that: "The smoke and steam seen by those outside are simply a bluff, as the men are doing nothing."[28]

The Amalgamated Association believed this and similar account. They thought that none of the steel turned out would stand testing, and that, while there were doubtless plenty of men ready to work, there were few, if any, skilled laborers among them.

Secretary Lovejoy claimed the opposite: that the work was of the most satisfactory nature, and that he had more applications for positions than he could fill.[29] As proof, Assistant Chairman Childs admitted a favored reporter to the mill, on his promise to report faithfully what he saw: that a large number of men were at work making steel and that nearly all machinery was in motion.[30]

Given the contradictory nature of the reports, contemporary Stowell said he himself was unable to determine the true state of affairs. And even Company supporters, Superintendent O'Mara, Inspector McKelvey and Captain Brophy of the Pittsburgh Police Department, visiting Homestead to determine if more deputies were required, found few signs of mill activity beyond the machine shop in the upper mill and the forge at the lower plant.[31]

Yet with the service of a few skilled and their direction of the unskilled, practically all the machinery, save the 23-inch plate mill, was eventually put in action. Only mediocre steel was produced and delivered.[32] But steel it was nonetheless, and contracts were again being filled.

The Homestead workers continued to resist, assuring the men on the inside of the stockade that only kindness awaited them on the outside. As the non-union men were too closely guarded to be reached by ordinary methods, the workers had prepared circulars in both German and English, stating the situation and promising help to get the non-union men home and to find other employment for them. These were thrown from the train as it passed over the mill yard on the trestle, the Company being unprepared for this move and unable to prevent the men from reading the tract. Thereafter: "Guards were stationed along the line of the trestle to prevent a repetition of this move," an eyewitness reported, "but the mischief was already done, and men began leaving the mills in large numbers."[33]

The Company also intensified its psychological warfare, issuing passes to several strikers to enter the Works; and they emerged to say that considerable progress was being made.[34] Without reservation, the Company now insisted it had 2,000 men at work in the mills.[35] And half a dozen strikers, who had been drawing minimum wages before the trouble could not withstand the temptation of securing maximum wages, deserting their old friends for new ones in the Works.[36]

The pressures were becoming unbearable, as the strikers watched their places being taken, the evictions of their families from Carnegie housing continue, and their efforts at self-protection being put down. It had been distressing enough for families to have to move out of Company-owned houses under notice that the cottages - where once hung above their doors of good luck – would be wanted for the families of new employees. But their tranquillity had its limits: Displacement from jobs and homes upset the balance of both family and community, setting the stage for further conflict.

Though before being forcefully evicted, the old workers had proceeded at once to find other quarters,[37] they nonetheless intensified their demonstrations against eviction – only to be dispersed by the soldiers under orders.[38] Their interference with the scabs moving into

Company housing also led to Sheriff and deputy complaints that the non-unionists were being mistreated, thus bringing back the soldiers to patrol the streets of Homestead.[39]

Under renewed Militia protection a handful of Pinkertons then invaded the town: "Three Pinkertons . . . searched a striker's house in quest of guns, of which they had been relieved upon a certain occasion," a witness sarcastically relived the recent events. "This trio of man hunters, as they said, were 'out for revenge.'" James Gibson, a striker, was arrested by these Pinkertons and charged with conspiracy and riot[40] – though their authority to arrest anyone was beyond the pale of law.

Meanwhile, the Amalgamated held out steadfastly that victory would be theirs. When in the third week of August, the editors of the local media asked the Amalgamated for an official statement about the lock-out, the Advisory Committeeits offered its firm belief that the Company would have to cometo terms with the skilled labor essential for production.

There were four major points to be noted, the Committee said. First, the mills were not turning out good material and that delivered to the government was largely from stock made before the strike began. Second, the scabs were not skilled workers, were from disreputable classes, and accepted quarters in such degraded conditions as filthy bunk houses. Third, the Company was operating at an enormous loss and was bluffing, though newspaper reporters had been taken in by Frick's "dress parade" of production success. Fourth, a local boycott of Carnegie products would be extended nationwide when necessary to prevent successful Homestead output from being sold. Under such conditions, the Committee stated, "there were only two things necessary to ensure our ultimate victory. These are the presentation of a solid front to the enemy and sufficient funds to prevent want. Of the latter we are assured, of the former we are certain."

But the Committee had now over-reached its underlying strength. Powerful precedent was with them, the Committee bragged, emphasizing that solidarity had been exhibited in the past, despite the employer's success in using scabs for production:

> In 1882, the strikers tramped the streets in snow to their knees, many with hardly clothing enough to cover them; and suffering with hunger and cold they watched the mill run successfully on double turn. But they never wavered and finally won their fight. Strikes of three, four, or six months duration are not unusual, and except when men are starved out, they are almost uniformly successful. With such an example as that of 1882 before them, men should even be ashamed to think that the

> struggle is lost, and a man who would endeavor to lead his associates to think so, should be ostracized from their society.

Such not-so-subtle intimidation was framed to enforce an allegiance by each worker essential to avoid being described or caste as an enemy. Using such transparent means, though, the Committee again displayed it arrogance, attempting to enshrine itself in union-righteousness – to label those from within its own ranks as being either with them or against them, with no middle ground for discussion or other course of action:

> The few men who have deserted us are those in whom we have placed no confidence. Some more of the same stamp are still among us, acting as emissaries from the firm. The most contemptible creature on the face of the earth is a man who causes others to commit a wrong act that he may have an excuse to follow their example. . . . It is those who are among yourselves, posing as one of you, who should be guarded against.

In the context of the Committee's weaknesses, this was more a piece of propaganda about the need for continued unity - with the threat of excommunication - than a statement of the comparative strengths of the workers and the Company. In truth, Amalgamated officers were quickly losing any influence over the rank-and-file.

In evaluating the availability of sufficient funds to prevent want, however, the Committee was more calculating, though still exuding unwarranted optimism:

> That we shall not suffer from starvation is certain. We have the sympathy of every union workingman in the world. The subscriptions so far received have come mainly from outside sources, although labor unions have also responded freely. The systematic collection of aid has just commenced. What may be expected can only be estimated, suffice to say that there will be plenty of money to keep the wolf from the door for months, yes, years to come, if it should be necessary to continue the fight so long.

The Committee convinced itself of certain victory based on its superior tactical maneuvers: "The fight so far has been little more than a skirmish, the heavy guns have not yet been fired. The one thing to be remembered by the men of Homestead is, that in union, there is victory. So long as they stand united they need not fear a dozen men such as H. C. Frick, who has met his Waterloo at Homestead."[41]

* * *

Under armies of guards, production in the mill had gone forward in fits and starts by the end of August. O'Donnell tried to minimize the danger, counseling that: "The Company [managers] are loudly boasting that they are now running their mill with experienced labor and producing just as good material as that they turned out by their old hands. This is entirely false. I know they have no highly skilled labor in their mills at the present time that are capable of doing the work of the locked out men. I feel confident in stating . . . that the Carnegie Company is not producing a merchantable product, and it will be impossible for it to do so until it treats with its old hands and swings inward the gates that are now barred against them."[42]

But it was already evident to the rank-and-file that many of the non-union men in the mill had come to stay. By August 20th, about sixteen families were already quartered in Company houses and others were prepared to come when the new houses being built were ready. Meanwhile, a large number of the locked-out men had found work elsewhere and were moving away from Homestead. The Company appeared confident that the new men could replace the old: Repairs on the new Bessemer mill were progressing favorably, the Company expected to start them up on August 22nd, and, by the close of the following week, to have some 2,000 men working![43]

Frick on the last day of August paid his first visit to the mills since the strike began, spent a couple of hours assessing the situation, said he was well pleased with the progress, and insisted with utmost confidence: "The strike is a thing of the past!"[44]

**Stalemate**

Steadfastly pursuing his original plans to defeat the Amalgamated and introduce an infrastructure to accommodate non-union workers, Frick's successes were enumerated by Homestead's *Local News* in early September:

> The prominent features of the Homestead lockout this week were the numerous arrests of Amalgamated men and others made by the Company on the charge of conspiracy and rioting. No disposition is manifest on the part of the Company to let up on these prosecutions - persecutions, the men say, and many additional arrests are threatened.
>
> The erection of new houses on Carnegie Hill is proceeding vigorously, and were the stranger to see no other part of Homestead, he would conclude the town was being favored with a boom. Some moving [by the working population] has taken place, but considering the large number interested, is not of material consequence as affecting the situation. With a singular unanimity the ranks of the locked-out men still present a strong

> front, and the leaders are as firm in their assurances of a favorable termination as ever.
>
> A representative of the Company circulated among the real estate agents during the week, seeking information about houses, a movement that excited much speculation as to the object. One circumstance is noted, favorable or otherwise, that there has been no reduction in rents in Homestead. Nothing unusual has occurred within the mill. The number of men is continually increasing, and the smoke and steam are indications of work in progress, but as to its probable character, the reader is left to conjecture.[45]

In mid-September, both the managers and the locked-out workers were uncertain – the first because the Works could not yet be operated by the "green hands," the second because the workers did not have sufficient means to live without work for more than a few months. "If ultimate failure is before the management," the *Local News* assessed the situation, "they will soon realize it, and either make terms or close down the whole establishment indefinitely as many fear. There has been an utter failure thus far to cause a break in the ranks of the locked-out workmen, but while few of them are willing to go back, many have given up hope of an amicable settlement, and are scattering far and wide. Thus it is that sure defeat is staring both sides in the face."[46]

* * *

The patriotism of the past was finally giving way to September's harsh realities, nostalgia being overthrown by the city fathers as well as by the work force. "One year ago today, Homestead was in her glory," the *Local News* pointed out the previous September 17th's devotion to pomp and slogans. "We were dedicating the soldiers monument and the town was wreathed in the stars and stripes, thousands of people thronged the streets and music and cheers filled the air. It was the proudest day Homestead ever saw. The contrast with the present is too apparent to every Homestead citizen to need one word as a reminder."[47]

If the end was nearing, it was nearer for the people of Homestead than for Carnegie. By mid-September, a hospital had been erected at the Steel Works proper to care for the increasing number of injured workers,[48] and new housing was also purchased for the non-union workers from attorney John F. Cox, who sold the Carnegie Steel firm his three blocks of tenement houses for the sum of $37,500 – a small fortune in the '90s. "This purchase," the *Local News* pointed out the consequences, "will give the steel firm twenty-two separate houses for their new workmen and it is expected that they will occupy them as soon as they are vacated."[49]

More company houses were planned as well, the paper reporting that absentee landlord, "A. J. Boreland of Reynoldtown, Pennsylvania, owner of a block of houses on Fifth Avenue near Dickenson, is said to have sold the property to the Carnegie firm for $3,000. It will accommodate four families."[50]

There were hopeful developments for the strikers, nonetheless. One was the conundrum of Roher Rattigan, a leading striker who had returned in mid-September to work in the 23-inch mill. After working a week, he contracted the dreaded mill disease, quit work on September 25th, but could not be replaced by the Company – which then closed down its 23-inch mill without explanation. The strikers declared jubilantly that the remaining unskilled hands could not keep the mills open.[51]

Strikers argued the Company was worse off than ever. Two months earlier, in July, when the Company attempted to run the mill with non-union men, the machinery was in good condition and the unskilled were able to do a little work. But, due to their ignorance, the new men had wrecked and smashed the machinery to such an extent that nothing could be done with it. Within the week ending September 30th alone, three lower rolls and one center roll had been broken, each of these weighing from fifteen to seventeen tons. Until a new roller was secured in the 23-inch mill, the crew of workers had to be laid off.

But manager Potter insisted other departments were operating as usual, and Frick asserted with his usual confidence that the workers were still operating satisfactorily.[52]

**Production Successes and Military Contracts**

Frick was more right than wrong: By mid-September the Homestead production curve was smoothly ascending, and it was still the nation's principal arms maker for battleships.[53]

On September 3rd, the strikebreakers had completed the last piece of armor plate for the government cruiser *Monterey*, making it plain enough that the Works were being operated in earnest. Expectation that the firm would be unable to execute its contracts with the government, as long as green hands were employed, was a false hope. When confronted with the circumstances, the answer of the strikers was that the material turned out was necessarily defective and must be rejected by the government.[54]

"We know positively that the mills are not turning out good material," the Advisory Committee insisted. "A government inspector stated a few days ago to a prominent official of this town that 'not one pound of material has yet been produced that was acceptable to the

Government.' The largest total daily shipment yet made, including plates, beams, slabs, billets, angles, channels, etc., is 108,600 pounds. The armor plate mill alone, when running successfully, can produce this much in one hour. And it may also be stated that a good proportion of the above shipment was from stock made prior to July lst."[55]

Having held control over the mill floor operations, the Amalgamated knew precisely. This claim was substantiated when the Navy Department complained about the quantity and quality of output, the deliveries being far below the government's contract requirements. Yet, despite the express prohibition of any transfer of work under the contracts, the Navy allowed Carnegie to exchange with Bethlehem a large part of its obligations on the basis that it could be accomplished better and more rapidly there. The Homestead mills were effectively relieved of their obligation to produce structural steel and thin boiler plate.[56] Thereby, the mills could continue to operate with unskilled, scab laborers unable to produce the high quality structural parts required by the U. S. Navy.

By this time all parties either claimed victory or disparaged the claims of others. The government, now taking official cognizance of the strike, still would not push the Company to meet the quality requirements of contracts for armor plate. Frick conveniently interpreted this as meaning that, as the strikers' cause was finished, the Secretary of the Navy could not understand why there should be any delay in delivery of a certain quantity, and thus production would go forward as planned.

Since the quality of the Carnegie plate delivered was still in question, the Knights of Labor, mixing patriotism and esteem for the Homestead workers, proposed to test the material designed for the great warships and to thereby learn whether it was of sufficient excellence – and incidentally to satisfy the country at large as to the claims of the Steel Company that it had ample expert labor to turn out good work.[57]

Yet all these self-serving antics proved nothing worth anything to anyone.

## Frick's Calculations, Committee Naïveté

Adept at outmaneuvering opponents, Frick laughed up his sleeve at the *naïveté* of the Advisory Committee, the leaders of the Amalgamated, and their supporters. They were misleading the rank-and-file, who were meanwhile becoming increasingly desperate.

At weekly mass meetings, the general officers of the Amalgamated had continued to inculcate a spirit of quiet waiting alongside moderation and respect for law. Other orators simultaneously surcharged their opinions with bitterness and resentment, unmerciful to Mr. Frick, the Sheriff and the common pleas judges. The general tenor assured the rank-and-file that the Company was only "making a bluff" and that, sooner or later, old hands would be reinstated, the Amalgamated Association recognized, and wages kept up to a decent figure.

Oratory along these lines kept life in the strikers' organization for an astonishingly long time. The loyalty of the workingmen to one another and their cheerful confidence in the power of union principles and unionist methods to overcome all obstacles were almost sublime.[58] For a full four months, the strikers held solid to their beliefs; and by late September few, if any, of the strikers and only six former Amalgamated men had applied for reinstatement and were back at work.[59] "The firmness with which these strikers hold on is surprising to everyone," Frick wrote Carnegie, adding his hope that there would be change after the election.[60] But a political solution was not forthcoming – this election being lost to the Republicans in part because of the occupation of Homestead.

Even when the Militia left, Homestead remained a central issue in the national political campaigns. The Democrats paraded in Pittsburgh and invited the Homestead strikers to participate in presenting the cause of labor. As similar demonstrations were carried out in all the manufacturing centers, the workers switched their allegiances to the Democrats add away from the Republicans in the November election.[61] Yet the combination of the Militia occupying Homestead and the victorious Democrats helped to weaken the Amalgamated and the Republican Party at the same time!

Still, Frick had not yet broken up the essential unity of the Homestead workers nor crushed their determination to resist. It was three months after the outbreak, but he had yet to acquire enough trained, highly qualified men to run Homestead near full capacity.[62] He found instead three sure-fire methods of breaking the workers' will: prosecution, bankruptcy, and starvation.

### Approaching the End

With production started, only a military force could continue to protect the scabs as long as the strikers held out. But the strikers' zeal was eroding, and General Snowden, sensing the approaching end, shifted authority to the Sheriff.

Under Snowden's original mandate, the Sheriff had to be obeyed in

Homestead; this was the single legal rationale for the presence of the Militia from the start. Accordingly, when the Militia planned its exit, the Sheriff began preparing for the day he would have to run the city under arms. "As the soldiers disappear, the Sheriff's deputies appear," the *Local News*[63] traced the shift of authority. During the week, July 24th-30th, the Militia departed day by day until the point was reached where the number of soldiers in the camp was reduced to about 1,50O.[64]

"That the steel company began to feel safe was apparent from the movements of the Militia," wrote a witness. "A consultation resulted in an order to the effect that four companies of the Sixteenth Regiment should prepare to depart, which was done without delay. This left companies A, C, D, and I, of the Sixteenth Regiment, to control the entire situation."[65]

These soldiers were to remain as long as it was deemed necessary by the authorities and Carnegie steel. Meanwhile new sheriff deputies began arriving in Homestead on July 29th, bringing the posse up to a total of 50 men. They planned to control the vicinity of the mill and the Borough day and night to prevent the workers from intimidating the scabs.[66]

County police powers thus mobilized to back the Company as the Militia prepared to leave,[67] the *Local News* ominously warned on September 17th: "During the week, the coal and iron police employed at the mill were deputized by the Sheriff. Since then they have been frequenting the town, and made several arrests. Their presence in regular police uniform is quite noticeable on the streets. The rumored withdrawal of the troops next Thursday has to some extent occasioned alarm among the citizens as trouble is thereby anticipated. It is said that the effect inside the mill will be greater than in the Borough and as a result a large number of timorous non-union workmen will leave."[68]

As the withdrawal of the Militia was to take place in step with the beefing up of Sheriff McCleary's appointments and Frick's iron and coal police during the week ending September 24th, the Fifteenth Regiment was withdrawn from Homestead, with the Sixteenth expected to go early in the following week. But as Sheriff's deputies took their place, they overstepped the law, handling the strikers roughly. 'Sheriff's Deputies Rakeshaw and Smith came to grief the other day on a charge of assault and battery," the *Local News* reported one altercation, "and were sent to jail by Squire Kuhn in default of bail for appearance at court."[69]

A week later, four companies of the Militia had gone; the four remaining numbered less than 150 soldiers, while the Sheriff's deputies, including the coal and iron police, were "doing service within the Borough."[70] Despite the fact that the situation at Homestead was just as critical as at any period, the last four companies of soldiers finally went home on October 11th, and the Sheriff increased his force of deputies, composed of non-unionists, to about 100.[71]

**The Great Chess Match**

The workers still did not lie down supinely and let the Sheriff's new deputies run the city.

The Company's attempt to mobilize the Hungarians and Slavs at a meeting was prevented by the locked-out men, demonstrating, the *Local News* said, "that they are still holding the fort in the Borough as certainly as the firm is holding it within the mill."[72]

It was a great chess match. The Company again induced first-class workman John J. Rattigan to return to work. The workers countered by exposing the death, sickness and injuries among the scab workers, pressuring the Company to put out funds to improve the sanitary conditions in the mill, and to create an infrastructure to improve health both within and outside the Works. Carnegie Secretary Lovejoy parried by leveling charges against the workmen; and the Grand Jury found[73] 167 "true bills" for prosecuting the several charges.[74]

A large number of those indicted were not apprehended owing to their convenient absence from home. As a few of them were seen on the streets daily, the *Local News* was nonplussed: "it is regarded as a little strange that they were not arrested."[75] But by this time it was difficult, if not dangerous, for the Sheriff to openly make arrests in Homestead proper.

In countermoves, Burgess McLuckie and the Amalgamated charged all Carnegie Company officials and their hirelings during the lockout and battle with riot and conspiracy, starting with H. C. Frick and including managers and stockholders George Lander, H. M. Curry, J. G. A. Leishman, Otis Childs, T. F. Lovejoy, L. C. Phipps, John A. Potter, E. A. Corey, J. F. Dovey, and Nevin McConnell. Included were also mercenaries and their paid agents: William and Robert Pinkerton, John Cooper, C. W. Bedell, P. Primer, W. H. Burt, and Fred W. Hinde.[76] They all provided bail money of $2,000 each.[77] Unlike many indicted workers, they were free.

Meanwhile the struggle to man the mill continued. A good many scabs deserted, but others settled in as permanent hands, and a half dozen locked-out men returned to work in the week ending October lst. "Large numbers of non-union workers frequent the Borough hotels and restaurants for their meals, and others are boarding and living in the houses lately purchased by the Company," the *Local News* reported Frick's successes. "Over a hundred houses are in the course of erection on the Company's property and will shortly be completed. It is thought that before cold weather comes on, all the non-union men will be housed."

"Notwithstanding these factors," the paper was quick to add, "the locked-out men as a body show a remarkable determination to stick to the fight. There is no indication that the Association means to call the strike off, and those who go back do so upon their own responsibility."[78]

**Keeping the Peace**

Besides the Sheriff, the Advisory Board continued its role as a force to keep the peace in Homestead. Thus against violent-prone workers, it urged restraint and trust in eventual legal victories, while the Amalgamated leaders also continued to condemn them. Though murder charges were brought against its leaders and its treasury was drained by legal expenses, the Advisory Board still offered rewards for the apprehension of "irresponsible parties" who were throwing bombs and setting fire to the big boarding houses where the non-union workers lived. Nor did the Board defend itself against the Pinkertons who were now back in Homestead, busily engaged in hunting for their rifles and manufacturing false evidence to sway public opinion against the strikers.[79] Following this contradictory policy, soon the entire Advisory Board was itself arrested and trying to make bail at $10,000 each.[80]

Over the main course of action to pursue, the workers and their leaders had been divided for months. Thomas Crawford, made head of the Advisory Committee in September, counseled moderation, condemning "the loud-mouthed, anarchistic screechers for gore," and moving to stop the attacks upon representatives of the press. But others in Homestead demanded militancy. The Federation's cigarmakers who had taunted the non-union men were periodically disciplined by the Militia, while Ward, an orator at the workers' regular Saturday meetings, returned the compliment with several inflammatory tirades against the occupation. The strikers repeatedly molested non-union men going for their meals and moving their household goods, and the steamer *Tide* was fired upon by unknown parties with a small improvised cannon. From outside Homestead, too, on September 19th, Alexander Berkman's written address to the court denounced Frick, the U. S., the American flag, the courts, and all other institutions preserving "peace" and promoting "patriotism."[81]

A Kangaroo court it was, without defense lawyer, proper translation of Berkman's Russian-language defense, under the duress of a District Attorney who physically stepped into the jury-box to pressure a verdict of guilt without the jurors ever leaving their seats. The sentence imposed left Berkman physically trapped and politically helpless.[82]

He was found guilty of murder in the second degree, and given 21 years in the penitentiary and one year in the workhouse. It would be many years before Berkman would again be free to join Emma Goldman in working for Anarchist and radical causes.[83] "It is no earthly use to try to do anything for me," he wrote Emma. "It will take years to accomplish a commutation, and I know that Frick and Carnegie will never consent to it. Without their approval the Pennsylvania Board of Pardons will not act. Besides, I cannot continue for long in this living tomb."[84]

Anarchists in the Berkman *coterie* were also put away – permanently out of the reach of Homestead. Bauer and Nold were convicted, put in separate sections of the same prison as Berkman, and only later established contact with him – together putting out their sub-rosa, politically-volatile, *Gefängniss-Blüthen* (Prison Blossoms).[85]

Back in Homestead meanwhile, lawyers and judges assumed Anarchist goals had infected the actions of the Amalgamated. These learned jurists seriously haggled over whether the strikers had engaged in "riot," penalized by imprisonment and fines, or "treason," punishable by death; and whether or not they had a treasonable "intent" to overthrow the State. No matter if the leaders of the Amalgamated all along preached moderation, limited their struggle to trade union objectives, and did everything possible to prepare peaceful opposition to the occupation by the Militia. Few now recalled Homestead Burgess McLuckie's careful evaluation in early July that: "Our people as a general thing, believe the Pinkertons to be a horde of cutthroats, thugs and lawless invaders, employed in compelling the submission of honest labor into accepting the demands of capitalists. This accounts for their dislike of the Pinkertons. There would have been no resistance to a lawfully constituted authority of the State. There has been, however, in the coke region and in many other parts of the country."[86]

Although McLuckie could clearly distinguish between Homestead's acceptance of the State authorities and the coke workers' efforts to overthrow the State and take charge of the means of production, relatively few jurists could do the same. The best legal doctrine could say of the Amalgamated was that it had led a riot and not an insurrection, declared ex-chief justice of the Pennsylvania Supreme

Court, Hon. Daniel Agnew, reasoning the distinction was obvious: "It lies in the purpose or intent of the traitor to overthrow the government or subvert the law or destroy an institution of the State. Riot is a breach or violation of the law, but *without a purpose against the State*."[87]

Major General Snowden claimed the honor of initiating treason prosecutions against the Amalgamated leaders, saying his duties at Homestead led him to suggest such charges to Carnegie attorneys, Knox and Reid. The lawyers had pooh-poohed the idea at the time, but three weeks prior to the issuance of warrants by Judge Paxson, Mr. Knox had met with the General in Philadelphia and asked his help in the preparation of briefs. Snowden responded that it would hardly be wise for him to appear as legal counsel for Carnegie while serving as commander of the State Guard. Nonetheless the brief was prepared, sent to Snowden to be passed upon, and received his endorsement.[88] The aristocratic Snowden wrote that the penalty for the Advisory Committee should fit the heinous crime of representing the union cause, criticizing the mild State penalty for insurrection: "The punishment, I believe, is a heavy fine and long imprisonment. It ought to be changed to death!"[89]

This was clearly nonsense, as the Amalgamated Association, seeking job security, had no goal to overthrow the State. It lacked as well any goal beyond trade unionism, distinguishing the limited vision of the leaders of the Amalgamated and the Homestead commune from the quest of the revolutionaries in the coke regions, the Socialists and Anarchist leaflet distributors at Homestead, and the Berkmans and Emma Goldmans.[90]

### Prosecutions and Failing Union Resources

Despite its moderate stance, the end was nearing for the Amalgamated. In August, Hugh O'Donnell had scampered among the cities on the Eastern Seaboard telling the tale of Homestead and trying to raise funds. But when he returned August 13th,[91] warrants were sworn out against the leaders to drain the Amalgamated treasury of its last dime. In September, Carnegie Secretary Lovejoy already had warrants issued against strikers for the murder of Pinkerton agents, and the defendants were already under $10,000 bonds on the charge of killing two of the workers who had clearly been murdered by the Pinkertons in the raid of the 6th. The new charges revealed a plan "to multiply the amount of bail until the resources of the union leaders were exhausted and confinement to jail could no longer be avoided," a witness judged.[92]

Murder charges instituted almost indiscriminately, the courts refused to consider self-defense an issue, and on September 21st, the grand jury had returned the austere bills for indictment against all 167 of the Homestead cases presented to that body.[93]

Frick then played his trump card, picking the very court to try the strikers. In Pittsburgh the Supreme Court of Pennsylvania, headed by Chief Justice Edward Paxson, was initially pressured by Frick to take up the case against the strikers. Judge Paxson then conferred with the District Attorney and P. C. Knox, the principal counsel for the Carnegie Company; and he next proceeded to instruct the D. A. to bring charges against the strikers before himself (Paxson) and arrest the persons charged.[94]

The charges were for insurrection against the State of Pennsylvania "to the end that its constitution, laws and authority were defied, resisted and averted by the said defendants and their armed allies," with penalties under the Crimes Act of 1860 of a monetary sum not exceeding $2,000 and imprisonment by separate and solitary confinement at labor not exceeding 12 years![95]

"Homestead's striking steel workers have prepared to fight the last move of the Carnegie Steel Company, in instituting charges of treason against the Advisory Board, to the bitter end," the national Amalgamated Association publicly responded."[96] President Weihe served notice that, with the indictment of the entire Board, no stone would be left unturned to fight the great battle to a successful finish.[97]

Judge Paxson was now to act as judge, jury, prosecutor and jailer: He would hear in person the application of accused strikers for bail; he would instruct the grand jurymen as to what constituted treason under the statutes of Pennsylvania; he would sit on the bench and try the defendants if the cases were brought to trial; and his was the court of only resort! "The court of last resort," said an observer, "had been by a Frickian turn of the wrist, converted into a court of preliminary resort, immediate resort and all other known varieties of resort, and the strikers were led to understand accordingly that with Paxson armed to the teeth and ready to bring them to bay at all points, they might as well throw up the sponge at once and be done with it."[98]

Paxson instructed the Grand Jury that Carnegie had a right to hire non-union workers, that the Homestead commune and the Amalgamated organizers could not deny these workers the right to employment, and that the court would protect both the scabs and the employer's means of production. "If," said Justice Paxson, "we were to concede the doctrine that the employe [sic] may dictate to the employer the terms of his employment, and upon the refusal of the latter to accede to them, to take possession of his property and drive away others who are willing to work, we would have anarchy."[99]

Local efforts to usurp authority was the same as a design to overthrow the government, the judge concluded, and thus the Grand Jury should indict the defendants for treasonous actions:

> It is a state of war when a business plant has to be surrounded by the army of the State for weeks to protect it from unlawful violence at the hands of men formerly employed in it. Where a body of men have organized for a treasonable purpose, every step which any one of them takes in part execution of their common purpose, is an overt act of treason in levying war. Every member of such advisory committee, or by whatever name it is called, who has participated in such usurpation, who has joined in a common purpose of resistance to the law and denial of the rights of other citizens, has committed treason against the State.[100]

This was no abstract account of the status of the law; as Paxson stated that the specific acts of Advisory Committee were treasonous, he assumed the role of the Grand Jury in indicting the Committee members as well as of the jury in finding them guilty of the charges.

As the Kangaroo court marshaled the case through trial, W. W. Erwin closed the defense by stating that Frick had purchased arms and sent a foreign force to invade Homestead. In that the U. S. and Pennsylvania constitutions empowered the people to bear arms in self-defense, and the crowds on the shore were local and faced the Pinkertons who were outsiders, he argued, the real question was, "was this battle at Homestead a riot or an unauthorized invasion? If you answer that it was an invasion without excuse, that is the end of it, for the people never delegated away their rights to resist armed invasion."

Erwin then added the critical dimension: "I don't see how you can escape from this position, that whatever was done at Homestead was done in resisting armed invasion. . . . [I]f Frick had not sent armed men to Homestead there never would have been a riot."[101]

Since the people on the shore had the same right of self-defense as an individual, he continued, they could use dynamite and oil to rain "on this unman-fearing and unconscious crowd of Pinkertons. . . . The only reason why the Pinkertons did not wallow in the blood of the people," he fulminated, "was because of the magnificent resistance. There were only eight rifles behind the barricades and three hundred in the boat. Think of the absolute cowardice of men with three hundred rifles against eight! Such effrontery is only equaled by their cringing fear."[102]

It was Judge Paxson, the prosecutor, against the defense throughout the trial. Condemned by practically the entire membership of the

Pittsburgh bar and legal experts, the Chief Justice's arbitrary interference was seen as harassment. The consensus of legal opinion was that the disturbance at Homestead was purely a local affair directed against a private corporation and could not under any conditions be construed as treason against the State.

This consensus gave encouragement to the rank-and-file that they could somehow win the strike, yet it and the October 2nd oration of Amalgamated Vice-President Carney at the weekly mass meeting were episodic ventures in delusion. Carney had assured the strikers that the treason charge trumped up as a measure of intimidation would fail in a court test, and thus the strikers should stand their ground. Other speakers also confidently warned the Carnegie Company that, when all members of the Advisory Board were imprisoned, there would be other men to take their places; and that if all the men of Homestead were cast into jail, the women would still be there to keep up the fight for the rights of organized labor – sentiments received by the rank-and-file with clamorous approval.[103]

A heady show, but as bail for the indicted exceeded half a million dollars, one hundred of the indicted men judiciously skipped town, while four leaders, including Hugh O'Donnell, were jailed seven months without bail on murder charges. McLuckie and a few others put up $10,000 each to gain their freedom pending trial. But such judicial machinations raised doubts among the rank-and-file: "were they, after all, nothing but a collection of killers, rioters, and conspirators?"[104]

As legal expenses mounted and the Amalgamated increased its legal staff to seven, President Weihe nonetheless refused to surrender. The costs then cut into the Amalgamated's resources for sustaining the strike. Weihe appealed to the Federation of Labor in December for funds to cover legal costs, delineating that 225 men were under legal indictment, of whom five were charged with murder, thirty with treason and the remainder with conspiracy and riot. Some $400,000 had been given in bail bonds for the release of a number of the accused, and the cases against the arrested strikers had cost the Association $10,000 for lawyers and witness fees. There were still 2,200 Homestead workers who had not been re-employed, so that relief funds were also needed. Yet only $41,000 had been contributed during the strike.[105]

Amalgamated Attorney Brennan, previously busy in encouraging the opening of Homestead to the Militia, now sought to "bring suits against the Carnegie people for treason, for bringing an armed force into the community and attempting to usurp the power of the Government." But Judge Paxson refused to indict them, instead concentrating his venom on the workers.[106] "It seems to me," Brennan criticized, "that Judge

Paxson should have issued warrants for the other fellows."[107]

In the end, no Pittsburgh jury ever found any striker or Carnegie official guilty of the charges brought, but by mid-October the combination of legal costs and subsistence outlays had drained the treasury of all eight Amalgamated Lodges in Homestead. Even the Amalgamated national organization, supporting its Homestead local to the limit, was running dry. Contributions from other sympathizers and unions dwindled, as such donations seemed like pouring water down a bottomless well.[108]

Valiant efforts had been made. Funds were first raised in Chicago and other large labor centers to protect the Advisory Board.[109] Homestead Day in Chicago during late October was celebrated by 90,000 unionists from various crafts, each pledging a day's labor for the strikers' cause. $40,000 was contributed to the Amalgamated Association and then given to the Homestead Lodges. But this impressive financial boost only prolonged the agonizing day of reckoning.

The Amalgamated was soon broke, and the American Federation of Labor extended aid reluctantly. The American Federation had already provided a paltry $1,000 for the defense of Beatty and Dempsey.[110] When it then became unwilling to put up more funds, the strikers charged with treason were forced to make individual appeals to the Federation for funds to pay their attorney, Erwin. And Amalgamated President Garland then put Federation head Samuel Gompers on the spot to determine the scale of Erwin's legal fee.[111]

Would the Federation now turn its back on the 1,600 strikers and their 2,400 dependents in Homestead and Munhall?[112] After some relief was offered, the leader of the Slavs and Hungarians, Paul Oldshue, notified the relief committee that even more funds were necessary to keep his followers cared for and in line,[113] while to make matters worse, typhoid fever and dysentery were rapidly spreading in Homestead.[114]

Gompers hesitated, calling off the already ineffective boycott of Carnegie steel, promising little help to the major union in his alliance. When Homestead's Amalgamated head Thomas Crawford pleaded that the city's workers were "badly in need of shoes and wearing apparel," Gompers promised another Homestead Day to raise funds, but as the strike was clearly losing, the planned December 13th rally never took place. The Federation effectively deserted the Amalgamated.

The Knights of Labor, 980,000 strong, meanwhile offered brave words, to the Amalgamated to hold out against Carnegie for ten years, yet sent a total contribution of only $700 throughout the entire struggle! This was almost a calculated insult to the rival A. F. L., of which the Amalgamated was the most potent member.

Such drying up of labor support helped break the back of the union: Workers were desperate. More locked-out men left the Monongahela Valley for jobs elsewhere. More strikers began to quietly assess their own situation and look to the Company for their old jobs.[115]

After another thirty days, the lock-out had turned previously well-off workers into paupers, a first-hand report outlining the crisis:

> A majority of the families of the defeated strikers are suffering seriously for want of the bare necessities of life . . . . Homestead citizens are rendering every possible assistance, but are entirely unable to meet the fast increasing demands. . . . James Powers of Chicago, one of the committee sent by Chicago labor organizations to investigate the condition of the Homestead strikers . . . conducted a house to house canvass of Homestead and reports the condition of many of the once prosperous families as pitiable in the extreme. He says that not over 5 percent of the total number of strikers have secured work, and that there is much distress.[116]

By this time over 180 Homestead strikers had been indicted for the crimes of murder, treason, aggravated riot and conspiracy.[117] The leaders were charged and jailed on one legal indictment after another. O'Donnell and others were kept confined in jail until as late as March, 1893, first being charged with murder, then with riot, next with assault, and so on.

Yet after fifteen weeks, many people elsewhere had so wearied of the news on Homestead, they were actually surprised to learn the struggle continued.[118]

### The Dimensions of Struggle

On October 1st, the *Local News* outlined the dimensions of the struggle:

> The Lock-out is now three months old.
> There are signs of winter's approach.
> The Presidential election takes place on November 8th.
> Work appears to be a drug on the market in Homestead just now.
> It is said many of the striking coal miners up the river have
> obtained work in the Homestead mill.[119]

Frightened by such information, the skilled men were being enticed back to work by Frick's offer of their old positions. A good number of them had intended to return as soon as it was conclusively shown that the strike was lost. Yet, when the Company charged the entire Advisory Board with treason, it "made it almost impossible for any self-respecting man to go back," said a reporter. Rather, he wrote, the workers were drawn even closer together and became more determined "to hold out to the bitter end, notwithstanding the overtures of the Company to induce them to go back, or the prosecutions, which are judged more in the light of persecutions, for the purpose of forcing the men into submission."[120]

But could the men really hold out successfully? Winter had arrived and the requirements for heat, warm clothes and good food were evident in every working-class household. Yet people had neither wages nor sufficient savings to take up the offers of city merchandisers, some blindly advertising an untouchable cornucopia of wealth. McCune's on Sixth Avenue in the city center heralded the advance of zero temperatures: "Blankets! Comforts! . . . Warm Underwear, Warm Hosiery, Warm Gloves, ARE WHAT YOU NEED. Save colds and Money By Buying at McCUNES's".[121]

A solution to the wage question now seemingly beyond reach, many Homestead workers looked momentarily for an electoral upset that would resolve their material needs. A change in congressional and state representatives might somehow get them back to work and save their skins in the courts and elsewhere. Some thought Attorney John Cox was on their side, and he had now thrown his hat into the ring for Congress. "As the outcome of the labor upheavals and political mutterings," a self-described "Toiler" wrote the *Local News*, "our esteemed friend, John F. Cox, Esq., is out as a reform candidate in the 24th Congressional District. Mr. Cox is well known about Six Mile Ferry and will no doubt receive the almost unanimous vote of this vicinity. It is to be hoped that he may find a little spare time from his arduous duties in defending their cause in court to address the citizens of Six Mile Ferry and surrounding districts."[122]

Like so many politicians, John Cox had divided loyalties. He sided with the workers in defending their cause in court, but lined his own pocketbook by selling buildings to the Carnegie Company to house non-union workers. (No sooner, in fact, had Cox sold his block of houses at the head of Fifth Avenue to the Steel Company than they were thoroughly overhauled and refitted as a spacious hotel and boarding house for the scabs.)[123] As well, Cox hobnobbed with the local conservative elite; Republican Party friends close enough to celebrate his 40th birthday surprise party included Judge Frank F.

Baldwin, Postmasters McWhinney and Dodds, Dr. John Osborne, Judge John C. Oeffner, Esq., and former Homestead Burgess, Captain M. M. Wilson, among others. They joined in giving Attorney Cox a fine leather-covered library chair, which Cox said he hoped to take with him to Congress.[124] It was an open question, though, if in this chair he would ponder the political side to which he would lean in resolving the questions raised at Homestead.

**Schwab Moves In**

One question that needed an answer was the degree of workers' solidarity while awaiting political solutions.

Rumors floated wildly about this or that worker caving in to Frick's enticements of high-paying positions. And many workers became more suspicious of their fellows than of glib politicians handing out vague promises of future relief.

Rumors demanded a response as well. Like other strikers feeling the impending crisis of no work and too little money to survive, J. I. Kerr, residing at No. 25 Eleventh Avenue defended his virtue in the local papers, so that his friends would know, he said, "that he is not and has not been working at the Steel Works, since the strike began, notwithstanding all reports to the contrary."[125]

Despite many workers' holdout, in mid November, striker morale sank with the thermometer. Individual workers had come up against the harsh economic realities – maintaining their homes, paying their rents, stocking an adequate supply of food to carry them over the winter - and realized support resources were failing. Frick's reemployment invitations looked more appealing.

Frick also brought in Charles Schwab from the Edgar Thomson Works to replace hated Superintendent Potter at the Homestead mill. The back-slapping Schwab took to Homestead's streets, visiting bars and homes, using his gift of persuasion to induce department heads and foremen to take back their jobs. They had struck in sympathy with the Amalgamated, but were still excluded by union rules from membership. Returning foremen then began coaxing their friends back to the fold, since it was clear that matters were hopeless.[126]

The contrasts were stark enough to everyone involved: The Homestead mill was picking up the work pace, filling back orders and booking new ones. Yet the diehard workers roamed the streets, afraid to go home to their womenfolk, who nagged their men to go back to work, since the pathetic union handouts no longer helped them much. Only children ate a normal diet; fuel and clothing were lacking, and rents and mortgages remained unpaid. The three thousand workers who had joined the fray with confidence were being whipped by the approaching icy winter.

By early November, the men previously employed in the mechanical department made restless inquiries about return. Although not union members, they has sacrificed wages and security in solidarity with the strikers. Schwab had been busy among them as well; nearly a score now returned to work, assigned to their old positions in the mill. Others soon followed, breaking strike solidarity, crippling the Association plan, and signaling the general uneasiness and uncertainty of both the leaders and the rank-and-file.

The *Local News* read the tea leaves: As 200 former union supporters had already returned to work and there were 2,000 non-union men in the mills, the Company was gradually succeeding and the strike dying out.[127] Many Hungarians and Slav workmen returned on the 17th; half of the 300 who applied were given work immediately and the other half promised work sooner or later.

The pressure to return to work was becoming overwhelming: With 10,000 iron and steel workers unemployed in and around Pittsburgh and Homestead, jobs were hard to come by and rising destitution was clearly in evidence.

But what was to be done at Homestead? Both the Advisory Board and the Amalgamated Association convened secretly to discuss the mechanics' action and try to prevent their surrender.[128] The mechanics and unskilled laborers were of another mind. Meeting on November 18th in the open air and hotly discussing their predicament, they voted to be released from their strike pledge and permitted to return to work.[129] Though the Amalgamated tried to delay their action by voting that evening to reject the proposal by a vote of 106 to 75 (though too the Amalgamated decided to continue the strike by a 224 to 129 vote), it notified the mechanics and laborers that they could act as they liked. And in response, mechanics and laborers agreed unanimously to return to work, though under no circumstances to accept tonnage jobs that would "trespass on the rights of the Amalgamated men."[130]

Competition for jobs immediately began: 600 mechanics and laborers at once stormed the main entrance of the mill to put in their applications for reinstatement. Two clerks made out passes for them under the supervision of Schwab, who checked off each name against a long list as the men streamed through the office mill, rejecting some on the spot.[131] "They were given passes to enter the mill and directed to the various foremen of the different departments," the *News* reported the action. "In many instances, especially among the laboring classes, work was secured, but in other cases the old workers found their jobs in other hands and were turned away. Probably three hundred obtained work. A large number, however, were promised employment in a few days, or as soon as an opening should occur. There were a few though that were denied work under any circumstances, as they were objectionable to the company for obvious reasons. They turned away with a sorry tread."[132]

Since there were some 2,715 black and white non-unionists on the payrolls of the mill the day the mechanics and laborers went back to work, of the 2,200 men who had applied for reinstatement only 406 – almost all laborers – obtained immediate employment.[133] As the already-employed non-union limited the number of vacancies for mechanics, more than half of them were not given work for the moment.[134]

The count now favored Frick: With the English-speaking mechanics and a few Amalgamated men attempting to return on the 18th, a total of over 1,000 men had already reapplied for their jobs. Frick now able to pick and choose who would work, the *News* reaffirmed the defeat it had forecast:

> Today marks the closing scenes of the great Homestead strike. It is now virtually at an end. For a month or more it has been apparent that the workmen could not win and the prolongation of the strike was simply a useless sacrifice. This however was not acknowledged by the larger number, and they held out against fate. During the week a great change in the attitude of the strikers has taken place, and the conclusion seems to have been forced upon them that they must save themselves.[135]

This opening of jobs cracked the thin veneer of Amalgamated solidarity. The Association policy of holding out until the strike was officially declared ended had momentarily held the men in check.[136] But unity crumbled from the top, when Thomas Crawford, Chairman of the Advisory Board, resigned the next day, November 19th, and took a position at Uniontown at the Columbian Iron and Steel Works. Bitter debate quickly invaded the Homestead Lodges, as the drive for personal security exploded in charges and counter-charges of responsibility for the dismal state of affairs. A handful of union men then met at Munhall's Skating Rink, November 20th, casting 110 votes against 91 to end the strike and declare the mills open. Some 600 union men were too discouraged to attend, were in jail, or living and working in other places. As few Amalgamated men had expected to be taken back by the firm, many had left town since July; and on that Sunday,

November 20th, the few members of the Advisory Committee who were still at large voted to dissolve the Advisory Board, and announced that union men would be freed to return to the mills.[137]

"We surrender with both hands up!" exclaimed William Cashes, treasurer, on dissolution of the Board. "Great battles are rarely ever fought the way they are planned!" O'Donnell wrote the *Evening Leader* from the dark recesses of the county jail.

Early the next day, November 21st, the *Local News* summed up the situation:

> The last act of the strike drama was the official action taken yesterday at the meeting held in the Fifth Avenue Opera House by the Amalgamated Association. Although it was generally understood that at this meeting some kind of action was to be taken, so little interest was manifested that barely two hundred workmen attended. A few days ago such a meeting would have attracted 1500 or 2000 men. The indifference was, no doubt, due to the fact that of their own accord, a majority of the men had already decided the strike was off and did not care what was done officially.[138]

As the workers had already voted with their feet, the Amalgamated now served *formal notice* that union men would be permitted to return to mill work,[139] ending the lockout and the four-and-a-half months' strike.

"The strike having been officially declared off a grand rush was made this morning by the strikers to get back to work," wrote an eyewitness on November 22nd. "As early as 6 o'clock the rush began-and it kept the police busy handling the crowd. By 9 o'clock the rush was over, but they continued to go in in twos and threes, and this will doubtless be kept up for some time to come. A large number were turned away, some with good prospects of securing their former positions in the near future, while many were informed that they would be given no work under any consideration. It is now plainly evident that the black list is quite long, and those whose names appear on the list will be forced to seek work elsewhere."[140]

As the men were freed to return to work if they could get their places back, the Amalgamated Association was to take care of those blacklisted until they could find employment in other places. In all other cases, weekly benefits to the strikers were to stop immediately.

The *News* was optimistically cautious, editorializing that the old workers would quickly replace the new. It calculated that scarcely more than 200 workers would be blacklisted, 300 had already obtained work elsewhere, and the rest would find work in the mill sooner or later – repairing and cleaning up, making extensions or improvements, all of which would make extra work for extra men at least temporarily. "It is said too," the *News* emphasized, "that there are a good many incompetent workmen among the new men, who will necessarily be shipped as soon as their places can be filled with better men; there is also a large number who will leave voluntarily. In the end, then, nearly all the Homestead men may get back into the mills."[141]

The prospects for immediate re-employment nonetheless remained uncertain. "If the company is as good as its word and will stand by the new workmen of who there are about 2,500 in the mill," the *News* expanded, "there will scarcely be room for more than 1000 of the old workmen."[142]

Every man now for himself, hundreds of once active strikers thronged the mill office and besieged company officials with applications for positions.[143] Though more applied than the Company had places, the firm was definitely prepared to reinstate most. The Sheriff was contacted in case of possible trouble and he responded by transferring 15 deputies to the Works, leaving 17 on duty in Homestead proper.[144]

As more than a thousand men seeking jobs jostled one another, Schwab filed about 800 of their applications. Rigidly scrutinizing those re-employed under the new Company policy, Schwab admitted each round of workers in groups of five the next morning, checked them off his list, and hired a few who had apparently been inactive during the struggle.[145] "All day yesterday," the *Local News* recounted, "large numbers of workmen visited the mill office and made application for employment. Their names were recorded and passes given them to enter the Works and see the superintendents of the various departments."[146]

The new onslaught in motion was very different than in the months just past - unforgettable for those Amalgamated men who had insisted they would never give in to Carnegie's dictatorial ways. And the mill control mechanisms were now clear, as the mill hierarchy of bureaucratic managers now mechanically sorted the nearly 2,000 applications, awarded one-third jobs, promised others they might find work in the future, and held out the threat that the non-union men employed in the mill might be kept on if union workers did not toe the line.

Frick looked on, said little or nothing, savoring the fruits of his tenacity against those actively opposing his iron will. Almost all Amalgamated organizers were turned away. Though the Company required men with their higher level skills and their jobs were held momentarily by others with less ability, yet, regardless of skill and caliber, the blacklisted were viewed as troublemakers, to be permanently excluded from the Works.[147]

Those untarnished rank-and-file on wait also suffered Frick's cynical imposition of retribution. "Fully 2,500 men were left to keep the wolf from the door as best they could, without the assurance of early employment," a witness said. "Most of them were already embarrassed in consequence of their long idleness. Rent and taxes were unpaid, the endurance of grocers and butchers was exhausted, and with winter at hand and no money in sight to purchase the necessaries of life, what was to become of these destitute workingmen and their families?"[148]

More now sought to escape from Homestead while they could. "It is noticeable on the streets that many of the old workmen are leaving town and can be seen most any hour on the way to the trains with their trunks," the *Local News* reported, adding that within Homestead there was "also considerable moving going on."

Other Amalgamated members hoped to receive strike benefits until their jobs were reinstated in the wake of the increased activity in the mill. "Whatever relief is necessary for any one suffering through participation in the strike," the *Local News* editorialized, "they will be taken care of by the Amalgamated lodge to which they belonged."

With continuous production getting into swing, the rising possibility of future employment made itself felt: "The mill both day and night has the appearance of being in full operation, a fact no doubt due to the large addition of experienced workmen going to work within the last day or so."[149]

Within two weeks, this prophecy was being fulfilled: bureaucratic controls and concretized mill policies assumed their future form under Superintendent Schwab, who by January 4th had put the mills into full operation in every department. "The halls of the main office were crowded all day with men in search of employment and most of them got what they desired," a reporter explained the speed-up in the processing of the former workmen. "Superintendent Schwab received them in his private office during the whole day. While the several mills were closed down for repairs, a number of employees left, thus making room for the old men, many of whom were assigned places by Mr. Schwab."[150]

Jobs were now based on running the mills full speed, so that *continuous production* – the initial Carnegie goal in locking out the workers in order to stop Amalgamated interference – would become the rule. Such smooth running operations, without fetters, were, in the aftermath of the lockout, to become the benchmarks of Carnegie policy at Homestead.

The first steps towards this end were obvious: repairs and reemployment of the skilled, a reporter noting the progress:

> The repairs in the mills have been such that the Company will now be able to operate them without further intermission. This is a prophesy for a steady increase in the business of Homestead and vicinity. The mills are being fully equipped with men and in a very short time the work and production will be up to the old standard. . . . The Works presented a livelier appearance yesterday afternoon than at any time since the strike and it is safe to say that in the near future everything will again resume the usual smooth running order characteristic to the management of the Works.[151]

At this point the firm could not do without the skills of its old work force. "The firm is inclined to favor the old men as much as possible," wrote a witness, "and there has already been a large number of colored men discharged to make room for the old employees."[152]

Yet the momentary power balance struck between the strikers and the new workers led Company's Secretary Lovejoy to insist that "So far as the Homestead mill is concerned, places will be given to all the old men who wish to return, provided there is room for them, but none of the men now working and who are competent will be displaced to favor men hailing from the Association."[153]

Once Amalgamated men were re-employed, however, scab workers were laid off, either for "cause," incompetence, or of their own accord – it being, in a reporter's memorable understatement, "uncomfortable for them to remain."[154] The power balance had shifted once again, the firm wanting to replace the strikebreakers with skilled hands and the old skilled workers wanting their previous jobs at any price. Eventually the majority of old skilled workers were re-employed at Homestead – albeit only until the pre-strike machinery was replaced by the continuous steel making processes then coming into vogue throughout the industry.

For those who were rehired, meanwhile, wages would again sustain their families, a workman on the way to the mill reaffirming Homestead's traditional role-taking and division of family labor, telling one of the city's storekeepers: "If my wife calls to buy anything, let her have it. I have my job again." At which point he cheerfully proceeded up Eighth Avenue.[155]

## Notes

[1] Andrew Carnegie, "Cable to H. C. Frick," Sunningdale, Scotland, July 8, 1892.
[2] Stowell, 231.
[3] *Ibid.*
[4] Burgoyne, 184-185.
[5] *Ibid.*, 261-279; Wolff, 205.
[6] Stowell, 246, 249.
[7] *Ibid.*, 293.
[8] *Ibid.*, 292-295.
[9] *The Local News*, January 17, 1893.
[10] *Ibid.*, January 18, 1893.
[11] *Ibid.*, January 9, 1893.
[12] Stowell, 296-297.
[13] *Ibid.*, 298-300.
[14] *The Local News*, December 12, 1892.
[15] *Ibid.*
[16] *Ibid.*, December 12, 13, 1892.
[17] *Ibid.*, January 11, 1893.
[18] *Ibid.*, September 10, 1892.
[19] Burgoyne, 189-190.
[20] Wolff, 205.
[21] *The Local News*, September 10, 1892.
[22] Wolff, 205.
[23] *Ibid.*, 206; Stowell, 232.
[24] Jeremy Brecher, *Strike!* (San Francisco: Straight Arrow Press, 1972), 61.
[25] Stowell, 235-236.
[26] *Ibid.*, 236.
[27] Burgoyne, 190.
[28] Stowell, 232.
[29] *Ibid.*, 236.
[30] *Ibid.*, 237.
[31] *Ibid.*
[32] Wolff, 206.
[33] Stowell, 238.
[34] *Ibid.*, 240.
[35] *Ibid.*, 241.
[36] *Ibid.*, 243.
[37] *Ibid.*, 237.
[38] *Ibid.*, 242.
[39] *Ibid.*, 243.
[40] *Ibid.*, 243-244.

[41] *The Local News*, August 20, 1892.
[42] *The World* (New York, N.Y.), August 26, 1892; Wolff, 206.
[43] *The World* (New York, N.Y.), August 20, 1892.
[44] Stowell, 245.
[45] *The Local News*, September 3, 1892.
[46] *Ibid.*, September 17, 1892.
[47] *Ibid.*
[48] *Ibid.*
[49] *Ibid.*
[50] *Ibid.*, September 24, 1892.
[51] *The World* (New York, N.Y.), September 30, 1892.
[52] *Ibid.*
[53] Wolff, 205.
[54] Burgoyne, 190-191.
[55] *The Local News*, August 20, 1892.
[56] *The Sun* (Baltimore, Maryland), October 13, 1892.
[57] Stowell, 274-275.
[58] Burgoyne, 191-192.
[59] Stowell, 253.
[60] Wolff, 209.
[61] Stowell, 261-262.
[62] Wolff, 209.
[63] *The Local News*, August 6, 1892.
[64] *Ibid.*, July 30, 1892.
[65] Stowell, 253.
[66] *The Local News*, July 30, 1892.
[67] Goldman, 96-104.
[68] *The Local News*, September 17, 1892.
[69] *Ibid.*, September 24, 1892.
[70] *Ibid.*, October 1, 1892.
[71] Stowell, 261.
[72] *The Local News*, September 17, 1892.
[73] *Ibid.*, September 17, December 15, 1892.
[74] *Ibid.*, September 24, 1892.
[75] *Ibid.*
[76] *Ibid.*
[77] *Ibid.*
[78] *Ibid.*, October 1, 1892.
[79] Stowell, 260-261.
[80] *Ibid.*, 259.
[81] *Ibid.*, 249-251.
[82] Goldman, 108-109.
[83] *Ibid.*, 106-109.
[84] *Ibid.*, 110.
[85] *Ibid.*, 111, 127, 152.
[86] Burgoyne, 201-202; Wolff, 211-212.

[87] Stowell, 259.
[88] *The Sun* (Baltimore, Maryland), July 14, 1892.
[89] Burgoyne, 202.
[90] Goldman, 97; Alexander Berkman, *Prison Memoirs of an Anarchist* (New York: Mother Earth Press, 1912), Chapter 3.
[91] *The Local News*, August 13, 1892.
[92] Burgoyne, 192.
[93] *Ibid.*, 193-95.
[94] *Ibid.*, 196-97.
[95] *Ibid.*, 198-99.
[96] Stowell, 257.
[97] *Ibid.*, 258-59.
[98] Burgoyne, 199-200.
[99] Stowell, 271.
[100] *Ibid.*, 272.
[101] *Ibid.*, iii-iv, "Speech.of W.W. Erwin, November 22, 1892."
[102] *Ibid.*, vii-viii.
[103] Burgoyne, 201.
[104] Wolff, 209-210.
[105] *The Local News*, December 15, 1892.
[106] Wolff, 211-212.
[107] *Ibid.*, 213.
[108] *Ibid.*, 214.
[109] Stowell, 259.
[110] M.M. Garland, "Letter of July 8, 1893 to Samuel Gompers," *National and International Union Correspondence*, AA File, US, MSS, 117A, IIA, Box 44 (Madison, Wisconsin: State Historical Society of Wisconsin).
[111] *Ibid.*, "Letter of February 17, 1893."
[112] Stowell, 262; see also Samuel Gompers, *Seventy Years of Life and Labor: An Autobiography* (New York: E.P. Dutton, 1925, 1953), *passim*.
[113] Stowell, 250.
[114] *Ibid.*, 252.
[115] Wolff, 217; Burgoyne, 296-98; Stowell, 262.
[116] *The Sun* (Baltimore, Maryland), November 30, 1892.
[117] *Ibid.*, November 24, 1892.
[118] Wolff, 214.
[119] *The Local News*, October 1, 1892.
[120] *Ibid.*, October 8, 1892.
[121] *Ibid.*
[122] *Ibid.*
[123] *Ibid.*
[124] *Ibid.*
[125] *Ibid.*
[126] Wolff, 220.
[127] Stowell, 275.

**A Time for Charity**
**(Credit: *Harvest Quarterly*)**

# Chapter 11

# Deputies, Fear and Labor Tension

. . . [I]t is apparent that the matter could not be controlled by the peace officer of the township. It was therefore the duty of the sheriff to take action to preserve the peace and prevent the destruction of property. If this could be done by the ordinary force under his control it was his duty to use . . . [his regular deputies]. If this force was not sufficient, it was his duty to employ all other legal means necessary to effect the purpose. The only legal means within his control was the *posse comitatus*. . . . This he did not use or attempt to use. On the contrary, he deputized the men in the employ of the defendants [Carnegie and Phipps' Allegheny Bessemer Steel Company], and then appointed special deputies, – at first a few and afterwards in larger numbers, – as in his judgment the emergency demanded.

– Finding of the Court of Common Pleas, 1889 Strike[1]

Frantic, fearing the worst as the last soldiers marched out of Homestead on October 13th, Sheriff McCleary had doubled his force of deputies by swearing in new mill workers. The deputy sheriffs had already been making daily arrests in Homestead without first serving warrants.[2] Tension was mounting, as the local municipal authorities and the deputy sheriffs tried to hamper each other's operations, and Sheriff McCleary increased his force once more, as street and barroom fights escalated.[3]

Yet even McCleary's forces could not handle the emerging situation in October, when most of the scabs living inside the mill moved out to lodgings in the boroughs of Homestead and Mifflin, and a reign of terror against them began in the community. Ugly instances also revealed racism, as the considerable number of black strikebreakers continued to rely on public accommodations in the city, only to find themselves on the defensive. "The negroes, always dangerous in a fight," one writer revealed his own biases, "pulled the weapons they never should have been permitted to carry, and shot right and left into the crowd, wounding three persons. Then the crowd followed them into their lodging houses and threatened lynching. It was only by great effort that the officers succeeded in rescuing them and in locking them up."[4]

But so long as Frick employed scabs as a counterweight to die-hard union organizers, they did not willingly leave.

A month later, in mid-November, conflict and racial tension had not subsided. Two blacks had been jumped by the strikers at Fourth Avenue, the opponents drawing guns, firing fifty rounds, and leaving three men wounded and the blacks badly beaten before the police took them into custody. Deputies also had to rescue strikebreaker Martin Conrad, a particularly unpopular scab, who faced a large crowd gathered in front of his house with a revolver in each hand, warning he would shoot the first man who approached. This was but another instance in Homestead's tense atmosphere punctuated with innumerable street brawls, flying bullets and freely issued recriminations.[5]

Carrying revolvers had also become commonplace, a typical case on January 14th, involving two black men who got into fight with some whites in a restaurant; they "pulled ugly looking revolvers and invited all in the room to stand aside," the *News* reported the links between scab labor, institutionalized racism and violence: "The Negroes are working in the mill, and it is said had a hand in the row which occurred at the church on Tenth Avenue sometime ago."[6] Blacks, with their own church in Homestead, had entered but were refused prayer in one of the city's established churches – leading to a rowdy fisticuffs on God's watch.

Tension and gun-totting on the rise, each person was now on their own watch for the "enemy," a simple mistake like entering the wrong room of a boarding house at night bringing death for the "invader," in at least one instance.[7]

Nor was tranquillity a quality Homestead could boast, once the new Company housing for scabs was put to use. John F. Cox might well have earned a handsome profit by selling his Mansion House properties to the Carnegie Company. But as soon as the steel firm filled it with nonunion workers, it was dynamited.

"Between one and two o'clock yesterday morning a terrific explosion was heard by the inmates of the house, and others living in the neighborhood," the *Local News* reported. "The Amity Street front of the house contained large lights of glass which were completely demolished. A large hole was shattered in the floor. Beyond this there was no damaged done, yet the room and front of the building presented a wrecked appearance. No one was hurt. There are various speculations as to how the explosion was accomplished and who did it. The locked-out men repudiate all connection with it, as much as they repudiated the Berkman affair."[8]

Militancy, if not unsophisticated strains of Anarchism, were alive and well in Homestead.

The local Homestead police, determined to maintain order, meanwhile clamped down on the workers to keep them from upsetting the delicate balance of power. Street fights between scabs and workers were stopped by the Borough officers with force, despite popular sympathy for angry unionists; and the overtly bellicose were struck with maces, apprehended and locked up.[9]

Despite the new efficiency of the local police, the violence of the city workers continued; and at the end of October their battles with the scabs reached a new crescendo. Neither the Advisory Board nor the Sheriff and his deputies could stop the street encounters, and the *Local News* demanded either law and order in the city – or the return of the State Militia:

> The Sheriff is preeminently the one who should preserve the good order and peace of this community. His efforts all along have been signally futile and ineffectual. The presence of his handful of deputies in fact seems only to contribute to the disorder that prevails. It is not only the actual outbreaks that are so harassing to the people of the community, but the fear and expectation of a direful event like that of July 6th occurring, which will put the town in another state of siege for the next three months

> or longer. If the Advisory Board and the Sheriff cannot or will not improve the present condition of things, let the latter call upon the Governor forthwith, in time to prevent something more serious.[10]

The elite of Homestead joined the chorus, the six doctors of the borough joining the Schooleys of the *Local News*, the retailers McClures and McCunes, the realtors and financiers Coens, Bullocks, Bryces, and Giles, and the other well-to-do. A meeting an October 27th in Schuchman's Hall resolved that the prevailing lawlessness within the Borough had to cease. Dr. M. H. West presided, stated that the object of the meeting was not to discuss the right or wrong of the trouble between the locked-out workmen and the mill owners, but simply to express disapproval and condemnation of the prevailing lawlessness from whatever source, and to urge the authorities to take steps to prevent it. The fifty-six residents present then unanimously approved the following resolution:

> Whereas: The disorders, outrages and crimes have become so frequent and common in the borough of Homestead within the last few days, causing an almost complete reign of terror and bringing into disrepute the town and every law abiding citizen and injuring every inherent right of the individual and the common interest of the community.
>
> Therefore be it Resolved: That we the undersigned citizens of Homestead in Mass meeting assembled, stamp with our disapproval such disorders, outrages and crimes and call upon the civil authorities to take immediate steps to prevent all such further violations of law and if the civil authorities are unable to cope with and stay further outrages and criminal acts, we request the Sheriff of Allegheny County to call on the Governor of the commonwealth for sufficient military force to preserve the peace and good order, and we further pledge ourselves and call upon all law abiding citizens to aid the authorities in maintaining good order and government in the borough.[11]

## A Time For Charity

The dichotomy was clear-cut: Workers displaced by the strike and lockout had found protection in the union and commune. But the latter disintegrated, made them competitors for too few jobs, led to their unrest and violence, and brought forward attempts to control them by Militia, Sheriffs, local constables and the Amalgamated itself. Besides police power used as a stick to control, now good order and government in Homestead was also reestablished by providing for the families of the unemployed and needy.

As everyone was now fighting for jobs, and not every worker nor Amalgamated man got a job, needs had gone unmet. Thus on November 23rd, though there were few applicants at the mill, all were turned away. "The disheartened appearance of the men as they turned home through a blinding snowstorm made them objects of pity," said a witness. "There are many families among the strikers, while suffering from a scarcity of provisions, have not even enough fuel in their houses, and much more suffering is sure to result."[12] The situation changed for some a week later, at the end of November, when nearly 200 non-unionists were replaced by former Amalgamated men. The burden was shifting: The scabs now had to make do on their own.

Among those who had supported the strike and were not rehired, the situation went from bad to worse. Appealing for these workers' needs, the *Local News* proclaimed Thanksgiving Day a special time:

> What is the most imperative duty of the well fed and well to do portion of this community? it is to see that no family goes without sufficient food, fire and clothing. The wonder is that after such a long period of forced idleness, there are so few families in actual want. Homestead can easily afford to care for all who need and are worthy of assistance. It will only be a little while till all will again be at work here or elsewhere and the duty of the hour is to lend assistance where it is needed. on this Thanksgiving be generous.[13]

Not more than 500 of the 2,000 applications by the old nonunion workmen had been accepted by Carnegie up to November 28th, and Amalgamated rules stopped union funds going to them unless they were blacklisted by the Company. Since there were relatively few in this employer-designated category, most were left to their own resources and the good will that the Amalgamated and others could drum up in relief funds.

Most of these workers lived in the Second Ward, the heart of the proletarian district where the houses were closely packed and - with the fall in the level of nutrition and thus resistance to viral infection – the dangers of spreading disease were greatest. "While there is no general destitution prevailing at present," the *Local News* wrote at the end of November, "there are strong reasons for expecting there will be, and before long too."

Proof was not hard to find. The City Assessor John Dierken had canvassed the Second Ward and found a good many families living in privation, with the need for a few hundred dollars for coal being the most essential to prevent suffering before the harsh weather began. "Out of money and out of work stares a thousand or more people in the face on the approach of a hard winter," the *News* editorialized. "The strike is over and there is no longer any fighting to be done on one side or the other. It is simply charity now for each and all to do what can be done to prevent and relieve suffering."[14]

The leaders of the strike had meanwhile been making an effort to collect funds, as had the members of the McKeesport Select Council in arranging for a public meeting in that city for Homestead workers without jobs.[15] The *Local News* became the focal point for the collection and distribution of local aid, acknowledging contributions and turning them over to a committee for distribution to the needy.[16] The burden of collecting funds had also shifted to the community, the *News* viewing contributions from Homestead as a method of stemming hard times until the unemployed could secure work:

> As long as contributions were pouring in from the country at large, and the contest was undecided it was manifestly not the imperative duty of our citizens to give save as they were prompted by their sympathy with the locked-out men. That is all past now and it is the duty of every citizen to lend a hand to help the needy and restore the workmen to self-support. In view of the fact that many of our people were not affected by the depression and that many others are at work and that the prospect of good times just ahead are good, let Homestead provide for her needy ones so long as her means are not exhausted.[17]

The argument was infallible in the context of reestablishing Homestead's traditional prosperity: Yesterday's class struggle and unity of labor had ended with the workers' defeat; support for the wage system now meant all classes should contribute to keep the unemployed in a stable condition, so they could be re-employed at a later date. In this way, both day laborers and skilled workers, as essential production forces, could be maintained in Homestead pending the day Carnegie would re-employ them. Re-employment seemed just around the corner for some, and to get production in motion the foreman of the all-important Homestead Armor Plate Mill, W. L. Leadbeater, was brought back from Uniontown to take his former place in the mill.[18]

To receive relief, then, a moral standard was put in place: The families of working men would be maintained until they could finally get back to work. The Relief Committee itself set up the stringent rule that no money would be given out except for relief and that single men would not be given relief of any kind.[19] This standard had three requirements: that those helped had to represent the families of striking workers, that the family helped had to be from Homestead, and that workers temporarily out of work for department repairs alone could not receive aid.

Keeping the unemployed acquiescent – on tap for future employment – as the funds collected increased, these standards were more rigidly enforced,[20] donors setting the specific conditions for welfare distribution. The Pittsburgh *Press* alone had collected $2,000 in cash to be turned over to the Homestead Relief Committee.[21] But when banker T. J. Keenan, representing the *Press*, came to Homestead, he was accompanied by Superintendent Linderman, representing the City Poor Farm. Keenan would only turn over the funds if the Relief Committee was structured to his liking and used approved methods of aid distribution. Stating that he had been requested by some of the contributors to the *Press* fund that the paper have some representation on the Committee, he suggested the names of conservative Superintendent Linderman, bankers Louis Rott and himself, and lawyer, politician and real estate investor John F. Cox, all being gentlemen with whom he was closely acquainted. These four were then named as a committee to act in conjunction with the Relief Committee in the distribution of *Press* funds.[22]

The arrangement seemed innocuous at first. But maintaining a small body of Homestead unemployed might keep the workers anxious that their jobs would remain insecure. Pressure was also being put on union men to first reapply at the Homestead mill and, if they had to go elsewhere for work, to drain the Amalgamated treasury to finance their costs.

The new balance of interests on the Relief Committee thus led to a strict definition of "need" being enforced to maintain Homestead's unemployed labor pool. And the Committee now refused to hand out cash to enable former strikers and supporters to pay their railroad fare to go to Pittsburgh to find work. Instead, if men wanted work outside Homestead, the Amalgamated Association should assist.

The committee meanwhile became a policing body, working with the *Local News* and officialdom to control collections in Homestead in order to enforce conformity on the destitute and maintain surveillance over them. Considering its politics to reestablish law, order and regularity in both civic matters and mill employment, the *News* followed a similar approach, with a double edge, in all its appeals: "This is not to maintain and support a strike, but to help people till they can get work."[23] "The sooner a large number of the men now idle obtain work the better for them and the public."[24] But despite repeated attempts, there was no immediate inpour of funds: Homestead had just come through five months of idleness and imposing expenses.

Meanwhile, Homestead's unemployed were suspicious of any new obligations or limitations attached to handouts. The employed also understood the Relief Committee would not give them help when they needed it during the holiday season shutdown for temporary repairs at the steel works.[25] So, after repeated and unsuccessful appeals for funds by the Relief Committee, the workers set up their own committee to meet with other workers' committees from Chicago and Pittsburgh, as well as to set up a Homestead ward committee.

Workers again organizing and in charge of relief funds rubbed against the designs of the local elite. Businessmen were conspicuously absent from the meeting called by the new city Burgess, Hollingshead, though many well-to-do like lawyers Stenger and Cox, Judge Oeffner, and Principal Kendall took charge of raising funds in the City and themselves contributed generously. When Homestead coal mine owner W. S. B. Hays contributed 500 bushels of slack coal,[26] some businessmen judiciously defended their right not to give. One explained that they had extended credit to the strikers for many months and thus had done their part:

> It may not altogether have been an act of charity on their part, but it is a fact nevertheless that during the long months of the strike, they freely distributed goods and stores to their patrons who bought in very many instances without money. The storekeepers have trusted out thousands and thousands of dollars which [they] will never get back. Is it not wrong to expect these businessmen to contribute now to a relief fund? And if they decline or fail to respond should they be criticized?[27]

The same could be said for the merchants still offering goods, food and clothing, as well as for landlords who had lost from two to four months' rent: all of them had suffered losses and were unlikely to be repaid.[28]

When it became increasingly apparent that most Homesteaders were not yet back at work and stability was dependent upon a well-provided population, business contributions began to pour in. Cain Bros., well-known coal dealers in the city, contributed one thousand bushels of coal for relief of Homestead's needy, and delivered coal in small lots throughout the Borough.[29] The owner of the S. S. Marvin & Co. bakery of Pittsburgh, after investigating Homestead needs, committed his firm to contribute 3,000 loaves of bread to be delivered at such times and in such quantities as the Committee might suggest.[30]

Early estimates placed 48 Homestead families still in great need and 56 receiving a small amount of help amounting to an daily average of

$2.83. Even so, by December 6th, the Committee had expended all its resources. Since most of the strikers had not yet obtained work, the potential need was sure to increase and the labor members of the Committee undertook to stir up labor organizations elsewhere, to inform them that the end of the strike did not mean the strikers had obtained work and that contributions were still needed.[31] In fact, stores extending credit had reached the limits of their good will.[32] And though promises of fund-raising and Homestead contributions were made,[33] the growing need remained clear and present.

When the Committee finally surveyed the community, they discovered that by December 9th, only 800 of the 3,800 former workmen at Carnegie Steel had found employment in the mill city and only a limited number elsewhere; that there were 218 families in Homestead and vicinity in a state of destitution; that the vast body of unemployed men would be unable to secure work for many weeks; and that the public outside Homestead would likely have to help relieve the prolonged and increasing distress.[34]

Meanwhile, funds for the needy came from working class communities themselves, as the ward organization of the Relief Committee canvassed the neighborhoods for contributions.[35] The Federation of Labor meeting in Philadelphia also donated $1,500 for the relief fund.[36]

At holiday times charity almost swamped the needy. "Is there not some danger of overdoing the Christmas racket for Homestead suffers?" the *Local News* asked:

> According to the papers, McKeesport will send a thousand turkeys here on the day before Christmas, to be distributed among needy families. No thought seems to be taken as to whether or not there are so many families here. The number reported a couple of weeks ago was 218, but the Relief Committee has not yet found two hundred of them. Just who are to get the turkeys is not quite clear to us.[37]

Though the inflow of help was nor steady, once afloat, the news of Homestead's poverty had spread and drawn new rounds of supporters to aid the needy. The Pittsburgh *Bulletin*, which had opposed the strike and the strikers, nonetheless now announced: "The learning of a bitter lesson will not be hastened by letting the Homestead sufferers freeze and starve at their unfueled grates, and the big heart of Pittsburgh will not let this come to pass."[38]

So it went. So many turkeys were on hand at Christmas that the Relief Committee sent them not only to the needy, but to those "in straightened circumstances." As well, various persons outside the

Committee arranged to distribute large quantities of candy and useful presents to the children as well as to needy families. The *News* concluded the obvious: "none will go hungry for Christmas here, unless it be their own fault."[39]

By mid-January, though, relief provisions had again run low, and a $500 contribution of W. J. Bishop was much needed. "The gift was a timely one," the *News* explained, "as the committee has been doing all in their power to aid those in need and the fund on hand was rapidly decreasing."[40]

Beyond welfare, the key question uppermost in the thoughts of most Homesteaders was still aid for the workers' cause. Striker leaders were still in jail or on trial, and lacking its own press, the defeated Amalgamated again turned to the *Local News* to solicit any help it could get. But by November 29th, the *News* viewed the strike as a past event worthy of the least possible support, giving it scant press. It gratuitously acknowledge receipt of an Amalgamated ticket for the "Benefit Defense Fund of the Homestead Strikers" in Allegheny, an event attended by supporters from Homestead, Pittsburgh and Allegheny.[41] But its fund-raising days were over.

**Getting Jobs and Weeding out**

New production relations were meanwhile initiated under Carnegie managers, determined to introduce new work rules, technology and labor conformity. This was made easier by their power to play the various factions of workers and others against one another.

Too few jobs for too many hands meant that struggle, conflict and hatred would emerge from within the ranks of labor. And though a large percentage of the old men eventually got back to work, scabs, both white and black, were in the mill and city for many months before they were displaced. Some of these lacking skill were fired, while others were attacked and otherwise harassed by the old hands until they left. "The colored men are said to be leaving the Steel Works and seeking employment at the coal mines up the river," the *News* reported this realignment of jobs.[42]

The strike leaders were also weeded out by Carnegie's new policies: blacklisted or given the option of signing a yellow-dog contract, whereby the individual would agree never again to join a labor union while employed by the firm. As not one of the original 350 members of the Amalgamated lodges in Homestead had a chance of being rehired without signing and all refused in the immediate aftermath of the lockout, they were covered by the blacklist.[43]

Frick confused the public by declaring a general amnesty for the strikers. "The active leaders of the strike were, of course, excluded from the amnesty, and few of them have since been able to secure employment at their trade," one critic wrote in 1893. "They are the victims of a form of ostracism; blacklisted as dangerous agitators in every steel and iron mill in the country."[44]

Fear and conformity were built into the new hiring arrangement, and many former union men gradually gave way to the pressure in order to regain their jobs. Only 1,300 of the original 3,800 strikers had immediately gone back on the payroll.[45] Among them by early 1893 were some three-fourths of the former union men, most in their old jobs. And by January 16th of '93, of the 3,500 men working in the mills, some 2,400 had held positions before the strike.

"Room for more Homestead men is being made daily," the *News* optimistically reported, noting that inefficient workers and others who had not settled with their families in Homestead were leaving the Works.[46] Behind the mill fence, Schwab had meanwhile discharged the nonunion green hands and filled their places with ex-strikers, whose experience made them indispensable.[47] Old superintendent Potter and Chairman Frick had conveniently gone back on their promises not to replace the scabs, now heeding the momentous discoveries of foremen that these men were indeed incompetent. In November and December more had been discharged to make room for the experienced former workers; and replacements continued for many months to come.[48]

## Conditions of Labor Competition and Disunity

Without compassion or scruple, the juggling of workers, scabs and populations was now in full swing under Carnegie's new regimen of production. The scabs so recently deployed in the mills as pawns were speedily forgotten, with nominal payments at best made for those who had been injured or died. In a not atypical case, life itself was paid for to a widow and her child – free provisions and lodging in Company housing, $100 in cash to compensate for the death of her husband, accompanied by a ticket out of Homestead for Pittsburgh. "When she left. . . ," the *News* straightforward reported, "an officer [of the Company] was sent to Pittsburgh to see that she got safely started home."[49]

Former steel workers not rehired also left Homestead with little or nothing to show for their years of hard labor. In a not untypical scene in December, some half dozen steel workers who had failed to get into the mill left for the coal mines where they could find employment "digging the dusky diamonds."[50] Those blacklisted from the Homestead mills also left without a trace of their good work represented either by savings or monuments. Among those blacklisted, former Burgess McLuckie was caught between his legal prosecution for riot and murder and the necessity to earn a living by taking a job in Pittsburgh or somewhere else. "I cannot go to work because about the time I get in I may have to leave and attend my trial," he summarized his dilemma. "I have the offer of a position in Pittsburgh but I cannot tell whether I will accept it or not. I may still remain with my family in Homestead, in case I go to the city to work, or we may move to East End." McLuckie felt no uneasiness about his own case, because, he said, he had done all in his power to prevent trouble during the strike – a point amply confirmed by history.[51]

McLuckie was later seen in Mexico, some say having refused a Carnegie gratuity to support him for life. A few of those kept out of the mills also went into business for themselves, former steel workers John Miller opening a small grocery store on the corner of Eleventh Avenue and West Street, and John Hulse becoming proprietor of a variety shop opposite the P. McK. & Y. Station on Ann Street. Both relied on their popularity among the steel workers and residents to build up a good patronage.[52] They, too, depended on the old hands being re-employed in the mill, and thereafter could change the historic configuration not at all.

Skilled workers who marched back to their jobs as competitive, individual applicants now seemed to have no future potential for unity in struggle. Frick, on hand to watch their return to mill labor,[53] confidently cabled Carnegie in Europe that the strike had been declared ended, victory was complete, and the Carnegie Works would never likely "have any serious labor trouble again."[54]

"It teaches the workingmen this lesson," echoed Congressman-elect John Davis from Kansas, "they can't whip Carnegie's millions while their stomach is empty."[55]

The Amalgamated's national secretary, Kilgallon, agreed, saying that the workers' loss of the strike "did not reflect upon the power and influence of organization." Rather, he said: "The case was one where a vast accumulation of wealth in the hands of the Carnegie Company swayed the tide of battle in their favor." To this the grand master of the Knights of Labor, T. V. Powderly, added: "The Homestead trouble is as much the result of obstinacy and a desire to rule without question as anything else. Mr. Frick is overbearing and tyrannical. Mr. Carnegie is in Europe studying the conditions of labor, and reporting his findings to Mr. Frick, with instructions, no doubt, to go as far towards establishing similar relations in this country. . . ."[56]

Establishing the same relations of production between employers and workers as existed in Europe would permit the Company to successfully compete. By re-employing old skilled hands no longer protected by the Amalgamated, the firm could now pay lower wages, reduce the cost of production, and lower steel prices without squeezing profits. The way was now cleared to upgrade technology, do away with some skilled, high-priced labor tasks, and hire others as less expensive unskilled hands.

This was the future every steel competitor understood. It was the *raison d'être* of an industry as cutthroat in the market as it was towards the workers who created the steel to be sold.

"Let the Amalgamated exist and hold sway at other people's mills," Frick had cabled Carnegie when the strike was officially called off by the Amalgamated, bringing Carnegie's elated response, "improve works – go ahead clear track – tariff not in it – shake."[57]

The meaning was clear to Frick: cheaper foreign steel was no longer needed as the Company's excuse to lower wages in the aftermath of the great strike. With the Amalgamated defeated at Homestead, the firm could take away the elements of labor's control in the workplace, redesign the tasks of production, lower wages to any level it decided upon, improve its technical equipment, use cheaper unskilled workers to operate it, set price levels, and outflank competitors at home and abroad.

## Notes

[1] McCandless v. The Allegheny Bessemer Steel Company, Court Of Common Pleas No. 1, *Pittsburgh Legal Journal*, Number 32, March 9, 1892, 300.

[2] Stowell, 278.

[3] *Ibid.*; See *The Local News*, November 22, 1892.

[4] Stowell, 285. Spelling of "Negroes" in original text.

[5] *The Tribune* (New York, N.Y.), November 14, 1892.

[6] *The Local News*, January 16, 1893.

[7] Stowell, 285.

[8] *The Local News*, October 8, 1892.

[9] *Ibid.*, October 22, 29, 1892.

[10] *Ibid.*, October 29, 1892.

[11] *Ibid.*

[12] *The Sun* (Baltimore, Maryland), November 24, 1892.

[13] *The Local News*, November 28, 1892.
[14] *Ibid.*
[15] *Ibid.*, November 22, 1892.
[16] *Ibid.*, November 29, 1892.
[17] *Ibid.*
[18] *Ibid.*
[19] *Ibid.*, December 14, 1892.
[20] *Ibid.*, December 15, 1892.
[21] *Ibid.*, December 16, 1892.
[22] *Ibid.*, December 17, 1892.
[23] *Ibid.*, November 30, 1892.
[24] *Ibid.*, December 17, 1892.
[25] *Ibid.*, December 2, 1892.
[26] *Ibid.*, December 3, 1892.
[27] *Ibid.*, December 5, 1892.
[28] *Ibid.*
[29] *Ibid.*
[30] *Ibid.*, December 6, 1892.
[31] *Ibid.*, December 7, 1892.
[32] *Ibid.*, December 8, 1892.
[33] *Ibid.*, December 9, 1892.
[34] *Ibid.*, December 10, 1892.
[35] *Ibid.*, December 6, 1892.
[36] *Ibid.*, December 15, 1892.
[37] *Ibid.*, December 23, 1892.
[38] *Ibid.*, December 23, 1892. Quoting The Pittsburgh *Bulletin*.
[39] *Ibid.*, December 24, 1892.
[40] *Ibid.*, January 17, 1893.
[41] *Ibid.*, November 23, 1892.
[42] *Ibid.*, November 22, 1892.
[43] *The Sun* (Baltimore, Maryland), November 30, 1892; Wolff, 217-218.
[44] Burgoyne, 296-297.
[45] Wolff, 223.
[46] *The Local News*, January 16, 1893.
[47] *Ibid.*
[48] Burgoyne, 296.
[49] Wolff, 224.
[50] *The Local News*, December 13, 1892.
[51] *Ibid.*, January 14, 1893.
[52] *Ibid.*, December 7, 1892.
[53] Burgoyne, 229-231.
[54] Wolff, 225; *The Sun* (Baltimore, Maryland), November 22, 1892.
[55] *The Tribune* (New York, N.Y.), November 21, 1892.
[56] *The Sun* (Baltimore, Maryland), November 22, 1892.
[57] J. K. Winkler, *Incredible Carnegie* (New York: Vanguard Press, 1931), 218-219.

# Part IV

# The New Stage of Stabilization

> Labor people often have, it seems, a nice paradoxical loyalty to the corporations that they are pitted against.
>
> – John Gunther, Inside U.S.A.[1]

The new stage of stabilization of Homestead civic life and reorganization of mill production differed sharply from the past.

Carnegie's pre-strike design had been to concentrate on relations with the workers through the union, initially accepting the role of the Amalgamated because it put pressure on his competitors' mills raising their labor costs, enabling him to outflank, take over, or drive his principal competitors out of business. But he then reversed course, sharply charging that the union had placed the Homestead mill at a competitive disadvantage because it had organized there, not throughout the entire industry.

Once the Amalgamated had spread, helping to consolidate Carnegie's empire, culminating in the purchase of the non-union Allegheny Bessemer plant in Duquesne, the power-play at Homestead was decisive. Carnegie was now dominant in the steel industry; he no longer needed – and moved to destroy – the union.[2]

The owners of the entire steel industry in the Monongahela Valley then hung on Carnegie's victory over the Amalgamated Association and its allies. With the union defeat certain, the old relationship that had restrained the owners was completely undermined and restructured. For the new ties established between all-powerful employers and unorganized labor meant direction of production passed from the skilled workers to the managers; and the employers gained total control over the introduction of technology and the way in which laborers performed on the job.[3]

Carnegie now sought to establish an all-encompassing company town at Homestead. To upstage the old division of functions in Homestead he extended the role of his hierarchy of managers into the separate spheres formerly held by the union, the city's business interests and the Borough government – attempting to totally control the living conditions of the work force and manipulate the population. He thereby hoped to regulate not only work tasks and the wages he paid, but the city's commercial rhythms and price of basic amenities the workers required to support themselves and their families.

By lowering the cost of labor's maintenance, Carnegie planned to both cut wages and quiet labor protest over the sliding scale.

## Notes

[1] John Gunther, *Inside U. S. A.* (New York: Harper & Bhrothers, 1947), 618.

[2] Robert B. Brooks, *As Steel Goes, Unionism in Basic Industry*, (New Haven: Yale University Press, 1940), 26 ff.; Clarence Budington Kelland, *Dynasty* (New York: Harper & Bros., 1929), 178.

[3] See Wellington Roe, *Juggernaut* (New York: Lippincott, 1948); Selig Perlman, *History of Trade Unionism in the U.S.* (New York: Augustus Kelley, 1950); Carroll D. Wright, "The Amalgamated Association of Iron and Steel Workers," *Quarterly Journal of Economics* , July 1893.

# Chapter 12

# Homestead in the Aftermath

> The problem of our age is the proper administration of wealth, so that the ties of brotherhood may still bind together the rich and poor in harmonious relationship.
>
> – Andrew Carnegie[1]

In the grand sweep of emergent conditions in the Valley's mill towns, lower wages and harder work by themselves would strain labor and reduce the workers' standard of living. Yet once destroying the Amalgamated so labor could not advance on the job front, the employers in charge of the company towns quieted labor dismay by also attempting to lessen or alleviate the economic burden placed on the workers' families through lowering or otherwise controlling the cost of housing, food, gas, coal and other services.[2]

Consciously or not, albeit for far different purposes, the employers put in motion the program of several generations of Utopian Socialists. These socialists had theorized that control over the means of production and life would enable populations to secure inexpensive housing, free education, health services, and a well-rounded, wholesome diet, reducing the cost of labor's subsistence, caring for the working family, and energizing the work force. On the basic level of survival of the working class family – but not permitting their control – the employers led by Carnegie attempted to put this scheme into practice by cutting living costs with a view to make them correspond with the lower wages paid – as a means of both keeping down the cost of production and preventing labor's protest.[3]

Carnegie had long recognized the principle involved. Before the strike, he had periodically attempted to pacify his work force by lowering labor's cost of maintenance: providing company housing, loans at low interest rates for mortgages, inexpensive coal and gas from the mills, and such amenities as church organs, library and gymnasiums and other facilities that normally improved the bleak company towns.

Most of these benefits had been given to win labor's allegiance when the workers threatened to strike and/or were used to lock the workers into the interwoven structures of the wage system and company towns. Workers with a mortgage to repay to Carnegie, living in Carnegie housing, or receiving coal in the winter and free ice in the summer would be more reluctant to strike for their demands than those for whom such considerations did not enter the picture.[4]

These relations were tested in 1892, though. So long as the Amalgamated kept wages high at competitor mills, production at the Homestead mill remained relatively cheaper and limited the extent to which Carnegie tried to cut wages while offering benefits lowing living costs. But when the Homestead skilled demanded relatively higher pay, Carnegie moved to bust their union.

Still, at Homestead, taxes, the cost to clean up the city, make it commercially viable, train the work force, and provide essential amenities for the population and workers was largely borne by the workers themselves. If they lowered their cost of living by exerting their labor-time or spending their resources directly or in taxes, it all came from the same source and at best netted a small immediate lowering of their maintenance costs.[5]

The greatest beneficiaries were the commercial and business interests, especially Carnegie, for their costs of production were lower when streets were improved to transport raw materials, finished goods and a labor force to and from the mills. Their labor force was trained cost-free to them by the Homestead school system which the taxpayers (but not Carnegie) supported; and this work force was made healthy and secure by the voters imposing burdens upon themselves to initiate a water works, a fire department, and a system of streets, sewers and street lights.[6]

By these means the Carnegie Works and other Homestead

businesses benefited greatly. With the taxpayers footing the bill for the police department, the courts and the jails on the one hand, and, on the other, the same group financing the Borough government, its agencies, and the provision of water and gas, the burden of maintaining the city and family life rested squarely on the population – especially the mill workers - and little if at all on the owners of commercial and manufacturing firms.[7]

During the early '90s, though, these dynamics changed in the Monongahela Valley, punctuated after the Homestead events of '92, as the mill owners in charge of the company towns in the Monongahela Valley shifted some of these burdens to themselves. Having destroyed the Amalgamated Association as the protective shield of skilled workers, Carnegie and the other employers not only established control over what was formerly the union's domain on the job, but effectively invaded the workers' personal sphere involving family survival. For now they actively sought to cut living costs to make them correspond with the lower wages paid, as already explained, thereby also keeping down labor agitation. Thus many of the amenities the workers had previously paid from their own taxes and self-assessments were now paid for by the employers; though other charges, such as education, training the work force and sustaining their kin in case of injuries or death, were maintained by or carefully shifted back to the workers.[8]

Overall, though, the employers tried to lower the cost of the sustenance and reproduction of the working class family; and this was done in order to make these outlays correspond to the lower levels of pay, and thereby the lower cost of producing steel for competitive marketing.

Though this was the plan, given the commercial and social pressures in the industrial towns, the workers and their families could not and did not maintain their traditional living standards. Through the operation of the market, the independent commercial and financial interests sought to drain the working class family of the very benefits Carnegie wanted to award them. And though the working class population tried to improve the urban environment and strengthen their social, economic and political position in the community, they again imposed the burden of rectifying living conditions on themselves, so their payment of taxes and their exertion of self-labor was the main source enabling them to sustain, energize and reproduce their own families, attempting to improve the quality of every day life.[9]

In a comparable fashion, the cost of supporting the disabled sector of the work force was borne by those still able. And this paralleled the process of the socialization of the other costs of providing for the working class through the local government and public tax assessments – which supported and maintained the streets, the sewage system, and education.[10]

## The New Legerdemain

These overlapping pressures in the aftermath of the Homestead lockout not only filled out the new stage of proletarian legerdemain, but as well marked a critical juncture in working class life in America – with the post-strike dynamics of Homestead involving those of centralized wealth and control, quickened urbanization, city improvements, and the fiscal crisis of the local government.

Urban expansion had itself been given new life in the early 1890s, as growing industry demanded that government finance, organize, educate and deliver a healthy, fit labor force and an infrastructure for production consisting of bridges and roads, and access to raw materials, supplies and markets.

As the Monongahela Valley expanded, local governments created the conditions facilitating production and trade, as well as tried to oversee and regulate the population at large. For these purposes, city taxes were imposed for the police, a variety of social services, and construction of infrastructure. So though the earlier villages in the Monongahela Valley had been comprised of a population mixing farming and household handicraft with little government other than voluntary associations, the new cities were founded on regularized factory production and discipline, city charters and borough councils, tax collectors and allocation of city funds – and the emergence of bureaucratic controls over the community. The elaborate and intimate linkages between the rural and urban domain seemed irreversible. In response to these pressures, the architecture of the city was itself transformed – brick, steel and asphalt, box-like construction now dominated the urban domain.[11]

Nowhere were these changes more rapid than in Homestead after the defeat of the Amalgamated. For here the mills and its labor force were "restructured," and these readjustments in the sphere of production led to rapid alterations in the dynamics of urban expansion – efficiencies sought by constructing new square and rectangular factories and urban structures, new methods developed for marketing commodities and services, new pressures and drains imposed on the working class family, and new controls exercised through the market, the government and political parties.

More important still was the sphere of finance under the control of

Carnegie and the great bankers, for this financial power had overwhelming influence in shaping urbanization in Homestead, including the condition and demographic composition of the population, its system of taxation and finance, its Borough services in facilitating public health, sewage disposal, passable and safe city streets, efficient transportation, fire fighting, and adequate water supplies.[12]

**The Genesis of Post-Strike Prosperity**

Mobilizing his financial resources to lower the cost of housing, food and amenities so that the workers' daily needs would be covered by the lower wages paid at the Works, Carnegie attempted to build his company town anew. Without such finances, the facilities to house new workers could not have been built; and without stimulation from Carnegie wages, merchants could not have renewed their sales, nor could the city make improvements based on tax collections from residents on Company payrolls. The commercial life of the city and solvency of Borough government thus hinged on Carnegie employment, empowering the Company to attempt to establish direct controls over the work force and indirect controls over the city at large.[13]

With the initial bringing of non-union workers to live and work in the mills, the merchants and businesses of Homestead had experienced an increase in trade in the first two weeks of October. Scabs living in the Borough and handouts by the Amalgamated to members, mechanics and laborers combined to press more money into the market.[14]

Carnegie erected more than a hundred new houses and built a large storeroom to act as a food center for the "New Town" located on his poor-farm properties – placing orders with merchants supplying materials and labor, presenting, the *Local News* exclaimed, "such a scene of activity as has not been seen in Homestead since the building boom in 1881-2."[15] The old work force would now have to contend with the uncomfortable fact that some of the new workers would occupy better homes than they had lived in before the strike, that plenty of land remained for Carnegie to build more residences as he pleased,[16] and that low-price goods would be distributed from the Carnegie-run commissary operating as a cooperative.[17]

The January '90 opening of Company lodging houses, 80 rooms fitted with modern equipment, was in addition to the homes on the old poor-farm and the facilities available at the repaired Carnegie Hotel.[18] True enough, the non-union workers at the lodging houses were virtual prisoners under the eyes of clerks and watchmen. Electrical bells and call pipes regulated their activities and awakened them on time for work. But, viewing their health and strength as essential for the firm's survival, the Company established medical facilities in the lodge to care for those injured,[19] and built an emergency fire department on Carnegie's own property.[20] Here was built a stable to house 100 fire engine-pulling horses and large quantities of feed.[21] Together, these elements of security were designed to regularize the workers' daily schedules, lower their cost of living and needs below pre-strike outlays,[22] and provide the essential conditions to protect them and their energies for the next day's labor.

Secure conditions in Carnegie's New Town were also established by benefits provided on the job and in Company-owned homes. In part this was made possible by the financially-needy Borough government, which received subsidies from, and quickly became compliant to, Carnegie. Though the Company had built its own sewers and graded its streets, Carnegie homes were serviced by both the Homestead electric plant and the Borough Water Works.

At first, landlords and real estate brokers saw the new rental properties as a potential lever that would depress rents in the Borough. But the local media viewed the Carnegie houses as a precondition for a growing population in a larger city.[23] With more Carnegie Company living quarters, in fact, had come a rise in the activities of real estate financiers C. W. Meeks of Munhall and Schuchman & West of Homestead, both calling out the advantages of Munhall Terrace lots near the steel Works – available, they advertised, at low prices, accessible by board walk on a level road, and payable "out of pin money" in a low-tax district.[24]

Renewed commerce and wages at the mill soon drew more families to Homestead, too, leading traders to take up the vacant stores and stalls on Homestead's main streets,[25] also generating higher rents and valuations of real estate.[26] The old union men drove some non-union workers from Homestead, emptying the Carnegie's boarding house at a loss.[27] But the new employees pushed up rentals, preferring to occupy the old homes in town where they were close to the churches, schools, stores and city activities, leaving the Carnegie buildings largely unoccupied.[28]

With the return of mill production and employment, workers again sought permanent homes, brokers reacting to and shaping their demand by reaching to the outer boundaries of the Borough, carving up the remaining estates to sell small parcels to the growing body of wage earners.[29] Once these real estate sub-dividers resumed the place they held before the lockout and strike, maintaining a formal company town became increasingly unlikely.

Real estate speculation in the wind, new, costly transactions in city housing increased,[30] and the rise of realty values appeared certain.[31] The *Local News* heralded construction of a new urban infrastructure: "The improvements which the Carnegie Steel Company are contemplating, the building up of the New Town on the hill, the construction of a couple of bridges and electric street railways, a number of new industrial enterprises on the Hays Estate, are all things in store for Homestead's near future. Add to this the grading and paving of a few main streets, and no one will question the effect upon property values."[32]

With middling retail shops and business participating in completion of the essential infrastructure of the town, these commercial establishments catered to, linked – and acted as a partial buffer between – the working population, the soon-to-be-employed, and the owners of large enterprises already located in or entering the Valley. Petty commerce thereby oiled the gears of growing mill cities, themselves expressing proletarianization in the Monongahela Valley and its dozens of steel-making towns.

This division of functions gradually strengthened: The mills provided employment; the retail businesses, services and goods. Without local shops and small manufacturing concerns, those not directly involved in the mills could not have obtained employment, lived in the town, and, through the market, sustained the other commercial enterprises. Local financiers not connected with Carnegie also could not have begun their ventures without resources coming from the Homestead population and commercial sector. They, too, formed connections with, and a mutual support network for, the larger mills and factories, as well as for the railroads and tramways that were rapidly coming into the Monongahela Valley after 1892.

**Dividing Lines**

Through the city structures that emerged, the local elite and middle class also eased friction between the great enterprises and the Homestead population. In the post-strike milieu, these middling forces would again try to control the population through the marketplace for goods, services and jobs, as well as through the electoral process, the city government and the schools – also propagating assumptions and behavior useful to outside moneyed and political interests. The elite were such, only because they lived in the shadow of the Carnegie mills and the other outside financial interests, though still they had the pride of rulers of the town's economy, City Council and exclusive societies.[33]

The general belief was that they and others had earned their place by hard work, honesty and consideration of others,[34] a theory put in practice by the small-fry of enterprises,[35] which were part of the essential commercial infrastructure aiding big industry – also ironically reaffirming what Adam Smith had advised in 1776 about personal industriousness awarding the higher good of others.[36] Thus the strike had not gutted the foundation for smaller Homestead businesses to reassert themselves, and by January '93 a resurgence of the local market seemed certain.[37]

Middle-sized shops like the Noble & Neel mill[38] and the Bryce Higbee glass works[39] had expanded, introducing the latest technology. Smaller capitalists also rode the coat tails of the Carnegie's mills, drawing on the skills of mechanics to erect ancillary foundries to repair Carnegie machinery and deploy traditional means of transport by rail and river.[40]

A new dividing line was meanwhile struck between Carnegie's domain and Homestead's commercial spheres, with Carnegie's New Town providing only housing and amenities in close proximity to the mills, and the people living there also doing their essential commercial business in Homestead proper.[41]

The Company was now using the Borough in a different way than in the past. Before the strike, Homestead had grown helter-skelter to accommodate the labor force and meet the needs of workers while they were away from the plant. Shortly thereafter, the Borough commercial interests and city authorities realized they would be replaced by the Company if they did not efficiently perform their respective roles of serving the population and shifting the costs of city government to Homesteaders at large. After nearly six months of slack business and hard times, Homestead's merchants and shop owners were anxious to reassume the role of providing goods and services. As a demonstration of their new importance, the city authorities also renewed their vigor, replacing Burgess McLuckie as the representative of labor and making plans to put Borough finances on a sound basis.

The age of corporate liberalism dawned. The Company now recognized that it would have to provide, finance and lobby for essential infrastructure to sustain and reproduce a healthy, reliable work force. In the past, the Borough had clearly lacked sufficient funds to provide a sewer system adequate to city health needs. Homestead real estate agents and developers had too often concentrated on profitable deals, not on housing the work force adequately and securely with clean water and the other essentials. So long as a worker could not be housed properly, fed adequately, given safe water to drink and garbage and sewer services, so long as epidemics and sickness endangered the functioning of family life, the worker could not be expected to fulfill his potential for labor.

The logic of lowering the cost of labor's sustenance and raising its energy available for production seemed irrefutable to Carnegie managers. Thus were the New Town's 120 houses sited near the Works on a hill, so that the workers living there would be closer to the Works than those residing in Homestead proper. For the lucky potential residents, their accommodations certainly seemed better than many workers' homes in Homestead, with basement furnaces, hot and cold water, electricity, tasteful wallpaper and paint. The houses were also to be connected to the Borough's projected sewer expansion as soon as the city voted the money. And the Company planned to pay its part of the expense for connections provided the city built the lines. The Company meanwhile readied a grid of utilities for its Homestead properties and adjoining Company land.[42] As landlord, Carnegie's goal was thus to provide inexpensive housing in his hotel and boarding house, more expensive family facilities at the New Town,[43] and planned economical future rental homes as they were called for by Carnegie employees.[44]

Though he continued to dominate, Carnegie's full-blown scheme for a company town with living quarters and a single company store had limited success. He remained in charge of finance so far as steel-making and steel distribution were concerned. For in this sphere he efficiently integrated finance, production and distribution. Yet, in personal goods and services, in the million of petty transactions now needed to keep Homestead's thousands of households going, the local merchants and producers gained a victory over Carnegie and his company store. They were more sensitive to desires for variety, could make fine-tuned credit arrangements, and exerted personal magnetism on their customers. Carnegie's general store failed to keep pace and began selling out its stock at the close of November of '93.[45]

Homestead and Munhall merchants had joined the patronizing public in ensuring the death of competition from the great company store,[46] the *Local News* declaring that "such a store would never have been popular here," so that now Homestead would be "free to go ahead with business in the usual way."[47] Still, the Company store expired slowly in the hands of profit-seeking franchise holders, changing ownership a number of times, absorbing the economic shocks, and saving Carnegie from any loss.[48]

The contest with Carnegie's subalterns also provided a powerful lesson and, as a means of self-protection against any future threat to their domain, the merchants of Homestead pressured the City Council to pass laws requiring persons selling goods at public auctions and door-to-door on the streets to procure a relatively costly license, as the established Homestead merchants with steady business themselves had to do. Peddlers either obtained a license from the Burgess or paid a fine of $3 to $25 for each offense.[49]

Competition was also reduced, then eliminated, in the crucial areas of railroads, banking, and heavy-machine manufacturing. Though many still viewed smaller enterprises as a means of achieving economic democracy, was there even a glimmer of a way to actually attain the measures Homestead Rev. Galloway called for in the aftermath of the strike? "All unearned grants of land to railroad companies and other corporations should be reclaimed," he fulminated before his congregation. "All foreign capitalists should be prohibited from buying up these tracts of land in our country. These lands should be reserved for the poor of our own growing population, and the poor should not be brought into competition with the millionaires when they attempt to purchase a home. We need more small industries," he concluded, "not a centralization of wealth, but an equal distribution of it."[50]

Centralization was clearly the dominating motion of the times, and Homestead felt the grip of large financial institutions agglomerating local money pools around the Valley. Brilliant schemes put forward by merchants such as Pittsburgh's Solomon & Ruben sought to collect in a common fund the savings of the work force, to bring together $180,000 "in dollars for the poor man's dimes."[51] Here, too, the Carnegie and other steel works were central to the plans of the financiers, as the wages workers earned remained the life blood of the area.

A new center of power had been established: Banks collected the deposits of the community and became principal sources for business and real estate loans. They threw themselves into real estate and retail activity, enterprises both auxiliary to and benefiting the Carnegie mills.

A few financial institutions gradually took charge. The First National Bank of Homestead, having trimmed its sails to fit the economic storms loosened from Carnegie's mill, survived because it played the old work force against the new one, providing basic liquidity by discouraging withdrawals when the lockout began, then taking on the banking business of the Militia and the new, nonunion workers[52] – rolling up $194,000 in checking and time deposits on $63,000 in equity. Republicans Dr. George Gladden and Jacob Trautman were respectively bank president and vice president, both also part of the nine-man Republican board of directors comprised of Homestead's wealthy elite – including in the neat little circle, John F. Cox, D. R. McClure, Louis Rott, Thomas L. Parry, Joseph A. West and J. B. Neel.[53] Competitively, the Braddock National Bank had also accumulated Homestead savings, remaining a financial cut above, having nearly "a round millions of dollars on deposit," as the *News* reported.[54]

A Homestead financial dynasty was in the making. The First National had strengthened its position by interlocking directors with the only other important local financial institution, the Homestead Building and Loan Association, primarily involved in financing the city's real estate improvements. Five Republican gentlemen were directors and officers in both the Association and the Bank;[55] and the directors also linked

The next step in building this financial center was the rapid conversion of the old Hays Estate into a manufacturing center for the Baldwin Locomotive Works and several smaller industries, linked to the Borough by a railroad line. Joseph A. West of the First National negotiated a financial deal with moguls directing the Valley's railroads. The *News* leaked the December 28th clandestine business meetings of prominent railroad superintendents D. M. Watt, Robert Pitcairn, D. P. Corwin, and others, with Messrs. Joseph A. West, George W. Giles, Fred Schuchman, Reed Kennedy, Thomas B. Bridges, and Jacob Gillilland – all prominent Homestead bankers, businessmen, and proprietors. Together with the railway men, they surveyed the ground lying between Hays Station and the Homestead Borough Line, quietly planning to take by eminent domain a railroad right-of-way across the Hays Estate so as to link a new branch of the Pittsburgh-centered Pennsylvania railroad system from Six Mile Ferry to Homestead proper.[56]

The whole scheme seemed essential if Homestead commerce was to flourish, the Steel Works were to remain the center of employment, new industries were to be attracted, and the Borough's financiers were to prosper.[57] The new railroad project promised to stimulate the rapid development of the entire river front between Homestead and Pittsburgh, laying the foundation for a great boom in the area.[58] In January of '93, other investors also brought their resources together using the Hays Estate as a center, planning a brick works using modern handling machinery, a distribution system to market the bricks to Carnegie's New Town and in the Borough, and a real estate business.[59]

Smaller capitals barely survived the renewed struggle if at all, while to accommodate manufacturing firms located on the Hays Estate, a fight between the railroads ensued, bringing the Pennsylvania Railroad Company and the Monongahela and Street Run Railroad Company into a dominating position linking the Hays property and the Borough of Homestead.[60]

City residents could listen passively to the media spout the wonders of the financial legerdemain and the locally-funded Homestead advantages of these projects.[61] But there was an ominous aspect to the financial capital and unlimited economic power from which city residents were protected only by the Borough's austere-sounding ordinances compelling the railroads and street railway companies to construct, care for and secure its tracks within the city limits.[62] In the case of railroad negligence costing life or limb, it would take a survivor's pain, time and money to successfully pursue a fair settlement through the courts.[63]

Yet, one step simply followed another in creating the new economic complex enveloping Homestead. On December 30th, Joseph A. West of the First National Bank and former steel worker Reid Kennedy became partners as Kennedy & Reid to conduct a real estate and coal enterprise.[64] "Opportunities will be abundant," the *Local News* foretold, "but competition will be very great and therefore only the enterprising and far seeing ones will profit largely by the prevailing prosperity."[65]

A sign of the boom was a larger train station built to accommodate increasing numbers of passengers to and from Homestead.[66] Rail business also jumped as the great iron industries of Pennsylvania and Ohio boomed along with the demand for structural steel.[67] The whole region now noted the quickened pulse of the Homestead mills, and the *News* again published the Works' twice-a-month pay day schedule covering '93, giving an economic score card of these important days to debtors and creditors alike.[68]

With the railroads likely to provide a key industry in the Monongahela Valley in the '90s, the financial interests meanwhile scrambled to control regional politics and the component forces – iron and coal, steel, rail and other complementary industries. The McMahon Farms stretching 700 acres along the river were soon turned into a rail mill employing 800,[69] close enough to affect the noise and pollution levels of Homestead and the rest of the Valley.

After the intertwining of railroads, the Carnegies and their retinue extended control over the conditions essential for production and commerce, so much so that now they aimed to politically consolidate the industrially-centered population in Allegheny and contiguous counties.[70]

But barriers were quickly erected. For fearing the Carnegies and the railroads in Allegheny County, other counties resisted political mergers which would subserviate their local politics, the staid *Monongahela Republican* warning loudly that – "It is a rotten, corrupt, ring controlled, boss-governed county, where the worst possible political methods prevail."[71] But though each county strove to become the political center of the emergent industrial district,[72] Carnegie quietly prepared to take the greatest benefits – connecting rail lines from his Homestead Works to the convergence of the great railroads at Munhall Road,[73] thereby creating a link to connect, service and control the entire Valley and its economic center at Pittsburgh.

In the shadows of these overarching schemes, Homestead merchants enjoyed immediate benefit when the Steel Works started up again, the first full pay day on November 25th awarding the 2,500 new workers buying power. "It is expected that a larger portion of the money will be expended in the stores of Homestead than at any time since the work started up," the *Local News* announced, "and the merchants having pretty generally increased their stock of goods, will do a rushing business."[74] One merchant thus experienced "a regular old time pay day trade."[75] Even local laundry-man Bagley won Carnegie contracts to wash the 4,000 scab-used blankets, bringing more revenue into the city.[76] Outside peddlers now sought to enter the emerging market; five paid license fees to the ordinance officer on November 28th.[77] Revenue was further circulated by merchants placing ads and notices in the *News*, both new and old advertisers reaching out to a thousand potential reading customers. The *News* brightened its pages and coffers with new ads from McCune's dry goods; Wagner's clothing house, Erbeck Bros.'s hardware, Jos. A. West's coal and brick, and 0'Donnell's millinery joining the pre-strike advertising merchants, the Youngs and McClures – all being encouraged by the *News* to "spend not less than one fifth of their total expenses in newspaper advertising . . . in legitimate newspapers that go into the very house you desire to reach."[78]

Once again the wheels – and horses – of commerce moved, bringing on old conundrums. Officers now enforced ordinances demanding highways be kept clear, and citizens poured complaints on Officer Gessner about wagons and other vehicles blocking the streets and alleys. "This practice is in violation of the street obstruction ordinance," echoed the *Local News* in warning as it had before the lock-out, "and notice is hereby given that unless the practice is promptly discontinued, arrests will be made and fine imposed."[79]

Homestead commerce and city policies had clearly resumed their pre-strike postures, and production was again in swing.

## The Dynamics of Urban Expansion

In the past, Homestead's commerce had been a matter for individual storekeepers and service mechanics. They came to the city to serve and service mill workers and the people on the fringes of the rural Borough. Trade practices and hours had varied to meet a wide assortment of demands. But after the strike and lockout, 24-hour mill production using continuous-processing techniques demanded an increasingly regularized work force whose lives were nearly as mechanical and efficient as the changing industrial and commercial infrastructure.

The local newspapers recognized the needs for uniformity in the future. "Would it not be a good thing to revive the Homestead Board of Trade?" the *News* asked, answering, "It might be made a very useful institution in the near future. Very few towns with pretensions to enterprise are without a Board of Trade or chamber of commerce organization."[80]

By the end of January, business interests also showed a renewed interest in the Merchants Protective Association, their members voting expansion to represent, organize and secure benefits for all business interests of the town.[81]

The city's infrastructure was also expanding. Before the strike, the city had only partly regularized the cost of labor maintenance. It had improved streets somewhat – though even after the lockout dirt and clay under foot became a mass of slithering mud following the winter thaw.[82] The Water Works connected pipes to any household that paid installation and the monthly service charge, while the Electric Light Plant provided power, subject to the accidents of the day and difficulties with its latest equipment.[83] The routines of production and home life had come to depend on these conveniences, so much so that when the electricity failed, it disrupted the rhythms of the town.[84]

Carnegie desires for unimpeded production after the strike merged with these city developments and needs. Thus when the Company extended wires to more buildings at the Works, increasing its purchase of power from the Electric Light Company, utility stockholders matched demand with greater investments in new, improved technology.[85]

The new work patterns at Homestead also required efficient transfer of messages, people, materials and goods. In November, for the first time ever, the Homestead Post Office published hours of mail pickup and delivery. The city's railroads printed expanded, detailed schedules, meeting burgher and business requirements, often upstaging and twisting personal schedules.[86]

Carnegie's needs also imposed costly requirements on Homestead, adding to popular burdens and taking away previous benefits.[87] The Independent and Volunteer Firemen, for instance, had hitherto devoted their unpaid labor to construct fire houses, create and supply library reading rooms, and make city properties and the mills safe. For their services they had received payment-in-kind from the franchised companies providing services to the city taxpayers. Yet the Electric Light Company monopoly, secured by a fat contract with the Borough for arc lights on the streets, had second thoughts about furnishing free electricity to the fire company and its volunteers. Though the Electric Company had originally agreed to light all Borough buildings free, it now wanted to charge the Engine Houses, claiming they were not public institutions since each held a charter from the city and owned half their own property.[88]

So it went, each advance in city services, rewarding Carnegie in part, imposing new urban requirements, and taking away some of the benefits Homesteaders once received.

**The Fiscal Crisis**

The costs of the new ordering of Homestead life were meanwhile transferred directly to the city population and indirectly to the taxpayers in the form of moneys spent by the Borough to enlarge and expand services and jails. These were necessary to cope with the increased population, the ongoing battle between union men and scabs, and the extension of the urban infrastructure.

The enlarged population had impacted neighborhoods, shortening the fuse sparked by local conflict, also enlarging the need for Homestead outlays on city services and imposing taxes to pay for full-time policemen and city employees – both legitimating authority and sustaining the community. The existing jails had quickly proved themselves too small for the increased number of prisoners taken into custody.[89] Meanwhile other cities exhibited superior facilities Homestead needed, envied and ultimately would have to build with city funding.[90] City water connections, now legally obligatory for each family's survival, had to be applied and paid for.[91] Fixing the streets also required individual outlays of labor and money,[92] as the Borough treasury was temporarily depleted to the point where the city could not finance immediate improvements.[93] Other utility costs pressed upwards; the Philadelphia-controlled Low Pressure Gas Company raised its post-strike rates to Homestead customers by 10 per cent,[94] keeping households centered on earning wages and other incomes to cover outlays. And the Borough, lacking funds to maintain the essential infrastructure of the town, soon came to rely on Carnegie and other businesses for special project outlays and dispensations.

The Borough's financial bankruptcy meanwhile led to the refusal of the Birmingham Electric Railway Company to lay lines in Homestead.[95] Carnegie seized the advantage, taking up the project to link Homestead to his Duquesne plant by a direct rail route,[96] while the Homestead and Highland Avenue Electric Street Railway secured the Borough right-of-way.[97] Nor did the city have moneys to subsidize rail transport to the Bryce Higbee Glass Works or other businesses, so that in an effort to keep them from moving to other cities offering superior transport facilities, the Homestead Board of Trade considered a private subscription for funding the project.[98]

The *News* advocated that the facilities should be freely available for use by "any manufacturing establishments that may locate within reach . . . ,"[99] and that private "franchise" should not be granted by the Borough.[100] But without funds, the Borough Council could not finance, but only hope to purchase the required facilities, including land for the new fire department buildings.[101]

Indeed, the old Borough government was now bankrupt, could not undertake any new projects, and was unable to continue old ones. "All the street workmen have been laid off, except the street commissioner and the board-walk repairman," the *News* grimly reported the city's fiscal crisis.[102]

Given the complexities and demands of its enlarging urban domain, hereafter there was no way but deficit-financing to sustain Homestead with its inventory of 7,896 people, 4 bands, 19 churches, 9 public halls, 8 Amalgamated lodge buildings, 55 fraternal centers, 7 financial institutions, 2 parochial schools, and 3 public schools run by 26 teachers, 3 principles and a single superintendent.[103] The people could certainly have more services and an efficient infrastructure – more schools, police and jails, an expanded fire department, more government officials and bureaucrats – but only if they paid in taxes and interest on bonded indebtedness today, and more taxes and interest on subscribed bond debts tomorrow.[104] For now a perennial fiscal crisis, a product of costly city services periodically overwhelming insufficient sources of revenue, reflected the structured class relations of Homestead's new stage of urban, proletarian life.[105]

## Class, Caste and Education

Just as earlier family relations and community life had been regularized by the dominant rhythms of the Carnegie mill in the city, the pre-strike hierarchies of class, caste and power were perpetuated and rigidified after the lockout.

Aiding these ends, Superintendent of Homestead schools, Professor J. C. Kendall, had reached a new level of importance in Pennsylvania's pyramid of education, being cited for outstanding work at the County Teachers Institute held in Pittsburgh in August, and winning acclaim for the quality of Homestead's educational system from authorities in Cleveland and New England.[106] Through his efforts, Homestead schools were maintained as a technical treadmill for the boys entering the mills and the working class girls entering marriage, as well as a preparatory center paving the way to college for the well-to-do. Thus in September of '92, while most male graduates had waited for the strike to end so they could enter the mills for their life-time work, Homestead Court Squire A. J. Kuhn helped send his niece, Miss Bertha, to the State Normal School at Westchester, Pennsylvania.[107] So, too, others from Homestead's young elite also sought higher reaches.

Despite a continuing division of future educational opportunities by class and caste, after the strike Superintendent Kendall and the city fathers upheld the school system as a model. The local media also asked parents and guardians to impress on their children "the importance of the opportunity and the value of the excellence of the public schools," adding that on September 6th: "Could the outside world look in upon Homestead . . . and see our twenty-eight graded schools and three fine ward buildings and sixteen hundred bright people and trained teachers, they would have a better opinion of Homestead than they have formed from the things they have been hearing and reading."[108]

Reaffirming the belief that knowledge itself bred upward mobility, Pennsylvania's state authorities also took measures to turn public education from a partial service into a compulsory one, providing free textbooks and enforcing strict discipline.

"Experience has shown that wherever free school books and supplies are furnished, the attendance has increased, the system thus serving the good purposes of compulsory education and avoiding its objectionable features," Governor Pattison noted in early '93. Recommending such free textbooks everywhere, he added: "It will enable each district to purchase books best suited to the community," contrasting this with the then existing system under which the purchase of school books involved an expenditure of at least $800,000 to parents, some of whom could not sustain the burden.

"By the use of free textbooks," Pattison concluded, "every pupil, as soon as his name is known to the teacher, would be supplied with all the necessary books, and many of the excuses now advanced for the nonattendance of children in the public schools would disappear."[109]

Homestead education also held a wider scope. Libraries were initially established in the fire company buildings, not so much to win the people away from feelings of deprivation in the working class town, but to provide a genuine service by the volunteers staffing these companies.

The public collection of the works of George Elliott (aka Mary Ann Evans) and other serious books and interesting magazines and papers was begun for parlors of the three fire companies,[110] and Fire Chief Bryce and the heads of the companies supported the action to hold the fire department together and make it pleasant for members.[111]

Carnegie, shaping the tenor of the new proletarianization at Homestead and elsewhere, then built upon these initial efforts with grandiose libraries bearing his name and stocked with volumes stressing metallurgy, chemistry and technical sciences.

Such technical knowledge accompanied a renewed demand for obedience to orders and promptness of execution; and the structural liturgy of the classroom locked Homestead youth into the doctrines of the past. The *Local News* stressed:

> The training of all the children of the country in schools that teach the same elements of knowledge and inspire the same principles of patriotism and morality tends mightily to the development and fostering of great principles and doctrines that are of universal application and benefit. There is but one hope for the future of the Republic, but upon that we may confidently rest, namely through the medium of the public schools, the opinions of the rising generation may be so moulded that correct ideas of civil government may prevail a quarter of a century hence. The duty of this generation is to properly teach the youth of this land.[112]

Despite the strike and the brutal lessons learned by the Borough population, the schools continued their molding of youth into obedient servants of national patriotism, religion and the business of Homestead. Homestead schools, like those around the nation, pushed the belief that all students started at the same point, and only the best would attain the upper reaches of success. This theme, with its nationalist trappings of uniformity, was carried forward at Homestead in October '92,[113] wrapped in staccato euphemisms, trite shibboleths, and the myth of America's founding bound up with biblical lore. "The program will be nearly as follows," the *Local News* delineated the Columbus Day celebrations, October 21st:

> Schools to assemble at 9 a.m. Short stories about Columbus
> to be read or spoken.
> Displaying the flag with demonstrations and music.
> Prayer or reading of Scripture.
> Songs.
> Short addresses.[114]

Nor were teachers speaking to their peers immune from the pressures of conformity. At the Braddock new school building dedication in September, '92, invited teachers and directors of Homestead schools found a familiar framework waiting: "Elegant copies of the Bible and beautiful silk flags will be presented," the *News* foretold the dual exercise of faith.[115]

Homestead was also anxious that such education be universal. "Let all the children of proper school age at once be enrolled as pupils," the *Local News* insisted. "The present attendance makes a favorable showing for the town but there must yet be scores of boys and girls, especially among the older ones, not in attendance. Such an opportunity as our excellent schools afford should not be neglected."[116]

Many strikers sought to place their children in religious schools to keep them away from the children of the scabs and newly-arrived immigrant laborers. Yet, despite the turmoil and transfers from public to parochial schools, the *News* reported no "falling off on account of the labor agitation."[117] In fact, due to the enlarged family population of workers and new arrivals seeking jobs, the combined enrollment in both school systems outpaced the pre-strike roll call.[118]

The population remained segregated by class and caste, though, those directing the teaching and publishing of ideas themselves part of or beholden to the upper reaches of the social pyramid. They associated with their own kind, married within that circle,[119] yet exhorted those less fortunate to do hard work at small wages.[120] Conforming students would thereby advance under the eyes of an impartial elite, the media admonishing:

> Do the best you can where you are. Fill the place you are in more than full . . . .
>
> The world generally gives its admiration not to the man who does what nobody else ever attempts to do, but to the man who does best what the multitudes do well.
>
> The fortunes are not all made, neither are the good situations all filled. During the next ten years there will be five million first-class openings

for the young people who are competent to fill them. Be ready.[121]

Despite such promise of upward mobility for excellence in one's pre-assigned place, the elite of Homestead followed other advice to fill their aspirations, going beyond provincial limitations to take advantages where they found them. The McClures, typical of the local elite – doctors, landowners, businessmen and lawyers – did not rely on merely local training or faithful "servanthood" for their progeny in their reach for success. Abbie R. McClure thus took the two year course in architecture at Pennsylvania University at Philadelphia, with a "promising career before him."[122] Another of their upwardly mobile caste, Miss M. Hilda Kuhn, daughter of Magistrate Kuhn, kept up post-Homestead graduation studies at the Westchester State Normal School, now and then visiting her parents and friends at Homestead during holiday vacations.[123] The Barton family, business people known in the community, similarly sent their son Claude to the not inexpensive Shippensburgh State Normal School.[124] And other old, well-known families in Homestead similarly sent their children away to school – facts well publicized in Homestead – letting those who would never get to college know there were those from higher social posts who would.

Meanwhile the taxpayers gracefully paid for functional education of Homestead youth, as they had before the strike. Rated at about $14 a year to educate each pupil, Homestead's September '92 enrollment of 1,205 students cost state $2,400 and local citizens another $14,480.[125] Of this outlay Carnegie paid practically nothing save the small tax for the educational budget in Mifflin Township where the Works was sited, though in Homestead he had his entire male work force trained in the skills of mill production. Little had changed in terms of financing education or the purposes it served.

**At Risk in the Mill**

Nor had much changed in the mills, so far as burdens and dangers were concerned. Men familiar with the industry knew well that the machinery itself frequently malfunctioned and the workers' surroundings were often as likely as the individual to create dangers.

A misstep often caused an accident, usually compounded, as in January when Andrew Leibold at the McKeesport's National Rolling Mill fell from a car and another car passed over him, badly mangling a hip and an arm. An emergency amputation by Dr. H. Black resulted in the not unusual death by operation.[126]

In a typical week, Nick Rusher of Whitaker had his hand crushed while operating a crane in Homestead's converting mill on January 12th of '93.[127] Five days later, Jacob Rohe of Tenth Avenue had his left leg broken while working in the converting mill. The next day, a Swedish worker, name unknown, also in the converting mill, had his left arm crushed-off near the elbow. "The sight was very sickening and, it is said that several parties nearly fainted," the *News* reported of his fellow workers.[128]

All these were considered worker-caused "accidents" for which Carnegie had no responsibility so long as his machinery was in good working order.[129] "Yesterday about noon while Michael O'Tool was coupling two buggies near the converting mill," one factory report reads, "he slipped and when the buggies came together his head was crushed." The effective cause-effect relation made this an "accident" in legal terms:

> He was employed as a hookon, or brakeman, and had given the engineer of the dinky the signal to back up. It is not clearly known just how he happened to be caught, but it is supposed that when he attempted to make the coupling his foot slipped. His head was caught between the moulds and was smashed flat.
>
> The unfortunate man was removed to the office and a physician summoned, but it was evident that he could not live. He suffered in a horrible manner for about an hour, when he died. . . . The young man was about 17 years old and well known in town.

Magistrate Oeffner's inquest resulted in a jury verdict of "accidental death," witnesses testifying that, indeed, the young man attempted to make the coupling on a curve and that, when the moulds came together, he was irretrievably caught between them.[130]

Accidents accumulated without end in the post-strike period. "A man whose name could not be learned," began another story relevant to every Homestead mill hand, "had his foot caught in the cogging mill on Tuesday night and it was badly crushed. It is said, the foot had to be amputated."[131] True, a foot might be saved, a heel torn off,[132] but the accounts had become commonplace, almost casual in tone: "Robert Abbott a workman in the converting mill was badly burned yesterday afternoon but his injuries are not of a serious nature."[133]

Amputations were so frequent that parts were discarded in garbage bins, Black "Zack" Blake, who hauled waste from houses in town and at the Steel Works, shoveling up a badly crushed human arm, which, with his other garbage, he dumped in the river in his state of excitement over his find.[134] There were plenty of mangled limbs to be found, the shattered leg of 17-year old Henry Stahlhut being amputated at the mill hospital on January 26th after he fell under the engine of the mill transport department and had his leg badly crushed.[135]

John Peacock had his fingers smashed in the 40-inch mill on January 28th.[136] On the same day, James Morris had the toes of his right foot taken off in the open hearth mill No. 2, when the chain about the runner near the pit fell on his foot.[137] Countless Hungarian workers were also on the injury list, one laborer, John Hovedo, living on Eighth Avenue, having a finger of his right hand severed in the 35-inch mill on January 31st.[138]

There was an industrial war in progress and reform to patch up the work force was becoming essential to keep labor compliantly efficient.

To deal with comparable injuries before the strike, the workers had created their own insurance benefit programs. Yet, many injuries placed tremendous burdens on the families involved and led to suits by the workers against the employers. Though the Fellow-Servant law absolved the employers from liability if the injury was proved the fault of the injured worker or his co-workers using employer equipment that was not defective, the employers were anxious to see a discontinuance of these suits by offering nominal benefits for injured workers or the survivors of workers killed on the job.[139]

Towards this end, Carnegie initiated a new scheme, hoping to pressure the workers themselves to finance those injured or killed by withholding a portion of their pay each week. Notices were posted in all the Carnegie mills, at Beaver Falls for example, that 5 cents per week would be deducted from the pay of each employee, the proceeds to go into a fund to provide relief benefits for employees injured while at work in the mills. Thereby, any employee injured would have his doctor bills paid and would be given $3 for the first week he was unable to work because of the injury, and $5 for each week thereafter, for a term not exceeding 13 weeks. In case of death by accident while at work, the funeral expenses were also to be paid from this fund.[140]

Here then was Carnegie's new scheme to regularize the work force by providing for those injured who could not immediately return to work; to cut benefits to those who could return; and to remove momentary strain on families which had lost their principal wage earner. By these means as well as the building of public libraries nation-wide, Carnegie hoped to win back workers to a belief in his humanitarian and generous nature, to have them look upon him as a purveyor, if not promoter, of domestic tranquillity in the mill towns as a medium of peace in both steel markets and international affairs.[141]

## Notes

[1] Andrew Carnegie, "Wealth," *North American Review*, June 1889.
[2] *The local News*, October 29, 1892, 1. See also Carnegie, *Autobiography, passim*; Fitch, *The Steel Workers*, 213-216, 219-220;
[3] *The local News*, August 20, 1892, 2.
[4] *Ibid.*, September 17, 1892, 2; *The West Elizabeth Star*, November 22, 1892, 1.
[5] *The local News*, August 20, 1892, 2.
[5] *Ibid.*, September 24, 1892, 2.
[6] *Ibid.*, August 20, 1892, 2.
[7] *Ibid.*, August 27, 1892, 3.
[8] *Ibid.*
[9] *Ibid.*, August 20, 1892, 2.
[10] *Ibid.*, September 17, 1892, 1.
[11] *Ibid.* For the general process and theoretical premise, see William Cronon, *Nature's Metropolis: Chicago and the Great West* (New York: W. W. Norton, 1991),
[12] *The Local News*, September 17, 1892, 2. For the influences of financial capital, see J. K. Winkler, *Incredible Carnegie*; Harold C. Livesay, *Andrew Carnegie and the Rise of Big Business*; Jean Strouse, *Morgan: American Financier* (New York: Random House, 1999).
[13] *The Local News*, September 24, 1892, 2; see also *Ibid.*, August-November, 1892, *passim*.
[14] *Ibid.*, October 15, 1892.
[15] *Ibid.*
[16] *Ibid.*
[17] *Ibid.*
[18] *Ibid.*, January 20, 1893, 1.
[19] *Ibid.*, January 22, 1893, 1.
[20] *Ibid.*, January 30, 1893, 1.
[21] *Ibid.*, October 15, 1892, 2.
[22] *Ibid.*, November 22, 1892, 4.
[23] *Ibid.*
[24] *Ibid.*, November 22, 1892, 1.
[25] *Ibid.*
[26] *Ibid.*, 4.
[27] *Ibid.*, 1, 2, and January 16, 1893, 1; *The Sun* (Baltimore, Maryland), November 24, 1892; Stowell, 278, 285; *The Tribune* (New York, N. Y.), November 14, 1892.
[28] *The Local News*, November 30, 1892, 4.
[29] *Ibid.*, January 10, 1893, 4, and January 13, 1893, 3.
[30] *Ibid.*, January 27, 1892, 1.
[31] *Ibid.*, 4.
[32] *Ibid.*, February 1, 1893, 4.

[33] *Ibid.*, January 3, 1893, 1.
[34] *Ibid.*, December 31, 1893, 4.
[35] *Ibid.*
[36] Adam Smith, *An Inquiry Into the Nature and Causes of The Wealth of Nations*, (New York: The Modern Library, 1937), Book IV.
[37] *The Local News*, January 3, 1893, 4.
[38] *Ibid.*
[39] *Ibid.*, January 4, 1893, 4.
[40] *Ibid.*, January 30, 1893, 1.
[41] *Ibid.*, January 28, 1893, 1.
[42] *Ibid.*
[43] *Ibid.*
[44] *Ibid.*
[45] *Ibid.*, November 29, 1892, 4.
[46] *Ibid.*
[47] *Ibid.*, November 39, 1892, 4.
[48] *Ibid.*, January 6, 1893, 4.
[49] *Ibid.*, October 22, 1892, 1, 2.
[50] *Ibid.*, July 23, 1892, 4.
[51] *Ibid.*, January 19, 1893, 1.
[52] *Ibid.*, December 16, 1892, 4.
[53] *Ibid.*, December 19, 1892, 1.
[54] *Ibid.*, December 16, 1892, 1.
[55] *Ibid.*, December 28, 1892, 1.
[56] *Ibid.*
[57] *Ibid.*, 4.
[58] *Ibid.*
[59] *Ibid.*, January 28, 1893, 1.
[60] *Ibid.*, December 28, 1892, 4.
[61] *Ibid.*
[62] *Ibid.*
[63] Childs v. Pennsylvania R. R. Co., *Pittsburgh Legal Journal*, No. 50 (July 13, 1892), 466-468.
[64] *The Local News*, December 30, 1892, 4.
[65] *Ibid.*
[66] *Ibid.*
[67] *Ibid.*, December 31, 1892, 3.
[68] *Ibid.*, 4.
[69] *Ibid.*, January 4, 1893, 1.
[70] *Ibid.*, 4.
[71] *Ibid.*, January 10, 1893, 1.
[72] *Ibid.*, 4.
[73] *Ibid.*, January 5, 1893, 1.
[74] *Ibid.*. November 26, 1892, 4.
[75] *Ibid.*, November 28, 1892, 4.

[76] *Ibid.*
[77] *Ibid.*
[78] *Ibid.*, 1, citing *The Elizabeth Star*.
[79] *Ibid.*, November 30, 1892, 1.
[80] *Ibid.*, January 18, 1893, 1.
[81] *Ibid.*, January 26, 1893, 1.
[82] *Ibid.*, January 14, 1893, 4.
[83] *Ibid.*, January 23, 1893, 1.
[84] *Ibid.*, 4.
[85] *Ibid.*, January 24, 1893, 1.
[86] *Ibid.*, November 22, 1892, 4.
[87] *Ibid.*, October 15, 1892, 2.
[88] *Ibid.*, November 19, 1892, 4.
[89] *Ibid.*, October 15, 1892, 2.
[90] *Ibid.*
[91] *Ibid.*, October 1, 1892, 2.
[92] *Ibid.*, November 5, 1892, 2.
[93] *Ibid.*
[94] *Ibid.*, October 22, 1892, 1.
[95] *Ibid.*, October 15, 1892, 1; October 22, 1892, 2.
[96] *Ibid.*, October 22, 1892, 2.
[97] *Ibid.*, January 17, 1893, 1.
[98] *Ibid.*, October 15, 1892, 1.
[99] *Ibid.*
[100] *Ibid.*, 2.
[101] *Ibid.*
[102] *Ibid.*, October 22, 1892, 2; October 24, 1892, 1.
[103] *Ibid.*, August 27, 1892, 1.
[104] *Ibid.*, August 27, 1892, 3; September 17, 1892, 2; September 24, 1892, 1; October 29, 1892, 2; January 30, 1893, 4.
[105] *Ibid.*, November 18, 1892, 4; November 22, 1892, 1, 4; November 23, 1892, 4; November 25, 1892, 4; November 26, 1892, 3, 4; December 2, 1892, 4; January 10, 1893, 1, 4; January 18, 1893, 4.
[106] *Ibid.*, August 27, 1892, 3.
[107] *Ibid.*
[108] *Ibid.*, September 3, 1892, 1.
[109] *Ibid.*, January 26, 1893, 1.
[110] *Ibid.*, January 16, 1893, 4.
[111] *Ibid.*, January 17, 1893, 1.
[112] *Ibid.*, September 3, 1892, 2.
[113] *Ibid.*, October 8, 1892, 2.
[114] *Ibid.*
[115] *Ibid.*, September 10, 1892, 1.
[116] *Ibid.*, October 29, 1892, 2.
[117] *Ibid.*, September 10, 1892, 2.

---

[118] *Ibid.*
[119] *Ibid.*
[120] *Ibid.*
[121] *Ibid.*, October 15, 1892, 1.
[122] *Ibid.*, November 19, 1892, 4.
[123] *Ibid.*, January 4, 1893, 4.
[124] *Ibid.*
[125] *Ibid.*, October 29, 1892, 2.
[126] *Ibid.*, January 13, 1893, 1.
[127] *Ibid.*, January 19, 1893, 1.
[128] *Ibid.*
[129] See Tunney v. Carnegie Bros. & Co., Limited, *Pittsburgh Legal Journal*, No. 24, January 13, 1892, 219-221.
[130] *Ibid.*, January 26, 1893, 1.
[131] *Ibid.*, 4.
[132] *Ibid.*, January 28, 1893, 1.
[133] *Ibid.*, January 27, 1893, 1.
[134] *Ibid.*, January 26, 1893, 1.
[135] *Ibid.*, January 28, 1893, 4.
[136] *Ibid.*, January 31, 1893, 4.
[137] *Ibid.*
[138] *Ibid.*, February 1, 1893, 1.
[139] *Ibid.*, January 25, 1893, 1.
[140] See Tunney v. Carnegie Bros. & Co., Limited, *Pittsburgh Legal Journal*, No. 24, January 13, 1892, 219-221.
[141] Andrew Carnegie, *Autobiography* , *passim.*

**Sheriff McCleary Among the Workmen**
**(Credit: *The New York World*)**

# Chapter 13

# Homestead as Metaphor

Upon the sacredness of property civilization itself depends – the right of the laborer to his hundred dollars in the savings bank, and equally the legal right of the millionaire to his millions.

– Andrew Carnegie, *The North American Review*, June 1889

United States steel makers suffer from destructive price wars and excess capacity. The companies are susceptible to economic downturns because their profit margins are being squeezed.

The solution is clear: the merger of several companies so they can combine operations while cutting costs. The resulting steel behemoth would be a competitive force in world markets.

– President Charles Schwab of Carnegie Steel, the nation's leading mill, envisioning the dawn of the steel industry in the 20$^{th}$ century, speaking as the guest of honor to business executives gathered in New York City's exclusive University Club, December 1900

Pierpont Morgan is apparently trying to swallow the Sun.

– Henry Adams, amazed at the great banker's 1901 acceptance of Carnegie's offer for his steel empire, inflated at a $1.4 billion price covering equity of $480 million, debts of $920 million, and fees[1]

I met more socialists in Homestead and Munhall than elsewhere – enough to lead one to expect to find a local organization there. There may be as many in other mill towns, yet in the four more important steel centers, McKeesport, Duquesne, Braddock and Homestead, there was not in 1907 a trace of a socialist local. I wondered at this until a Homestead steel worker told me that they simply had not dared to have a local within the borough. To be known as a socialist the men thought would be to court discharge, and most of the Homestead socialists held membership in a Pittsburgh local. I have since learned that in the campaign of 1908 a local was organized in Homestead.

These skilled workers['] . . . extreme caution, the constant state of apprehension in which they live, can have but one cause. It is the burnt child that dreads the fire. It would hardly be expected that the loss of one strike, disastrous though it was, should have so disheartened the workers that they would never attempt to revive collective bargaining. It has been through a series of such attempts that the men have learned to respect the vigilance and power of their employers, and they have learned the cost of defiance.

– Investigator John A. Fitch, 1907[2]

Now is the time to break down the open-shop slavery that has cursed the steel industry ever since the defeat of the heroic Homestead strikers in 1892. Now is the opportunity to build the Amalgamated Association of Iron, Steel and Tin Workers into a great union, powerful enough to bring a happier life to the steel workers and their families.

– William Z. Foster, leader of the 1919 Steel Strike[3]

Though this union we shall win higher wages, shorter hours, and a better standard of living. We shall win leisure for ourselves, and opportunity for our children. Together with our union brothers in other industries, we shall abolish industrial despotism. We shall make real the dream of the pioneers who pictured America as a land where all might live in comfort and happiness. In support of this declaration, we mutually pledge to each other our steadfast purpose as union men, our honor and our very lives.

– Declaration of Independence, Homestead, June 1936[4]

This is an industry that has significant national security implications. It's important from the whole basic concept of maintaining manufacturing in this country. It's important to have a vibrant domestic source of steel. Allowing the industry to be destroyed would put this country on a path to over-dependence, similar to oil, on foreign sources.

– Pennsylvania Republican Senator Rick Santorum in December 2001, echoing H.C. Frick's 1892 point of view.[5]

Homestead is an America's epic – a model and a metaphor of the general development of class relations at a point when monopoly capitalism was strengthening itself in the steel industry and other sectors of the national economy.

Great capitalists held court in the marketplace. And Andrew Carnegie's holdings were unequalled, a horizontal constellation of mills, vertically integrated with an empire of suppliers and customers. Carnegie's massive accumulation of capital and takeovers of other steel companies equipped each of his mills with the latest blast furnaces and giant rolling plants. So too Carnegie held a significant financial stake in Henry Clay Frick's immense deposits of western Pennsylvania coking coal, as well as a substantial interest in the rich, Lake Superior iron-ore area. And Carnegie bought outright his own fleet of carriers to bring the ore across the Great Lakes to Lake Erie ports, linked by rail to his mills.

These became critical pivots to outflank competitor mills, which either had too few supplies for expansion or were forced to pay a higher price for their coking coal and ore. Personal ties with railroad executives also brought Carnegie large customers, supplying their lines with rails, bridge steel and other structural shapes.[6] And the scale and capability of his mills ensured lucrative government contacts for warships; influence over lesser mill owners; and effective lobbying power with Congress over tariffs, as well as with Pennsylvania politicians over the deployment of sheriffs and Militia.

With such conglomerated resources and political puissance overarching industrial towns like Homestead, the workplace and the surrounding community together became an expression of a new stage of proletarianization, though the work force was still unable to effectively use these focal points to unify itself against those in control of integrated capital, company towns, and far-away governments.

Labor's weakness also stemmed in part from the way workers organized themselves. For unions in the late 19th century usually structured themselves by separate crafts in order to command sufficient strength for protecting their members' wages, controls over production, and veto power over the employers' attempted introduction of new work methods or technologies.

These craft unions had emerged at a time when hand tools were widely in use in the mills, and inanimate steam and other power had not yet displaced the need for the workers' exertion of great physical strength. Given such power and specialized, manual techniques were demanded by and ran parallel with the technical structure and production methods of the nation's factories and mills. And thus skilled workers organized themselves along these specialized, narrow craft lines.

Periodic strife revealed the downside for such unions in dealing with a common employer of several different craft specialties as well as those without skill. A single craft union might call a strike demanding better wages or improved working conditions for its members; but if the output of this craft was the raw material for the next production stage, the strike often shut down the entire production line, idling and angering members of other craft unions who would gain nothing from the strike. The union structure and periodic demands of each craft thus tended to undermine any lasting solidarity that might have developed between the various skilled crafts, as well as between any union of craft workers and the unskilled.[7]

After separating the members of various crafts into different unions and union subdivisions, then, each set of skilled workers pursued their own goals. Each craft kept their members' wages high by limiting its number to an employer's specific requirements; and this exclusion of the unskilled was achieved by craftsmen refusing to share their knowledge with non-union members.[8]

This exclusiveness stemmed from the underlying logic of craft workers' control throughout the 19th century: to protect and keep to themselves the know-how and training associated with each craft specialty. By linking this knowledge of the skilled with their own trade unions, they developed and codified a body of instruction and work rules that excluded the semi-skilled, unskilled workers and day laborers. Yet, locked into a written code of conduct excluding all others from their lodges, craft union members neglected to upgrade or extend their circle of relations or methods of organizing to other crafts, so that each union with its fractionalized lodges had less strength than if all were unified in a common movement with the same mission.[9]

Thus, at the very time when all the unions hoped to gain and maintain control over the immediate production process for the benefit of their own members, the craft unionists' principle of exclusivity and organizational separation from other workers in the same workplace made each union and each lodge more vulnerable to the power of a single owner of capital employing the various branches of skilled, semi-skilled and unskilled labor.[10]

At Homestead this narrow vision, practiced by skilled workers, was changed in part after 1889-90, when Carnegie forced labor's initial solidarity – solidarity in common exploitation - by extending the technical revolution into his mill operations calling for continuous, uninterrupted production. At first he needed the union workers' specialized knowledge to make the new system run smoothly. But as the new technical equipment embedded in its working mechanisms what were formerly the skills of the craft workers themselves, Carnegie effectively restructured the work process, demanding a new level of efficiency and labor productivity that threatened to break up the old craft specialties and their members' exclusive organizations and subdivisions. In turn, this pressured Homestead's Amalgamated to attempt to shield the skilled workers, though their demands were limited to protection of job places for their members alone; to standardization of craft functions to stop Frick's speedup; to established hours to regulate the work week; and to minimum wage rates to maintain living standards.[11]

Though collective union power was used tactically to win the solidarity of the heads of mill departments, the semi-skilled and day laborers in the course of the strike of '92, it did not last. With defeat, the workers were forced to deal with Carnegie as individuals selling their power to work; and the remnant locals of the union retreated to other mills in bargaining for the workers' collective labor-power, no longer dreaming of an influence over all production in the industry, much less control over their members' lives.

In pursuing these limited union demands both before and after the strike, Utopian goals were largely put aside.

There were obvious ideological limits to confronting the state and its authority structures through collective union action or the mobilization of the working class community. The path to secure broader, social demands were even less clearly defined, came from those outside the Amalgamated lodges, and called either for Utopian reforms within the wage and political system or for an Anarchist revolution which was based neither on Homestead workers' own organizations nor on their knowledge or capability to bring it about. Entering Homestead with such plans, Anarchists like Alexander Berkman "had waited for this sublime moment to serve the Cause to give his life for the people," as

Emma Goldman later put it.[12] But the cause was seen otherwise by Homesteaders and the Amalgamated, and even that was decisively lost to a better-organized, employer alliance, resulting in what was later castigated "open-shop slavery" by steel-organizer William Z. Foster.

The crushing of the strongest union in the nation followed the defeat at Homestead. Carnegie completely eliminated unionism at his Homestead Works. Similar measures were taken in almost all the steel companies in the Pittsburgh area. Blacklisting and firing of workers who spoke out at Homestead were nearly guaranteed, as Carnegie created an insidious espionage system, aided by Pinkerton agents, scabs and frightened or compliant former union men.[13] Then the panic of 1893 set in, unemployment escalated, and steel workers who would work for next to nothing were "a dime a dozen." It was in this *milieu* that the new stage of proletarianization clearly appeared at Homestead and within the steel industry at large.

**Proletarianization & Its Consequences**

The overall consequences were these: Without solidarity among all workers in the Carnegie mills, the emerging mass labor force was controlled on the job by Frick's policies, hemmed in by borough councils, and pacified by a cash and credit marketplace that provided subsistence for those who earned wages and could buy. So long as labor remained tied to the existing economic sphere, and workers' political desires could be confined to the financially-subsidized, two-party political system, Homestead's power elite could handily regulate the conditions of working class life. And, in fact, Homestead's advance in steel-making methods and technology took away skilled workers' controls over the work process, destabilized the previous mode of working class life, undercut living standards, and set the stage for a new civil and political arrangement, fueling the reorganization of class relationships.

Yet, somehow, human dignity survived. Despite all of Carnegie's efforts, large numbers of steel workers were not pacified and as early as 1893 began to clandestinely reorganize their union against what they viewed as "industrial despotism."

Their determination now had outer limits, though, as it was clear to most that the Amalgamated Association had failed in its organizing powers, its lodge-societies stressing saloon standards and male-virility, its myopic militancy, and its weak union alliances with others to win demands within the wage system. It was also apparent that the Amalgamated leaders had mistakenly relied on Republican party promises to maintain steel tariffs, the assurances of local authorities and sheriffs, Governor Pattison's double-talk,[14] and the Militia as a shield protecting them against Carnegie's impositions.

These leaders had closed a blinded eye to the ties that existed between the government and the large mining, industrial and railroad interests, though these were the very forces that were central in mobilizing the Pinkertons and Militia, defeating the strike, undermining the commune, irreparably weakening the union and – with the Farmers' Alliance and People's Party drawing off votes from Republican nominee General Harrison – leading to a landslide Democratic party "free trade" victory for Grover Cleveland in the November 8, 1892, presidential election.[15]

Nor did the employers' measures flag after the strike. Led by Carnegie, the great steel barons again outflanked their mill hands by extending their relations with labor from the narrow realm of production and wages to the point where they tried to encompass the entire life condition of the workers and their families in the Homesteads, Braddocks, McKeesports and Duquesnes, Carnegie not only eliminated unionism from his plants, but took greater charge over and shaped the relations of labor with the commercial and political forces in each mill town – effectively taking away their means for self-determination – swaying, using and dominating all of them.

These efforts creating industrial fiefdoms, "mill protectorates" and company towns were quickly followed by almost all the steel firms in the Pittsburgh district, where wages were cut, the lives of working class families worsened, and the cost of producing steel was lowered.

## Plight and Struggle

The Panic of 1893 only added to the plight of the already disoriented steel workers, who saw no purpose in remaining members of, or paying dues to, the weakened union.

Union solidarity went from bad to worse. While Amalgamated membership plummeted to ten thousand by 1894, three years later in 1897 the largest independent steel producer in the Pittsburgh area, Jones and Laughlin, finally expelled the Amalgamated.[16]

The steel owners in the Pittsburgh area meanwhile standardized and regimented their labor policies – dividing the workers ethnically and degrading the various mill positions and pay scales of the Calvinist-English, Presbyterian Scotch-Irish, and Lutheran-German workers, putting them in desperate job competition with Polish Roman-Catholic and Orthodox Eastern European laborers. Relying on the fierce ethic of work as the driving force sustaining the varied ethnic and family bonds of the Valley's male-dominated households, [17] the mill owners' slight wage differences created nearly insurmountable jealousies. Yet, the owners controlled overall costs and made wage levels more or less uniform, enabling the Pittsburgh region to compete favorably with other districts still under Amalgamated pressures. For now the union had successfully lodged itself in the iron mills lying west of Pittsburgh, as well as in the biggest steel mills of Illinois and at a large proportion of the sheet, tin and hoop mills of the nation.

Repeating Frick's dictum that the Amalgamated should organize at other people's mills, the Pittsburgh mill owners savored their favored competitive advantage. And throughout the Pittsburgh region, the union was continually buffeted and beaten back by the extension of Carnegie policies, with the final battle line being drawn in 1901 by the formation of the United States Steel Corporation.

* * *

J. Pierpont Morgan, the "great Financial Gorgon" in the eyes of the laboring classes, had set his sights on buying Carnegie Steel as a stepping stone to merging eight great companies, including Morgan's own Federal Steel, forming United States Steel as the largest corporation in the world.

An epigram of the banal simplicities of high finance, Morgan had told Andrew Carnegie to name his price, Carnegie scrawling $1.4 billion on a single sheet of paper – the Titan's offer bringing the master financier's simple response: "I accept this price."

Apart from such rose-colored fictions, Morgan took $62.5 million in service fees;[18] and out of the total $1.4 billion price part was supported with I.O.U.s., as the company took on nearly $570 million in debt, largely in bonds – including a high-risk, speculative issue of $150 million of 60-year, second-mortgage paper.

Congressmen fumed at the financial agglomeration of monopoly power, while the *Wall Street Journal* expressed deep "uneasiness over the magnitude of the affair." Though some shareholders promptly sued to block the floating of speculative securities, New Jersey courts took two years of wrangling to allow their issue.[19]

Within months, the United States Steel Corporation became the first billion-dollar corporation through the merger of these ten steel companies. The new company controlled over 60 percent of the domestic steel market and 29 percent of the world market, with the domestic steel market still being protected by high tariffs.

The agglomeration owned hundreds of factories, employed 168,000 workers and controlled six sizable railroad lines. It oversaw the steel-making process from beginning to end through a combination of furnaces, ore deposits and transportation companies.[20] From that point forward it empowered Morgan as the purported "field marshal of industry" against a sundered working class nationwide – enduring the life of sweatshop labor at the furnaces and rolling mills, on the railroads, in the factories and mines.[21]

* * *

As the century turned, the disastrous defeat at Homestead and Morgan's foothold in finance and steel was now on almost every steel worker's mind. Financial panics had already led to immemorable failing cycles of production in 1873, 1884, and 1893, in part brought on by the deliberate manipulations and corrupt practices of large moneyed interests. Steel workers feared the worst at Morgan's hands for their job security, their slender, remaining controls at the place of production, and their wage levels. In defense, while Morgan's new corporation was trying to market its huge issues of stock in June and July 1901, the Amalgamated went into action – concentrating its forces on three U.S. Steel corporate subsidiaries, demanding that all their branch plants, not merely those that were organized, sign agreements. Foolishly or not, the union rejected the corporation's initial compromises, the committee of banker J. P. Morgan, old Carnegie manager Charles Schwab, and Steel Corporation head Judge Gary then offering less than company-wide recognition of the union.

At that point union solidarity shattered. The no-strike clause in the Amalgamated contract at the Illinois Steel Company spelled a weak and partial labor stoppage; and expected support from Samuel Gompers' American Federation of Labor and the United Mine Workers under John Mitchell failed completely. As the strike lost support, U. S. Steel offered even less acceptable terms to the union, then in September forced the Amalgamated to capitulate – taking away fourteen mills, forcing the union to agree neither to extend its influence nor to welcome new members in United States Steel plants. This left the union with a frail presence, remaining only in the tin plate, sheet steel, and steel hoop companies.[22]

An overwhelming victory, U.S. Steel was meanwhile accumulating revenues hand-over-fist. In 1901 alone its profits were $84 million, paying dividends to both common and preferred stockholders. Secure and awash in lucre, it extended its control over 75 percent of the entire steel market by 1907.

There was a hole in the bucket, though. For the merger of United States Steel had left it saddled with debt draining huge sums in annual interest payments. To meet these debts, the company ran its underlying enterprises and mills as "cash cows" – neglecting to upgrade its technical equipment to keep up with domestic competitors and international rivals, especially after World War II. Though United States Steel lost market share throughout the 20th century, by the turn of the 21st century it still remained the nation's largest steel producer – technologically backward, inefficient and producing only slightly more steel than it had the first full year of its existence!

Yet, during the first half of the 20th century, its underlying profit-source remained the mill workers, laboring and living under conditions of employer surveillance of their activities. Many thousands again worked as virtual helots, suffering privations and extreme mill dangers at near subsistence wages, constantly in debt to grocers, retailers and local banks. Especially at Homestead, where wage-dependent workers held bank loans on their frail equities in ever-vulnerable, row houses climbing the landscape's irregular, muddy slopes – it appeared industrial despotism would succeed.[23]

## The Degradation of Work Skills [24]

With the employers positioned to introduce technologies and management direction that undermined old skills, cut wages, and prevented labor organizing throughout the industry, the steel workers were clearly uneasy, vulnerable and on the defensive

The old methods of mill life were being undermined in step with the advance of machinery and the degradation of specialized labor, as specific skills once required by the employers were now being replaced by simplified task-work performed by all-purpose workers and generalized laborers. Hereafter a specific job task made each worker a machine "tender," an appendage of a particular process or mechanism fitted with a tool, tethering each operator to the specialties of output and, in the long run, blocking each worker's physical and mental growth.

By so detailing the labor force to the new technical production methods, work tasks were also divided into office and mill functions. And office workers themselves soon became a species of intellectual, mental laborers whose roles in preplanning and designing mill operations were variously designated and allotted under the unified direction of the mill's top managers.

These were no small matters for workers in the steel industry. In the mill, the tools now had been removed from the hands of the workers, fixed to machines, with the machines themselves driven by inanimate steam, electrical, and heat power sources. Propelled by such central-power mechanisms, the machines then put labor in motion, concentrating each worker's efforts on a single detail of production to the exclusion of all other functions. Not only were the old craft specialties carved into their respective operational steps and assigned to different workers in the mill, then, but the new technical instruments and processes tended to displace some specialized workers, leaving those remaining to operate machinery that moved faster and forced each worker to exert more energy in less time in order to turn out the same or a greater product.

Thus did the story of the exhaustion of the Valley's steel workers unfold. For the subtraction from the steel workers' physical force in running the machines and dexterity in operating them did not lead to less work tension; rather, the speed-up of production and the workers' loss of control over the pace and previous details of labor made them more vulnerable to increased exhaustion. With the employers' control over the work process and enforced wage cuts, the old concept of "A Fair Day's Pay for a Fair Day's Work" was thereby undermined. For in a very practical sense, the workers were increasingly exhausted each day, at the same time they had to make do with a smaller wage. And they had no way of contesting what they received. As at Homestead, they had lost control over both the process of work and the pay they received for their now specialized labor functions.

## The Job Ladder and Internal Labor Pool

The specialization of labor and the creation of a hierarchy of managers were also elaborated to the point where the relationship between workers and corporate owners was increasingly tension-riven in the workplace.

Within this contested terrain, tactical maneuvers and psychological gamesmanship meant the employers' rules for quality output and continuous production were constantly being broken. The employers tried to block the workers from slowing the pace set by the mechanical production processes, as well as labor's attempt to organize across the functional lines assigned to various sectors of the workforce. But the steel workers continued to do everything possible to reduce the speed-up and interrupt production.

The employers then devised a new hierarchy, deploying different gradations of managers to oversee and control the entire labor process that, in the steel industry, now required chemists, metallurgists, and office workers, as well as tiers of skilled, semiskilled, unskilled, and undifferentiated laborers. Together these workers performed the step-by-step technical functions of production along a pre-designed plan delineating their designated, assorted roles and wage categories.

At the upper tiers of management, corporate officers were positioned to track overall operations, assigning detailed controls to underlings who, in turn, designed and reassigned functional tasks to those below them. These demi-managers also established and oversaw an internal labor pool out of which each worker was encouraged to move upward on the firm's job ladder – this treadmill for upward mobility itself dovetailing with the introduction of the latest automatic processes of production setting the rhythms of work.

Such management techniques effectively divided functional tasks on various levels, acting as rungs of the hierarchical ladder, each relying on specialized training of specific groups among the old and new sections of the labor force. Management thus charted and controlled – advancing detailed schemes for upward job-mobility on a technical or functional basis, as well as using psychological techniques for behavior modification and oversight.

With such work categories established, pressures applied, and jobs assigned, labor's efforts to combine, organize and try to reassert control over the specialized roles of production were repeatedly blocked.

## The Mass, All-Purpose Steel Worker

Both Carnegie and the United States Steel Corporation that bought his assets elaborated this scheme of bureaucratic management – stressing job training and assignments to the new universal laborers, who could learn any job task in a relatively short time. With the job ladder as a central mechanism for control, the mass, all-purpose steel worker made its initial appearance.

The advanced technology that required these new, all-purpose workers also enabled management to extend and elaborate the job ladder based on assorted titles, greater or smaller differences of functions, and a progression of wage categories. Though technology itself required that the labor force be recast and workers specialized to operate particular machine processes, the functional division of labor was now extended by arbitrary categories and rules purposely set up by the employers to divide the workers and keep labor from contesting the managers' control over the details of work. Labor was thereby trained and specialized by management in order to dominate the work force as a whole, keeping the workers compliant and pacified.

What, though, was actually happening to workers in the mills? The new machinery turned large numbers into lever-pullers, monitors of technical instruments, isolated technicians in the laboratory, managers in the front office, and pencil pushers in the back room. The mental labor of the managers was thereby increasingly separated from the physical labor in the mills, in the laboratory and elsewhere. Not only was the intellectual preplanning of work separated from carrying out physical, manual labor – so that different workers performed each function, then – but the old skills were undermined and replaced by new tasks giving the operators a smaller degree of discretion over the immediate positioning, variance and movements of their own muscles and hands. The workers' previous self-direction was thereby being undercut by advanced processes and technical equipment that both regulated their physical motions and set narrow parameters for utilizing their intellectual capacity.

The consequences for the work force were two-fold, then: Technology led to the detailing of job functions in a way that restricted labor's natural, coordinate mental and muscular motions, so that the workers lost control and direction of both the work process and their intellectual ingenuity for production; and management took advantage of this detailing of labor to design a specialized job ladder that formally separated workers by categories, some of which were arbitrary, in an effort to motivate all workers to climb its rungs as emblems of security, marginally better wages, and prestige.[25] Thereby management successfully planned to keep labor contented, pacified – and from organizing to reassert control over the work process.

## Finance Capital and Labor

Throughout the nation, financial capital and labor had crossed swords at the turn of the century. [26]

Campaigning as "trust-busters," Republican Presidents Teddy Roosevelt and William Howard Taft had raised the nation's conscious of the sins of the monopolists, cartelists, and organizers of business trusts and finacial "pools." Roosevelt with his Attorney General and other key cabinet members had mapped out antitrust suit against John D. Rockefeller's Standard Oil empire to break it into 34 companies, prompting Henry Clay Frick's crude, breathless retort on behalf of his fellow oligarchs supplying corruption funds – "We bought the son of a bitch, and then he didn't stay bought!"

Other trusts were also scrutinized, none so great as the House of Morgan that had become the United States central bank, dominating finance, credit, stock markets – and, by their extension, production and employment. Morgan's fame stemmed in part from his intervention during the panic of 1907, when the infighting between the financiers for control over corporations and banks had caused the wheels of industry to slow and widespread unemployment and hardship to follow. Recognizing the crisis could devastate the economy on which the banks thrived, Morgan took command in creating a cartel among the largest banks in the nation, using the New York Clearing House, which the administration of President Roosevelt permitted to create out of thin air some $100 million in paper money. The cartel's banks then loaned this money at heavy interest charges to borrowers, so that while the bankers extended a species of fictitious credit, the borrowers were forced to repay these loans plus interest in labor-earned dollars. Then for his alleged yeoman work in cooling the panic, Roosevelt permitted Morgan's interests, centered in the U.S. Steel Corporation, to acquire and merge the southern steel industry emanating from Birmingham, ignoring the anti-trust law prohibiting such a merger.

Nonetheless, five years later, the trusts were again in the cross-hairs of Washington's cartel hunters. In November 1912, the Democratic party reformer Woodrow Wilson won the presidency, the Congressional Pujo investigation of the money trusts was launched, and the House of Morgan got the heat. Through interlocking directorates, Morgan men sat on the boards of the nation's major industries – railroads and steel companies, coal mines and public utilities. By 1913, a handful of Morgan-friendly New York banks held 341 seats on the boards of 112 corporations worth more than $22 billion. No one who needed a big loan or a new stock issue dared offend the New York cartel of bankers. And as part of Morgan's holdings, awash in seemingly unlimited funds, U.S. Steel had no apparent worries. Able to switch funds from speculation to production, the great banker took the raw edge off his methods of accumulation.[27] Only big steel's workers at the bottom of the financial pile were living a nightmare.

## Organizing Across Task Lines

How, then, could steelworkers pull together – magically drawing renewal out of Morgan's verger wand and financial hat?

The sorcerer's sword had been riveted at Homestead, where Carnegie held court: to eventually eliminate all aspects of skill; to raise technical capital to the point where, to use it, all detailed tasks would be simplified and controlled. Such modern methods were planned to progressively split old tasks into their component parts and motions – accommodating the technical machine processes as well as simultaneously creating new job functions undercutting the skilled workers' previous tasks and controls over the details of work.[21a] But, as Carnegie and his heirs lost the pulse of labor, it also opened the way for unanticipated consequences as these workers again organized attempting to exercise some measure of control over the new functional tasks they had to master to carry out production itself.

* * *

For eighteen years after the Amalgamated defeat in 1901, steel workers had centered their efforts on winning a larger relative share of the value of their output, recognizing that, as mass workers in an industrial apparatus that specialized their labor and made them functional parts of the machinery they used, they would have to organize across craft lines as a group - on an industrial basis.

They still feared on-the-job controls held by the Morgans and other steel employers, as well as the espionage of the coal and iron police in the company towns. But they were equally outraged by their low wages and the pall of incivilities and dangers they faced in the mills. And their collective voice was not stilled.

*The Amalgamated Journal* espoused their views as representing the near-universal opinion of the Amalgamated Association of Iron, Steel and Tin Workers of North America. Far less militant, the American Federation of Labor's *American Federationist*, *Labor Weekly News Service* and *Metal Trades Department Bulletin* all leaned towards accommodations with the labor needs of big industry in times of crisis. More radical views appeared in the pages of *Slovensky Obcan*, representing Slovak workers, as well as in *The Labor World* speaking from Pittsburgh to workers' international interests and worldwide solidarity. From Pittsburgh also came the heated call of the *National Labor Tribune*. Pittston had its *Industrial Advocate* representing the interests of the Central Labor Union, Shenandoah its *Anthracite Labor News*. Uniontown's *Working World* called labor to organize, as did the *York Labor Advocate* that came on-stream just as the U.S. entered World War I in 1917.

Yet, the pro-labor press quieted or otherwise bent their demands as the pressures of the war opened jobs. Slightly elevated wages also made coal, coke, steel, and other manufacturing workers bide time, standing in wait as the conflict consumed the products of their labor.

Steel-profits beckoned. To help the Allies secure armaments against the German-led Triple-Entente, in 1915, heir of the House of Morgan, Jack Morgan went into action as purchasing agent for Great Britain and France. He meanwhile set up a corresponding U.S. export division; issued a $30 million loan to the French; put together a syndicate of 1,570 banks to float a $500 million, five-year bond issue to finance

Allied munitions purchases; raised it to $1.5 billion by the time the U.S. entered the war in April 1917; and helped switch American industry to a wartime footing. Between early 1915 and April 1917, Morgan's agents had shipped the desperate Allies $3.5 billion worth of military hardware and armaments made at U.S. steel and other mills – recalling Andrew Carnegie's scheme, milling warships in 1892-93.

Morgan arms-profiteering, the country still at war, steel and coke workers fully employed, patriotism running high, workers remained unorganized to protest their miserable conditions. But as the war ended, these melders of the nation's brawn silently regrouped, then openly began mobilizing their forces to protest as one, nationwide.

Steel workers now faced employers who were ever more powerful than before. For the U.S. had moved from the status of a debtor to a creditor nation, taking in the gold supplies of other countries paying their war debts, and placing the House of Morgan at the apex of financing conglomerated steel. Though the steel workers tried, the odds in 1919 were not good for a strike to create an industrial union; and, indeed, they were stopped cold by the wealth and power of the great steel employers.

Once again two near-insurmountable barriers surfaced: The drives for *men*'s labor supremacy and skilled workers' control, along with religious and ethnic divisions, interfered with lasting union solidarity. And the rank-and-file were again outmaneuvered and blocked by the United States Steel Corporation working with the other giant steel companies, deploying police, Pinkertons, and scab workers.

As the strike shortcomings surfaced, strike leader William Z. Foster was taken back, forced to spell out needed changes for successful future union action. "The organization movement must be industrial and national in character," he recoiled. "That is, (a) it must include every category of workers in the steel industry, not merely a thin stratum of skilled workers at the top; and (b) the drive must be carried on energetically and simultaneously in every steel center, not simply here and there spasmodically in individual mills or steel centers."[28]

Given the great employers' positioning, though, was it really possible for the Pittsburghs and Homesteads, Youngstowns and Buffalos, Baltimores, Clevelands and Chicagos to simultaneously rise up to contest and throw off the yoke of the steelmasters led by the House of Morgan?

At first, the answer seemed decidedly *no*. But the course of the 1919 strike had offered a clue, helping to solidify a new awareness of the need for both an industrial union and a democratic organization based on mass participation. 1924-29 industrial expansion without a rise in wages kept the great employers' whip hand firm until the mid-1930s, when tremendous overproduction of steel in all industrial nations created a worldwide market glut leading to unrestrained dumping below production costs. As elsewhere, mills in the U.S. tried to remain open and competitive by cutting wages and imposing work speed-ups.

Craft-organized, union labor meanwhile gained an unexpected ally. President Hoover's "do-nothing" policies had helped bring on an economic depression, as markets, output and employment plummeted. The public incensed, their election of President Franklin D. Roosevelt quickly led to so-called "New Deal" measures that, however inadequate, attempted to reignite the economy, restraining the financial and industrial monopolies, while awarding labor the right to organize. Steel and other industrial workers jumped at the opportunity.[22a]

## A Historical Watershed of Working Class Life

Would this presage a historical watershed of working class life?

The vicissitudes of capitalist society and "industrial serfdom" had brought massive unemployment, desperate poverty and uncertainties. One-half of the nation's labor force was jobless, wages were down one-third, and the depression was threatening the survival of the inhabitants of every mill town in the nation. Laborers of every category were desperate, angry and on the defensive.

In steel, the companies continued to disregard labor's degraded condition. Facing unbearable poverty, to organize was to choose the only path most steel workers had out of idleness and desperation, opening the way for zealous communist field organizers and shop floor activists to agitate in the name of the Steel Workers Organizing Committee for an industry-wide union.

The risks to labor organizing in the mill towns were as great as ever. But there was weakness on the other side, too, as, in the midst of a depression and extremely vulnerable, the vast steel empires – imperiums linking finance, coal mines, coke furnaces, iron ore supplies, industrial production, rail lines and shipping – were operating far below capacity, suffering losses.

Given their own dismal situation, in 1936 the steel workers again prepared to shield themselves, this time behind the Committee for Industrial Organization, headed by John L. Lewis. Organizers began agitating and recruiting in the steel industry under Philip Murray. And the Steel Workers Organizing Committee was the central force, representing the insurgent wing of the old A.F.L. craft union – the Amalgamated Association of Iron and Steel Workers, handed down from its 19th century past long before the great battle at Homestead.[29] Still a leading steel organizer, now a zealous communist, Foster again recognized the advantages of the new campaign, didactically, if critically, analyzing the material base for finally building an industrial union structure:

> The industry is increasing production, the political situation is more favorable for maintaining the civil rights of the workers to meet and organize, the workers are in a more militant mood, the right of the workers to organize is more generally recognized, the campaign is being carried on upon the basis of one industrial union instead of 24 crafts, the illusions about company unionism are less now than-ever, the campaign has the solid support of a dozen powerful trade unions, there are ample funds for the organizing work, the language problem is not as severe as in earlier years, the radio now enables the message of unionism to evade the employers' censorship and to be carried directly into the steel workers' home. And, lastly, there is now in the field a strong Communist Party (which was not so in 1919) that is lending all its support to the success of the campaign.[30]

True, for a full year the giant mills fought back stubbornly, breaking strike after strike with Pinkertons, company-paid "goons," police support and scab workers. But after each confrontation, the employers again faced renewed labor militancy.

Workers withholding their labor were finally creating an unprofitable downside for the employers: production ceased, capital assets were being lost or destroyed when the furnaces were put out, and markets were being taken by foreign competitors. For the employers, it was a grim picture, as between 1929 and 1936, European steel held 70 percent of world exports.[31] True, the 1930 Hawley-Smoot Act sealed off U.S. markets, protectionist legislation keeping out foreign steel and manufactures, so that European countries could never pay their Wall Street debts, deepening the crisis.[32] But, after stonewalling and an all-out battle with union agitators, some employers sued for the stability of unimpeded domestic production. They discovered that the better part of their intransigence was a concession to recognize labor's right to organize.

After years wallowing in a Great Depression, U.S. Steel and the other giant steel employers sought to sidestep a threatening, catastrophic strike – on March 1, 1937, Big Steel signing a contract agreeing to collective bargaining, awarding steel workers a forty-hour week and a substantial wage raise.[33] This break point reached, the success of industrial organizing in the steel industry led to the formation of the United States Steel Workers Union, which was empowered to bargain for all steel workers on a national level.

But even this was ultimately modified by war, creating a fund out of which wages would be paid by the large steel and other industrial employers securing their own profits. Just as the sliding scale of wages was once capped by Carnegie and Frick bringing on the Homestead strike and the predetermined setting of wages in the steel industry, the U.S. government facing World War II imposed controls over wages, secured labor's no-strike pledge, set output levels and allowed the employers to set the pace of work through speedups of arms production.

The historical setting was much different than the feared military engagement with Britain and Chile in the 1890s. Then the struggle for domination of the Pacific depended on the size and viability of the U.S. naval fleet largely forged at Carnegie's Homestead and by Bethlehem. Now the global stakes were higher, as President Roosevelt secretly plotted to supply arms to the British and French under "lend-lease" in the wake of the 1939-41 German *Anschluß* with Austria, the Nazi invasion of Czechoslovakia and threats to Western European territory.[34] But pressures placed on steelworkers were similar, as out of the total production, the state-aided, military-industrial complex again established a cap for wage rates.

Two years earlier, during most of 1937 the smaller steel giants – Bethlehem, Republic, Youngstown Sheet and Tub, and Inland – refused to deal with the Steel Workers Union, contending they could not "afford" labor wage boosts. But several independent companies then faced militant strikes; and at the gates of the Republic mills at South Chicago on Memorial Day the police fired into protesting crowds, killing ten and wounding forty. In the wake of a massive public outcry, the National Labor Relations Board then ordered Republic to bargain collectively. And by the end of 1942, all of "Little Steel" had followed Big Steel into labor contracts.[35]

To win the war, steel production and employment reached all-time highs from 1940 to 1945. Yet, government wage-price regulations set wages in relation to the basket of goods and services workers could purchase - largely undercutting any rise in the steelworkers' real wages.

Though the war had fueled a rise in national income of 132 percent between 1939 and 1943 – from something over $72 billion to $168 billion – the lion's share went to the great corporations. Corporate income before taxes rose 302 per cent from $6 billion to more than $24 billion. Farmers' income rose 179 percent from over $4 billion to $12 billion. But labor's income rose only 72 percent, largely a result of overtime, from an annual average of $1,372 to $2.360.[36] War left workers with the short end of the rise in the nation's wealth they themselves help create.

**Union Hierarchies and Employer Defenses**

Though the United States Steel Workers Union had successfully won central bargaining power representing all workers in the industry, the reorganized union was also weakened by its own internal, structural cleavages – frequently undermining the democratic needs and equities sought by rank-and-file workers.

The union hierarchy had been very effective in dealing with the workers' overall demands in the face of the inequitable policies of the companies. But its leaders opposed the direct resolution of mill grievances by rank-and-file committees that carried out local strikes and/or presented other hindrances to continuous production. Such local, shop-floor actions were seen by the union overlords as threats to their authority as spokesmen for the great majority, effectively weakening their hand in dealing with the companies for industry-wide concessions in return for promised no-strike, labor peace.

As these leaders moved to purge communists and like-minded organizers who had once strengthened union ranks, sustaining their power, disaster loomed within the union itself. For over the years, these grass-roots organizers had created a democratic base, creating a groundswell of participation in local union affairs. But as the union's bureaucrats repeatedly issued drastic orders enforcing top-down union discipline, local participation flagged – empowering the employers to take the offensive, maintaining harsh work conditions that fell outside the local unions' grievance provisions set by the master contract between the United States Steel Workers Union and the mills.

The effect was the establishment of a national hierarchy of bureaucratic union leaders who acted as the employers' first line of offense that the rank-and-file had to overcome!

The second line of offense was the steel companies themselves. As smaller mills repeatedly regrouped into larger economic units whose operations and employment rolls extended into other industries beyond the scope of the Steelworkers Union, their was no effective way for labor to fight back in the context of industrial unionism. Especially after 1970, corporate efforts to concentrate industrial, commercial and financial capital also reorganized technology – throwing labor off balance by shifting financial resources from one branch plant, field of operation, or geographic area to another, displacing some working populations and creating unemployment in traditional steel-making centers.

To overcome these two lines of offense, steel workers were called upon to overturn the union bureaucracy and build a conglomerate union that would encompass all workers hired by a particular, transnational employer operating branch plants and facilities on a conglomerated basis worldwide. To do this, though, they would have to critically assess, counter or accommodate the owners' global base of economic, technological and political power – an effort not yet on labor's drawing board.

**Prospect for the Future**

As in 19th century Homestead, America today faces another transitional stage with old technology being replaced by new methods and machines.

Now, as before, great numbers of steelworkers are being forced into unemployment or lower-paid jobs as their skills and the methods of production become outmoded. As in the past, the workers made jobless initially respond with anger and militancy, rather than foresight. And since at least the early 1980s they have been demanding that the mills keep their job positions and production levels as they were; that the mills maintain the current technological *status quo*, rather than introduce innovations that would undermine their remaining skills. Yet, as happened in 19th century Homestead, by shutdowns, lock-outs, and/or refusals to bargain, once more the employers have attempted to introduce new technology and processes that requires fewer and less-skilled-workers to operate, thereby permanently displacing some remaining skilled steel workers while creating a new group of machine operators employed at significantly lower wages.

The story of Homestead's past thus partly explains America's current labor experience and technological dilemma.

But today there is also a shift of corporate focus away from basic production industries into new fields. This shift was accentuated in the early and mid-1980's, when high costs, inefficiencies, and a glutted market plagued the U.S. steel industry. Once again, giant steel companies closed one mill after another, firing thousands of workers. From 1982-86, a fifth of the domestic steel companies filed for bankruptcy protection, in part throwing the burden of unpaid workers' pensions on the federal government.

Steelworkers then joined in a call for higher tariffs to save their jobs, as they did in 1892 and 1929. But by then, the largest steel companies had already moved to diversify their operations in search of profits. Many steel companies took over chemical and oil corporations, while some went out of the steel business altogether. And diversification meant a cut of employment in steel, leaving workers with only a frail hope of future jobs in low-paid service "industries" – or in job-training, possibly for employment in finance, computers and micro-technology.

Where big steel mills remained open, new technology was introduced, jobs were simplified and workers were paid less under the terms of detailed union contracts signed by each plant with its local labor force. Such contracts weakened both local union strength and national labor unity.

American steel mills nonetheless remained vulnerable to lower-cost, foreign competitors, for new U.S. technologies of the late 1980s were not yet perfected.

Since its inception, the U.S. steel industry had essentially been using the very same technical methods. In a series of operations, coal was first converted into coke in giant batteries of ovens. Next, iron ore was fused into pellets. Then the coke was combined with the iron pellets and limestone in a blast furnace the height of a 12-story building to make pig iron. And finally the raw material was purified by a basic oxygen furnace into steel – synthesizing iron with a two per cent carbon base.

In the late 1980's, the new "bath" smelting process attempted to make steel ready for fabrication in one reaction in one big vessel. That was the 1892 claim of Carnegie's one-big-blast, Bessemer converters. To accomplish this in a modern giant hot-tub, a small amount of molten metal was topped with iron ore and cheap low-grade coal. Once the molten metal dissolved the coal, a catalytic combustive reaction was begun and continued by forcing a stream of oxygen into the vessel – the resulting high-temperature gases like carbon-monoxide also being channeled back into the vessel.

In theory, the process should have worked metallurgic wonders. But some chemical pyrotechnics repeatedly went askew.

During 1989, the experimental project to change age-old steelmaking techniques was launched at a USX Corp. facility near Pittsburgh, where the U.S. Energy Department contracted with the nation's steelmakers for $19.6 million in free-money. But the technique faced problems for a full ten years: Coal and ore poured into the furnace tended to cool the molten metal, undermining the balance of combustive reaction. And ash and undissolved metals left behind by the coal after it dissolved in turn left slag layers interrupting heat transfer for continuous reactions.

The technological question also boiled down to the practical, competitive one of revolutionizing steelmaking to compete against Japanese and European rivals. By 1989, the Japanese had invested twice as much money as the Americans to make molten iron which still required purification in basic oxygen furnaces. And American mills were at a competitive disadvantage, as the then most-recent, all-new U.S. facility was Bethlehem Steel's mill at Burns Harbor, Indiana, which opened in the late 1960s![37]

To outflank competitors by cutting production time, unit costs and prices, new technologies were constantly being tried, those coming on stream recurrently creating a world market glut in steel. The blindspot was not technological advance, which most large competitors pursued; but rather the enlarged output that necessarily followed.

And what would happen, though, if technological innovations increased output on a transnational or joint venture basis?

This happened during 1991, when Japanese and U.S. companies entered a joint venture that used a new continuous, cold mill process to cut manufacturing time to less than an hour from about 12 days! Inland Steel Company and Nippon Steel Corporation built the I/N Tek plant at New Carlisle, Indiana, spending $112 million plus start-up costs to use Nippon's latest technology to weave steel coils through a mill seven-football fields long. The giant mill took unfinished steel coils from another Inland plant, processing them so they could be cut and shaped to customer specifications for such things as washers, dryers, file cabinets, and automobile doors and fenders.

It was the technological wonder of the moment – promising to increase the world supply, despite the fact that global markets were already saturated and steel prices were bound to fall, returning too little revenue to cover cost outlays for the new technology itself.

## Computer-Controlled, Uninterrupted Production

This latest try at uninterrupted production is different from past hot or cold methods. For the key to modern continuous processing of cold steel is the computer-controlled, seamless movement of steel coils through miles of plant operations, rather than the traditional process that physically moved the coils in batches from one site to another – a process which could take as long as two weeks and require as many as 500 workers, compared with the 200 working in production at the entire I/N Tek plant.

The journey of cold steel production begins by cleaning the coils in a bath of weak hydrochloric acid to strip off rust and dirt. Then giant rollers reduce a half-mile strip of steel an eighth of an inch thick to a 64th-of-an-inch-thick sheet two miles long. Next the sheets of steel are threaded through a ten-story annealing furnace that reheats the metal at 1,470 degrees Fahrenheit, ensuring uniform heating and a more consistent finished product – properties that make the steel malleable, instead of brittle. This establishes continuity in the finished steel – unlike traditional annealing where the coils are enclosed and heated by large cones that cause the edges of the coils to be hotter than the interior, leading to a lack uniformity in the finished steel. The sheets are finally flattened and smoothed in a temper mill, then inspected for quality by a computer – designed to produce the world's highest-quality finished steel at tremendous savings in energy and labor costs.[38]

This technological wonderland has its social drawbacks, though. Not only does continuous processing introduce technology that eliminates former job functions and skills, but in the quest for cheaper costs and energy savings to win customers, it takes away the need for human labor itself – literally tearing away the means of employment, wages and purchasing power from the hands of former steel workers.

Besides some men being displaced, if not destroyed, in one domain by others who make the latest technical Cyclops in another, there is a downside, too: the capital invested in ever-depreciating, plant equipment has usually been borrowed from banks at high interest charges that offset savings in operating costs, so despite the elevated technological promises of production, the price of finished steel remains relatively high with little savings to consumers.

J.P. Morgan's dilemma of profit-sapping borrowed capital to buy Carnegie's steel empire thus reappears.

Worse still for those steelworkers in the smaller mini-mills using scrap steel (which is a cheaper raw material than turning iron ore into steel), each team of workers is paid on a piece-rate scale, so that wages rise or fall with the number of tons produced in a workday at ever-variable prices per ton.

This reintroduction of the piece-rate system in the mills drives the workers harder, so that, as in the Homestead era of the 1890's, sped-up labor again works on a sliding wage scale.[39]

Is this the future for steel workers in our era of global overcapacity? Perhaps so, if we may reasonably expect that, as cheaper foreign steel is dumped on U.S. markets, domestic production will periodically be cut and wages scaled down.

## Management, Labor and Two Great Barriers[40]

Today corporate control over the introduction of machines that perform what were once human functions has also led to the evolution of earlier management techniques that concentrate on assigning narrow, ever-limited tasks to each worker – in part following Andrew Carnegie's lead in seeking maximum profits by using new technology to simplify and divide work tasks and cut wages.

Whereas Carnegie broke the union and reorganized Homestead's labor force and technological base in a few years (1892-98), a similar realignment of labor that began in the late 1970's continued into the late 1990's and early 2000's. In the 1970's and mid-1980's, the most modern technologies were first introduced to industries other than steel, and only later surfaced at a few of the giant steel mills in the Monongahela Valley and elsewhere. In the interim, old steel mills were closed, their workers fired. And because new steel-making technologies were not immediately introduced, the steel workers were almost helpless – unable to protect their existing jobs, yet equally unable to switch to other jobs in technologies they did not understand.

Recurrent steel workers' efforts to protect themselves faced two great barriers: the competitive plight of the steel companies; and the nationalism of the steel workers' own union, leading them to ally with U.S. employers for tariff protection from foreign steel mills, steel workers, and steel imports.

The first barrier was handed down from the last century. Heir of the Carnegie interests, U.S. Steel Corporation (USX) remained the nation's number-one steelmaker, yet was unable to fully compete with foreign firms operating abroad or exporting to the U.S. market. From Russia, China, Japan and Germany to Mexico, Brazil and Argentina, foreign steelmakers were subsidized by the nation states of their origin – dumping their steel in the U.S. at a lower price than charged within their own national markets, periodically forcing many U.S. mills to close down.[41]

In the 1980's, U.S. steel companies diversified their investments to effectively upgrade their technologies, or left the steel business altogether; some sought a foreign refuge where wages and other costs of producing steel were lower. "Everybody must be competitive," the chairman of the U.S. Steel Corporation, David Broderick, insisted in the mid-1980's. "And if any of us fail to do that, any steel-making facility is in jeopardy."

Those that failed in the Monongahela Valley were shuttered,

Alexander Cockburn and James Ridgeway, reporting the *dèmarch*, March 8, 1983:

> We saw the snow collecting on top of the steel mills up and down the Monongahela Valley last week, mills whose names are synonymous with industrial America and a century of labor history: Homestead, where workers shot Andrew Carnegie's Pinkertons out of the water in 1892; Braddock; Duquesne; McKeesport. In the good times, the snow would have melted on the hot metal. There is no more pointed symbol of economic depression than these mills with their cold furnaces. Most of these workers have lost their jobs for good.[42]

The Homesteads and McKeesports closed, were dismantled and shipped away for scrap over the next three years. By the winter of 1986, little remained of the mills, the mill towns, or the working class that once magically fueled their fires, on-the-scene investigator Morris Edelson penned the Valley's bitter epitaph.[43] All that reminded of the strike of '92 were the grave markers, the memorials, and recall by the Society for the Preservation of Murals of St. Nicholas, a church in Millvale whose interior is deftly painted with scenes that bring together memories of rural Croatian peasant life, biblical stories, and episodes from the battle at Homestead.[44]

Other steel towns continued to resonate with a knowledge that technological competitiveness would have to be upgraded. Yet, in the early 1990's, 4 percent of the nation's steel was still made in open hearths of 1940's and 1950's vintage, 60 percent in Linz-Donawitz and other types of basic oxygen furnaces of the 1960's, and 36 percent in electric arc furnaces of the 1970's.[45]

The latest SMS Schlömann process was only in use in the mini-mills: designed to use less expensive scrap, as opposed to the ingots used by large steelmakers, and to form molten steel into slabs much thinner than conventional casters produce. As the slabs were flattened into steel sheets one-tenth of an inch thick, less energy and labor-power were required and less material was wasted. And the possibilities for continuous production without the intervention of human labor were enormous, despite the fact that so far the casters operate as a single unit – unlike the series of stations in a typical large steel mill – so the entire production line must shut down if one function fails.[46]

Thus the failing sequences of technical capital interrupt continuous processing – as separate and distinct trade unions once did – putting down their tools and interrupting work tasks of all other trade unionists on the production line! Employers may periodically displace labor with machines, then, but they have yet to escape the pitfalls of their own technical wonders.

* * *

Rather than upgrade domestic technological competitiveness and earn the wrath of displaced labor during the late 1980's, many large steel companies chose the point of least resistance – exporting their plant and operations to low-wage nations, abandoning mills in Pittsburgh, Chicago, Gary, Cleveland, Baltimore and the like. Intense competition to attract foreign capital meant newly industrialized nations like Taiwan, South Korea and Hong Kong opened the door to U.S., British and Japanese steel companies. Further pulls on foreign steel firms then appeared in Eastern Europe not far from Western European markets. Eastern wage rates in manufacturing ran at an average of $2 an hour in 1989; and Polish rates were closer to 50 cents an hour after devaluation of the zloty – cutting them below effective rates in the Philippines.

Though the typical steel mill in the East produced 25 percent less steel and used more labor than in the West, some types of steel made by these inexpensive workers were qualitatively competitive in Western markets. And this drawing card for Western capital meant less investment in U.S. steel mills.[47]

In response to the dual danger of technical upgrading and capital exports doing away with the need for domestic labor, United Steel Workers' treasurer Frank McKee returned to the union's traditional goal of job-security, guaranteeing U. S. steel workers "a fair day's pay for a fair day's work." He saw the futility of union concessions in the face of massive layoffs. He sought, rather, to secure jobs by pressuring the Federal Government to protect American steel companies against lower-price foreign imports with tariffs. If the steel industry and job places could together be saved, he argued, the union's traditional demands of fairness in pay and work would be won.[48]

As a sign of possible future victories under pressure of the union lobby, Pennsylvania's 1978 Buy-American state law required documentation that all steel used by contractors on state government projects be American-made and not foreign-produced.[49] But even this could hardly stop the industry's decline.

Symbols of the decline and fall of steel production at the heart of Three River Valley's fame were everywhere. The skyline of Pittsburgh was no longer hidden in smog and the grays of belching smokestacks. High-tech ventures, health services and banking had become the anchors of the city's steadily diversifying economy. New technological companies were extending their domain; 19th century factories became residential lofts and condos. Vacant lots and hulking shells of long-abandoned steel mills were also being tuned into parks, residential developments, manufacturing plants and technology centers. The old train stations were transformed into shopping malls. And the struggle to modernize the central city and build two downtown stadiums, a modern convention center and a "Cultural District" of theaters and galleries was the talk of taxpayers not only in Pittsburgh and Allegheny County, but in nine surrounding counties as well.[50]

Crime rates and real estate costs were among the lowest of any big city in the country; and the waterfronts of the Three River Valley became places of leisure and pleasure – bringing Pittsburgh the 1990's renown as the most livable city in the nation!

The grime of mill operations could now be transferred elsewhere – to Japan, Taiwan and South Korea, to Poland, Russia and Brazil, to the rest of the industrializing Third World.

Yesterday seemingly vanished. In the wake of his operation of mills and coke works, Carnegie had once left behind refuse, ash, slate and slack, all dumped into streams, overflowing farm land that was made unfit for cultivation.[51] For the next century, the iron ore, coke, steel, oil, and chemical industries indiscriminately destroyed terrain and polluted air and water. Congress reversed course for a moment in 1981, starting a program to clean up the nation's most seriously polluted toxic waste sites, reaffirmed in 1990 by a "national policy . . . that pollution should be prevented or reduced at the source whenever feasible. . . ."[52] But Congress then strayed as the Federal Government stopped collecting a tax on polluting steel, oil and chemical companies on December 31, 1995. Fiscal gridlock and political indecision ensued, leaving the program in turmoil and the violating great oligopolies and conglomerates, steel manufacturers among them, off the hook.[53]

Proposed 1999 legislation by New York Senator Daniel Patrick Moynihan then envisioned extending to a nationwide scale the 1990 restrictions on airborne pollutants – including sulfur dioxide and nitrogen generated by coal-fired power plants and other industries. These, he insisted, caused acid rain, destroying ecosystems.[54]

Lacking such legislation and its enforcement, some dismayed America would return big steel and other industries to Carnegie's polluting ways. For there was no assurance that Congress would ever restrict large-scale industry that gave vibrancy to national production, employment, consumption and voter reaffirmation of Congressional representatives' right to serve. And besides, the national budget was woefully inadequate to finance the rehabilitation of over a century's destruction of the natural domain.

## The Crisis of Globalism [55]

For nearly three decades overproduction of steel, insufficient markets and periodic crises threatened the survival of competitive mills and their workers on an international scale.

In the late 1970's and early 1980's, a worldwide recession and high interest rates led to weakened industrial investments and demand for steel, heightening competition for U.S. steel makers from more efficient producers abroad.

By 1980, American steel mills were operating at little more than half of capacity, steel industry employment was at its lowest level since the depression year 1933 – and the industry's trade group, the American Iron and Steel Institute, demanded strengthened government restrictions on imported steel, contending that European and Asian steelmakers were "dumping" at artificially low prices, possibly below their cost of production. From 100 million tons shipped from U.S. mill docks in 1979, sales fell to 84 million tons in 1980 and hit a postwar low of 75 million tons in 1982.[56]

"The industry's fundamental problem is that global steelmaking capacity far exceeds steel demand," economist Robert J. Samuelson noted the continued downturn in 1986. "With a worldwide steel trade, this imbalance created downward pressure on prices, profits and wages. Through protection, labor and management sought to insulate themselves from these pressures. As a political venture, the labor-management alliance succeeded. Beginning in 1969, when Europe accepted 'voluntary' restraints, the industry has enjoyed various forms of protection. Formal and informal quotas now cover European and Japanese imports."[57]

As in the U.S., European production also turned down sharply – with a devastating effect on steelworkers. "Deficit-ridden big companies increasingly say they can't survive if they can't cut their bloated payrolls," the *Wall Street Journal* reported the crisis in France in the late '80s. "But with unemployment over 9% and rising, saving jobs has become the most pressing concern of the government's working-class core constituency."[58]

* * *

The underlying themes that were seemingly new in the late 1980's also faced the steel industry in 1892-93, and Andrew Carnegie with the Amalgamated Association of Iron and Steel Workers could not have failed to recognized them: There was too much steel capacity in relation to prospective buyers; the steel companies were upgrading technology, simplifying labor, slashing costs, cutting wages, closing down or shifting production from inefficient, high-cost plants, and lobbying for duties, import restrictions and other limits on foreign steel dumped or otherwise entering U.S. markets.

Of all protection-seeking U.S. industries: "Most important is steel, which had already – in 1982 – used countervailing duty suits to force a negotiated quota arrangement with Europe, and is currently using the same path on heavily subsidized shipments from Brazil," said L. M. Destler of the Washington Institute for International Economics. "This is part of a three-pronged industry campaign, which includes a quota bill cutting imports by one-third, and Bethlehem Steel's petition for import relief on grounds of injury to domestic companies."[59]

Just as during earlier moments of the dominant positions secured by Bethlehem and Carnegie with government complicity protecting markets behind high tariff walls, big steel companies were again pressuring agencies of state to secure rewards for themselves and make labor shoulder the burdens. And to some degree they were successful: By 1994, steel production surged, equaling its 1980 level, making the U.S. the second largest steel producer in the world after Japan.[60]

But the crisis returned with the unrelenting 1997-2002 financial and market crises raking Asia, Japan, Russia, Mexico and Brazil. The earlier problems of overcapacity, competitive technologies, and dumping reappeared with greater ferocity and devastation – putting some big mills on less than full-time, creating unemployment for large numbers of the remaining steel workers.

* * *

Facing a classic dilemma, the large mills with older technical and labor methods were again being squeezed between four major forces on both a global and domestic plane:

* Falling prices of imported steel reducing the prevailing price they could charge customers and maintain production at full-tilt, thereby making them unable to spread their fixed costs over a maximum number of tons, slabs, billets or other units of output;

* The need to constantly upgrade their competitive, cost-saving, production technologies with capital borrowed at high interest rates, reducing profitability;

* The introduction of new technical methods and mechanisms that displaced old work tasks, creating a surplus of labor that had to be reassigned and ultimately discharged from the mills; and

* An ultimate fall in capacity production levels, with profit-margins again requiring that surfeit labor be shown the mill gate.

Clearly, in meeting global competitors dumping on the U.S. market, both the investment of borrowed funds in new processes and equipment and the cutting of prices and productive capacity released and demeaned labor.

The dilemma remained. Throughout most of the 1990s, steel imports accounted for more than 20 percent of the domestic steel market, and U.S. steel companies themselves bought a quarter of this foreign steel – most in the form of raw slabs that they turned into finished steel.

And here was the rub: The large, integrated mills like USX, heir of Carnegie and Morgan, as well as Bethlehem, the LTV Group and Weirton had pumped $50 billion into modernization from 1989 to 1998, yet failed to install the latest thin-slab casting methods. Even recent modernization programs were made with borrowed capital, so that almost every added dollar in revenue earned was paid out in interest to their debt holders! Securing lower borrowing-interest-rates was blocked by the complete absence of a market for steel industry bonds, the likes that Henry Morgan once created. And this was undermining Morgan's megalomanian dream: of ever more finance capital devoted to added technical means, in order to enlarge production and employ more workers at low wages, to outflank all competitors by lowering steel prices – the better to create an all-encompassing domestic monopoly and global emperium.

The mini-mills, so-called, were momentarily outflanking Big Steel methods with thin-slab casting that they introduced in 1989 and extended to capture nearly half of entire 1998-99 domestic steel output.

And there was a second sign of death for the large mills: fixed costs were increasing despite mills running nearly flat-out at up to 95 percent of operating capacity to supply the surge in domestic market demand during the 1990s. The delirium of U.S. steel consumption that reached a 1998 record of 138 million tons – overshadowing 96 million tons in 1987 – could not save the big mills living on massive debt-to-revenue ratios to finance the installation of technologies that displaced one round of workers, only themselves to be outmoded and replaced by other costly techniques and methods that, in turn, undermined another tranche of labor. Ruination threatened both mills and workers.

The shortsighted answer of the American Steel Workers Union was that cheaper foreign and non-union, mini-mill labor were to blame. "It's not that we haven't done things right," recounted Mark Roach, a Weirton union man, workers themselves having bought the threatened plant from National Steel in 1984 for $386 million, then sold off 76 percent of the stock to raise money for new technologies. "We just can't compete with slave wages."

Destruction of these integrated mills would overwhelm both plant and workers in stages. Even Weirton would destroy its own mills and lay off its owner-workers. In 1998, it dropped its prices to meet those of foreign-dumped steel, shut down one of two blast furnaces, and laid off 1,000 men, nearly a quarter of the work force, just before Christmas. Worker-managers would calculatingly impoverish their own kind, just as Carnegie had done to his workers in the winter of 1892.

"We no longer need these giant facilities to make steel," Brookings Institution steel expert Robert Crandall assayed. "Mini-mills are building sheet plants and making life miserable for the big guys. We're talking about a long, drawn out endgame for Big Steel." To which the Cato Institute's trade expert Daniel T. Griswold ominously added, with a grimace Andrew Carnegie would never recognize or concede: "There is nothing that Congress can do about it."[61]

That was not the thinking of steel workers handed down from the past. As in early 1892, the union would agree with Carnegie that Congress could secure the domestic market, jobs and wage levels. "The real question is, do we want a steel industry in the United States?" United Steel Workers Union president George Becker begged the question more than a century later in March 1999. "We know that there's an overcapacity of steel in the world. If we do not act decisively in the present crisis, there will be no American steel industry in the 21st century."[62]

## The Restructuring of Capital [63]

Now in our moment during the late 1990's and early 2000's, the Monongahela Valley has nearly completely dismantled its old steel industries, attaining a state of seemingly permanent high unemployment for the blue-collar work force, and exhibiting an increase in social and health problems typically associated with extreme poverty and insecurities.

Homestead has long been closed down as a steel-making town, likely forever. Other steel-making towns have followed its lead. And throughout the Valley many old mills have been melted down into scrap.

What alternatives are left for steel workers in Homestead and throughout the nation?

Steel workers now argue that most of their remaining jobs can only be protected by devaluing the dollar to cheapen exports and establishing quota and tariff barriers against cheaper foreign steel. They believe the politicians and financial groups that oppose these measures are undermining steel markets, production, and employment – and the steel workers say they must be stopped.

Though their United Steel Workers Union has long lobbied for higher tariffs to keep out foreign steel products to protect jobs for American workers, tariffs would only protect some jobs for the moment. Certainly tariffs would keep competitive pressure off U.S. steel companies that might otherwise expand production by replacing their old equipment still in use and discarding technical labor essential for operating these ancient Cyclops.

Yet, by calling for tariffs, the union has failed to consider the possibility of solidarity with foreign steel workers, who are faced with the same problems as American workers and are fighting to keep their own jobs. And if the U.S. Steel Workers Union wants to protect future American jobs, why doesn't it organize a public campaign to break new ground – by demanding that the federal government nationalize and modernize American steel mills?

Instead, the United Steel Workers of America now see foreign steel mills and steel workers as the "enemy." It is an opinion much like that held by the Amalgamated in the late 19th century. And thus today's steel workers have again entered into a loose alliance with the American steel industry lobby in Congress, to pursue tariffs as part of a nationalist policy.

The common goal, supposedly, is to rebuild a domestic steel industry that can offer high paying jobs, leading labor to agree with management that some aspects of steel production and some jobs can be salvaged if a wall of U.S. protection is built quickly enough against foreign sellers. The result is that both union officials and the labor rank-and-file are establishing solidarity with the great U.S. steel corporations at home, rather than with other workers employed by steel companies abroad.

Have today's steel workers learned nothing from the past, then?

Union leaders still support corporate capital's "renewal" of uncompetitive and technically backward mills. They reason that new investments in the old technologies for steel production can provide jobs at high wages and sustain local retail markets and communities.

Yet such a local union strategy is shortsighted – a determination to rebuild yesterday's industrial system with its old steel-making techniques. The very same demands had surfaced in the steel industry of the 1890's when unions sought to hold back harsh, corporate decisions eliminating old methods and technologies, outmoding high-paying skilled jobs.

## Industrial Decay and Labor Aspirations [64]

The dilemma the steelworkers now face is also similar to the situation confronting most industrial workers in the nation.

Over a decade ago, in August, 1987, Commerce Department analyst Rudolph DePass concluded, "We are seeing a fall-off in the industrial, agricultural and mineral heartland of the country while services, defense and high-tech industries have prospered."

According to DePass, regional economic health in the Mid-Atlantic states, which include Pennsylvania, dropped from 36 percent above the national average per capita income in 1929 down to 6 percent above the national average in 1979, and to even less by 1986. With continued industrial declineby the turn of the century, it was lower still.

Against such elimination of jobs, labor has so far developed few successful plans. Neither the smokestack industries nor the new and powerful science-based, high-tech industries are today training or hiring the industrially-discarded workers at fair wages or salaries. And union steel workers being displaced see no purpose in maintaining their union membership – just as Homestead steel workers gave up their Amalgamated affiliation after 1892.

Still, in the Monongahela Valley, some workers have organized to try to save selected mills from closing by buying their outmoded facilities and then upgrading their techniques and output. One local union's plan to buy out U.S. Steel's Duquesne facilities, for example, was based on the workers borrowing huge sums on which interest would be payable for decades while they reduced their normal wages. There have also been unending demands for quotas on foreign steel – and even schemes for cities to gamble and win in the state lottery to save the likes of West Homestead and Carnegie's old Works.

Throughout the Monongahela Valley, a few organizers still felt it was important for the many out of work to go on demanding full employment, good wages and job security. By the mid-1980's, a number of liberal organizations in the Valley had taken up, and modified, radical demands that employers show social and economic responsibility. Yet, new demands of the 1990's had become respectful requests – some offered by those appealing from a position of near helplessness. The crisis of steel had consumed tomorrow's hope.

## Domestic High-Tech and Runaway Production

As industrial jobs were being lost in Homestead and elsewhere in America during the 1980's and 1990's, U.S. firms focused elsewhere:

expanding employment either in upscale technologies at home or by exporting capital for investment in new mills abroad.

This left remaining U.S. steelworkers segmented into two categories: better-paid, highly-trained, technical workers, and relatively low-paid, semi-skilled or unskilled laborers, who could learn their essential tasks in a few minutes or hours.

The division recalled the one Carnegie imposed in Homestead after the defeat of the Amalgamated. But now another dimension had also been introduced: an increasing proportion of the labor force was being made dependent on capital's initiative to start or stop production wherever factory managers want to locate. Steelworkers were being departmentalized at various functional levels of expertise and unemployment on a global scale. And as more U.S. steel workers reached these dead-ends or points of no return, a permanent underclass without modern industrial skills was building up – a body of unpaid jobless, unprotected from the vicissitudes of the markets for both steel and consumer goods.

Indeed, America today is in the throes of de-industrialization through .the reorganization of its heavy industries and the permanent displacement of part of its labor force. For Big and Little Steel alike, the writing was on the wall: there were 500,000 steelworkers in the U.S. in the 1950's, a steady fall leaving only 170,000 in early 1999, 160,000 by the year 2000;,[65] and by the end of 2001 only 150,000 remained![66] And what would happen to these steelworkers if the Federal antitrust authorities winked at the potential merger of giant companies like the U.S. Steel Group of USX, Bethlehem Steel and Wheeling-Pittsburgh?

Markets for steel had worsened in 2001 and 2002. In order to stay afloat, in the face of below-cost-dumping by state-subsidized foreign steel makers – and fearing their own impending bankruptcy – the big mills mobilized their congressional lobbying power, trying to shift their internal crises to the Federal Government. They sought to secure $2.4 billion in government funds over a three years period in order to pay 80 per cent of their retired workers' health and pension costs. They also sought to shield themselves from market forces, mobilizing the state as their first line of defense for what can only be described as an overarching, state-secured monopoly, one that would put even the likes of Andrew Carnegie's Homestead to shame:

> The steel makers want protection from low-priced imports. They want the United States to put pressure on other nations to scale back their steel-

> making capacity. They want billions of dollars in taxpayer money to help pay health and pension benefits. And they want assurances that their proposal to merge many of American industry's biggest companies into a bigger, more efficient competitor will not run into antitrust problems.[67]

Underlying pressures were meanwhile being elaborated by six interrelated developments impacting U.S. steel mills:

* the rapid, competitive advance of technology, which was being applied abroad even though it was not adequately used in America's big plants at home;

* the lowering of labor and other output costs overseas;

* the export of U.S. capital in the form either of technology to activate foreign-located production, or of finance to buy existing manufacturing plants;

* U.S. corporate use of overseas plants to produce low-labor-cost-parts to supply and service other foreign-located manufacturers, who then sent their goods back to the U.S. market or elsewhere for sale;

* the late 1990's and early 2000's overseas financial crisis, undercutting domestic production and markets, fueling massive exports from Asia, Russia and Latin America - with cheap steel flooding U.S. markets and driving some integrated American mills to operate part-time or to completely shut down early in the 21st century; and

* the competitive merger and consolidation of steel companies on a national and global plane, Big Steel in the U.S. calling for U.S. tariff and quota protection against foreign-made steel.

These overlapping forces lashed industrial nations, bringing shrill, if contradictory, pleas for expanding ever-illusive "free trade." Yet the possibility for such barrier-free trade had come to an end with the 1998-2002 collapse of demand for steel from Asian economies. There was now a worldwide glut of steel; and low-price steel began flooding the United States and other markets where demand was still relatively strong.

Foreign steel makers were desperate to find new buyers because Asian demand had virtually vanished overnight, while their production costs had fallen because their workers and local suppliers could be paid with depreciated currencies that had plummeted against the dollar. By late 1998, low-cost imports from Japan, Russia, Brazil and other economically troubled countries soared to nearly four million tons a month, making up nearly 40 percent of the domestic U.S. market, and driving down steel prices to 25-year lows. Giant U.S. steel mills simply could not compete with cheaper Asian steel. "With 10,000 steelworkers already out of work, three major steel producers in bankruptcy, another on the brink, and virtually every steel producer experiencing huge losses, action [on tariffs] isn't an option; it's an absolute necessity," fumed George Becker, still president of the United Steelworkers of America in March 1999.[68]

As the U.S. faced a record 1999 deficit in its balance-of-payments, the issue of dumping steel hit a political button, the House of representatives voting to impose import quotas - limiting steel imports to pre-1997 levels – on all foreign steel, not just exports from nations whose producers were selling below their costs. Though there were 40 times as many American workers in industries that used steel – and benefited from lower import prices – than were employed in the steel industry itself, protectionism was seen as a way to simultaneously protect the remaining domestic steel companies and encourage foreign companies to invest directly in U.S. production facilities – effectively recycling the dollars the U.S. paid for the flood of cheap imports. But there was no assurance that foreign investments would continue, one columnist foreseeing "if that changes, the dollar could come under pressure and the Federal Reserve might raise interest rates to support the currency."[69]

By early 2002, that had not yet happened. The global economy faced a recession. Foreign dumping of steel on the U.S. market continued, threatening steel-making by the largest mills in the key states of Pennsylvania, West Virginia and Indiana – their congressional lobbyists raising a storm of protest.

"This is an industry that has significant national security implications," fulminated Pennsylvania Republican Senator Rick Santorum, echoing H.C. Frick's point of view in 1892. "It's important from the whole basic concept of maintaining manufacturing in this country. It's important to have a vibrant domestic source of steel. Allowing the industry to be destroyed would put this country on a path to overdependence, similar to oil, on foreign sources."[70]

### The New Legerdemain: Mergers, Dumping and Tariffs

No one can foresee the future of foreign investments or domestic merger or consolidation of competitive, nationalist steel companies seeking quotas and tariffs to keep out foreign steel. For by early 2002 there was no international steel cartel to direct how production might be allocated, exports regulated, and quotas and tariffs permitted. Rather, nationalist fault lines had already emerged, being sharply drawn across the mantle of free trade competition in steel and its products.

In theory at least, mergers and consolidation of steel mills in competitive nations could lower costs and limit the supply of steel produced, while pushing up prices to buyers of raw steel products. But industrial nation states do not see eye-to-eye on how such combinations should take place. "The simple solution, of letting markets weed out the least efficient producers, is not even being considered," wrote seasoned U.S. financial analyst Floyd Norris at the close of 2001 "The Bush administration, which seems to be a strong believer in free trade so long as it does not cost any votes in swing states, is pushing for an international deal to cut [steel] production and thus raise prices. In other words, it thinks an international cartel would be just fine."[71]

But a cartel was not what smaller steel companies in the United States were after. The Steel Manufacturers Association – the trade group for the highly-efficient, so-called mini-mills – were battling the big steel makers, pointing out to the Bush Administration and Congress that its members made steel in 126 of 435 Congressional districts and employed 120,000 people – most of the steel workers in the nation! That was a lobby that politicians dare not ignore.

Besides, the giant mills had become dinosaurs in an environment that could no longer sustain their consumption of costly, borrowed capital to upgrade their technological base. "The majority of our industry is not in favor of bailing out a few steel companies that mismanaged their assets," stiffed Thomas A. Danjczek, president of the Steel Manufacturers Association. "Combining several [giant] companies that failed to earn their cost of capital doesn't mean the new company will be able to do it. It's like two garbage trucks colliding." [72]

Though one or two giant mills could easily provide all the steel plate needed for U.S. armor or shipbuilding, the consolidation of these companies would not likely revive them. By late 2001, they were operating at half the capacity they held in 1975. And, without state intervention to allow their merger in order to shield them from competitive foreign steel makers and domestic mini-mills, some would go to the wall.

This was partly because overwhelming market pressures had put the world's five biggest giant mills – that account for 20 percent of world output – in a cost-price squeeze between their three major global suppliers of iron ore, setting their cost outlays, and those setting steel prices – the principle users of steel, the world's five biggest auto manufacturers, accounting for seventy percent of global automobile production[73] – themselves enduring a cost-price squeeze in 2001-02.

Still another way of viewing the world crisis, 2001 steel prices reached a 20-year low, falling about one-third as demand plummeted on the heels of a severe global downturn in manufacturing industries. U.S. steel makers lost more than $1 billion, 25 companies declaring bankruptcy as the $260 average cost to produce a ton of steel exceeded the market price of about $200.

U.S. steel makers railed they had been lashed by the dumping of lower-price steel by Russia, South Korea, China and Japan – the U.S. administration claiming these foreign mills were "inefficient," government protected and subsidized – adding to global "excess capacity." Lower-priced steel produced in Europe was also the cause, the U.S. insisted, as it tried to win agreement for a reduction of "over-capacity" and the need to shut down European mills.

Only a cartel among the largest steel companies could accomplish this. But even the best of cartels would be tenuous, as worldwide steel production was fragmented, with the largest five companies producing less than one-fifth of global output; and – unlike U.S. steel companies' costly health care plans for retired employees – European public health care for such workers was directly paid for by many governments.

By the close of the year 2000, the leading United States steel corporations had substantial revenues, yet had left many of their existing and retired employees with inadequate pension funds and retirement benefits – a shortfall totaling over $8 billion:

| **Company** | **Revenue (Billions)** | **Employees** | **Unfinanced Pension** | **Other Unfinanced[74] Postretirement Employee Benefit Obligations (Bill.)** |
|---|---|---|---|---|
| USX | $6.09 | 35,000 | Fully Covered | $1.54 |
| LTV* | 4.93 | 16,500 | $642 million | 1.61 |
| AK Steel | 4.61 | 11,500 | Fully Covered | 1.45 |
| Nucor | 4.59 | 7,000 | $32 million | combined |
| Bethelem* | 4.20 | 14,700 | $442 million | 1.78 |
| National | 2.98 | 9,300 | $106 million | 0.45 |
| Totals | 27.40 | 62,000 | $1222 million | 6.79 |

* In bankruptcy, with partial disclosure of liabilities.

The companies insisted these shortfalls should be covered by the Federal Government, in order to put them on an equal cost-footing with Europe's giant mills.

Yet, even before the issue was resolved, in 2001 the steel unit of the Organization for Economic Cooperation and Development tried to create a cartel among these and other steel makers to reduce capacity by mergers. "A merger of U.S. companies would be positive," advised Christian Mari, director of the European Confederation of Iron and Steel Industries."[75] But the U.S. insisted European steel companies do the same – to limit capacity output and redundant workers, the better to increase market prices.

Though a tenuous cartel to cut production by 10 percent over the next decade was agreed to in Paris by representatives of 39 nations on December 18, 2001, the agreement was not binding and did not eliminate the risk of a serious trade war between the United States and other big steel exporting nations. For the proposed cuts were voluntary; there were no detailed allocations of world markets by either steel companies or nations; there were no agreed prices for the various categories of steel sold either domestically or in global markets; there were no explicit prohibitions against dumping steel below the cost of production; there were no agreements over how fixed costs would be allocated by each company over all units of steel sold; and there were no designated, binding cartel penalties to be imposed for violation of proposed cuts in output, though these cuts were supposedly designed to eliminate nearly 150 million tons in surplus production each year:

| Nation | Million Tons a Year | Million Tons in a Decade |
|---|---|---|
| United States | 14-17 | 140-170 |
| European Union | 13 | 130 |
| Japan | 4 | 40 |
| South Korea | 3 | 30 |
| Other Nations | 113 | 1130 |
| Totals | 147-150 | 1470-1500 |

Failing this goal to cut global production, the United States rattled an economic saber, threatening to impose tariffs of up to 30-to-40 percent on all imported steel – cutting off foreign steel from the U.S. market.

It was clearly Big Power brinkmanship. And European officials were furious, claiming the U.S. was reverting to protectionist policies, while failing to close down its own inefficient mills, merge the remaining steel producers, and limit output. But even this seemed unlikely so long as the largest three U.S. steel corporations were at odds with the more efficient min-mills.

A lasting cartel also depended upon the degree to which each nation and each company was willing to close down mills, merge and reduce output. The United States held only 12 percent of global steel production of some 850 million tons in December 2001, though worldwide steel consumption hovered around 700 million tons. In this buyers' market, prices plummeted. And nationalist interests bred demands to secure domestic markets and self-sufficient steel production.

All nations wanted others to close mills, merge and cut production capacity – effectively demanding the forced idleness of workers elsewhere. And in this competitive *milieu*, the largest United States steel companies sought to cut off foreign steel imports that made up about one-quarter of all steel used domestically.

U.S. trade representative Robert B. Zoellick countered Big Steel, insisting the Bush Administration would impose protective trade barriers only if its industry consolidated to reduce its own overcapacity. The USX-U.S. Steel Group, Bethlehem and four other majors steel makers quickly agreed, hoping to generate some $10 billion in profits behind high tariff walls, saving 7,300 or more steel workers' jobs – and placing a $7 billion burden in higher prices on U.S. consumers in the four years 2002-2006! And so it was done, the Bush Administration imposing a 30 percent tariff to save Big Steel – and secure votes for Republican congressional candidates in key steel-making states.

Still, not everyone was pleased. "The domestic steel industry has been a corporate welfare queen for 30 years," criticized David Phelps, president of the American Institute for International Steel, an organization representing both steel importers and exporters. "The U.S. steel industry claims that if the government doesn't give them more aid, the whole industry will die. That is just nonsense."[76]

Nonsense or not, steel giants in jealous nations sketched a larger portrait of their place setting and influence in government affairs. By the close of 2001, steel company mergers in all the leading competitor nations were being subsidized by their respective governments. Their legislative lobbies were more powerful than the influence held by the purveyors of electronics, plastics and synthetics that, together, had not yet replaced steel as the essential ingredient of industrial power.

"If we want to save the steel business, it needs to be more concentrated" to limit output and push up prices, confirmed Pierre Bourrier, a spokesman for Unisor of France. "We are glad the American companies are starting to reorganize."

This advice was based on galvanized experience. For European steel had already undergone its own painful reorganization in the 1980s and 1990s, as part of a privatization program that abolished many state subsidies, closed inefficient mills, reduced capacity through mergers – and led to massive layoffs. Three of the top five European steel producers had merged, including Corus, which was formed by the 1999 merger of British Steel and Royal Hoogovens of the Netherlands, and ThysenKrupp the same year.

Europe was also readying for economic warfare. The February 2002 three-way merger of Usinor of France, Arbed of Luxembourg and Aceralia of Spain was heralded as the world's largest steel maker, being readied to cut overlapping facilities and redundant workers in order to reduce costs to supply steel throughout the European common market at prices no other competitor could equal.

Not to be upstaged, on December 11, 2001, Japan's Sumitomo Metal Industries pre-announced its proposal for "deeper ties" with its rival, the Nippon Steel Corporation, in order to combat a glut of production that was dragging down prices. The nationalist aura of the deal was clear, as Sumitomo Metal offered to exclusively supply Nippon Steel with semi-processed slab from its mill in central Japan – reducing the cost to Nippon to get the material from its own, more distant mills, the better to exclude all foreign steel from the Japanese market.

Meanwhile, struggling with bankruptcies and hurt by foreign imports, in December 2001 the largest U.S. steel companies pressured the Bush administration and the Congress to back a plan allowing up to half a dozen steel companies to merge into a single company and to shift more than $8 billions of dollars in retiree health care costs to the federal government (see the chart above). These companies insisted that Big Steel was vital to a manufacturing-based economy, that its linkages with other industries would continue to frame the development of new alloys, new production techniques and new uses in directly making machine tools, ships, tanks, bombs – and indirectly making the equipment to produce, or experiment with, computer-linked hardware and software. They insisted some among them were globally competitive, especially the likes of U.S. Steel and AK Steel, Nucor and Ipsco.

But what about the ailing half of the industry handed down from the 19th century? These 30 giant "dogs of production" devoured coke and iron ore in highly inefficient blast furnances. They held fast 15 or so percent of the nation's 100 million tons of annual steelmaking capacity. Yet sooner or later they would face bankruptcy – headed for the scrap pile if the government refused to protect them with tariffs and subsides.

So, then, should they live on forever as state welfare clients, or be allowed to gracefully die with the steelworkers who built the nation's industrial might? Then again, was there not another force that could somehow pump new vitality into these relics of the past?

Even the largest U.S. mills were small by comparison with those in Europe, Japan and Korea; and to try to equal their size U.S. Steel proposed a merger with Bethlehem, National, Weirton and Wheeling-Pittsburgh – trying to replicate Carnegie's late 19th century conglomerate quest:

**Raw Steel Production in 2000 (In millions of net tons)**[77]

| | |
|---|---|
| Usinor (France, Luxembourg & Spain) | 54 |
| Nippon Sumitomo (Japan) | 44 |
| NKK Kawasaki (Japan) | 32 |
| Posco (South Korea) | 31 |
| Corus (Britain & Netherlands) | 22 |
| ThyssenKrupp (Germany) | 20 |
| U.S. Steel | 12 |
| Bethlehem (U.S.) | 11 |
| National (U.S.) | 6 |
| Weirton (U.S.) | 3 |
| Wheling-Pittsburgh (U.S.) | 2 |

The logic was evident. Facing a 2-year low in steel prices, the template to reduce 2001 U.S. steel capacity of 125 million tons by 14 million to111 million by the end of 2002 could only be negotiated by a cash-less mergers of bankrupt mills. But to merger the five largest old-style mills into a profitable combination, the overhang of as much as $12 billion in health and benefit costs of retired steelworkers would first have to be shifted to the Federal Government. These costs were greater than the value of the companies, the steel industry's overall 150,000 workers effectively supporting some 600,000 retirees!! The only other alternative for the bankrupt mills was to liquidate – deserting their labor force and laving them without benefits – forcing U.S. Steel to compete with the remaining European, Japanese and American giants to buy their assets.[78]

And, yet, despite the consolidation of the great steel corporations, the issues of steel capacity and dumping steel revolved around much deeper roots.

## Deeper Roots

For twenty years or more, U.S. corporations had taken the lead in making strategic decisions about the geographic locale of new investments, factory production and employment, while U.S. unions remained relatively weak and mass joblessness in steel and the other basic industries were not reversed.

Big steel companies had refused to invest in the latest, domestic technology unless labor was partly discarded, partly retrained, and significantly cheapened. These companies sought higher returns by exporting capital in the form of production and buying foreign mills, hiring cheaper labor abroad.

At the start of the 21st century, the great steel employers continued to deploy their operations internationally, demanding investment-and profit-security through complete control over the reorganization of the workplace and wage levels in every nation where they operated. Since a least 1983, they had calculated where to place production by comparing relative U.S. labor productivity, costs and the size of domestic market with foreign labor and production outlays, price levels and nearness to markets[79]

Company managers had developed cost-saving strategies, which included upgrading their technical equipment and methods – or deserting one set of workers to employ another at lower overall costs. Clearly, when these companies upgraded their technical operations, they often displaced skilled labor; and if they moved operations abroad, they also deserted domestic labor. Trapped within the void of production in the old mill centers, U.S. steelworkers lacking mobility remained nearly helpless.

Steelworkers meanwhile had bide their time by seeking an alliance with their employers for tariff protection. But the future of any alliance was uncertain and would likely last only until the next downward price-adjustment of hot-or cold-rolled steel, when the employers would again move to protect their profit-margins by lowering wage rates on the sliding scale.

## Hollowing-Out

By pursuing profit goals without responsibility to their domestic employees, the great steel employers have undermined America's potential industrial power – a power handed down from the 19th century industrial revolution that once set labor afoot, demanding a better life.[80]

These companies' overseas adventures have already left behind nothing to replace lost jobs or the opportunity to fulfill the potential and aspirations once offered labor by steel and other basic industries – that for more than a century called millions of skilled and unskilled workers into various degrees of industrial servitude.

Few sing the praises of the low wages, dangerous work, and desperation suffered by the steelworkers. But the fact remains these workers of decades past built and sustained the nation's industrial base and made its economic prosperity possible. To hollow-out this heartland and desert those who followed yesterday's steelworkers – all in the name of lucrative opportunities elsewhere – seems morally banal, socially inequitable, and historically unwise for the nation's future industrial positioning.

This is not to say that one species of technology will not replace another in steel-making. In the long, drawn-out endgame, the old-line, big steel companies will have to compete with both lower-cost mini-mills and foreign mills, with far cheaper labor costs, periodically dumping on the U.S. market. Big steel will surely remain on the run. And by the late 1980's, the steel masters had already moved their industries away from communities in the Great Lakes and Midwestern states comprising the Rust Belt, as well as from the red hills and gulf lowlands of Texas, and from Alabama and the Monongahela Valley, its nexus in Pittsburgh no longer suffering beneath smoke-riven skies.

Questions remain, though. Are steelworkers and steel companies together caught in an all-but-unstoppable cycle of "creative destruction," as the overall U.S. economy periodically withers, constantly remaking itself in order to prosper by giving up once-critical, industrial operations that other nations can undertake more efficiently and cheaply?[81]

What will steelworkers do if and when the great employers again desert them, setting up operations in places where labor is less expensive, raw materials are cheaper, markets nearer and prices lower? If they protest or withhold their labor, will the employers lock them out, close down, and call out their private guards[82] – or their newly-hired armies under management of the Swedish Securitas, the European detective powerhouse that has taken over Pinkerton's Inc., and hopes to leap to prominence in the United States, offering guard services as well as security for multinational companies?[83]

The past may offer a clue to what is to come. After all, was not

Homestead in its death throes an epigram of the fate of the domestic steel industry – an about-face away from basic production, mill employment and essential self-sufficiency – that spelled doom for Carnegie's old mill and its workers' security?

**Rationalizing the Past**

The answer remains clouded by nostalgia that itself breeds justifications, as today the heirs of Carnegie and Frick rationalize the past, elevating the acts of their forbears to socially moral righteousness and the supposed natural evolution of our species.

Thus Frick's great-granddaughter, Martha Frick Symington Sanger, boldly insists that his purchase of expensive art, now grandly exhibited at the Frick Collection in New York, is an expiation of grief over losses of loved ones in his family and a recall, as in Rembrandt's "Polish Rider," of his early life journeys on his gray horse, Billy, through Pennsylvania's coal bluffs; that, too, El Greco's "Purification of the Temple," in which Jesus drives out the money-changers, commemorates Frick's violent victory over the intractable union at the Homestead mill.[84] And yet, it was not the workers, but Frick and Carnegie who inhabited the Temple of Great Wealth that housed their money-changers.

In expiation, a few historians similarly view Carnegie psychologically, as impelled to success by his mother, fueling his unscrupulous, brutal and bloody victory over the locked-out workers at Homestead; this despite his supposed empathy for labor rooted in his own father's ruin by steam-powered looms – that put Britain's hand-loom weavers out of work in the mid-1800s – initially driving the family to the United States.

Through this narrow prism, Carnegie seemingly sought to chase with the hounds of gilded wealth and run with the desperate hares enduring working class life. And yet the great steel-master carefully tried to avoid the public appearance of the Gorgon of labor's suffering, delegating the whip hand to others. True to his scripted calculations, cheering on the iron-willed Frick determined to win at Homestead, Carnegie retreated to Scotland, leaving Frick as his willing alibi – as if Carnegie himself need not face the logic that crushing the Amalgamated undercut his supposed belief in the so-called "principle" that workers have a right to organize for decent pay and protection of their jobs.

Still Carnegie's deviousness was transparent; and his image was once, forever shattered as a radical reformer – supposedly acting fairly towards his workers – despite his generosity, giving away expensive church organs and building grand libraries and fine opera houses to elevate the condition of workers in mill towns enclosing their lives.[85]

Why, though, was Carnegie's soul so confounded? Perhaps because, at the core of his beliefs, lay a deeper logic, enveloped in the concept of Social Darwinism, that was largely grafted from others holding eminence and standing in the *milieu* of late 19th century thought.

Edifying his creed that men of talent and ability should go into business and avoid the trivialities of politics,[86] Carnegie gregariously financed mill-town tours by famous philosophers and social thinkers of the day – one of the most eminent being Lord Herbert Spencer, advocate of Social Darwinism, a concept relying on the principle of "Survival of the Fittest," and from whom the eminent naturalist Charles Darwin himself took the phrase as more accurate than his biological term, "Natural Selection," in describing the struggle of species for existence.[87]

Darwin gratefully acknowledged Spencer's academic treatment concerning crossing domesticated species of animals, when Darwin wrote of the unknown bond essentially related to the principle of life: "that life depends on, or consists in, the incessant action and reaction of various forces, which, as throughout nature, are always tending towards equilibrium; and when this tendency is slightly disturbed by any change, the vital forces gain in power." [88]

Spencer respond in kind: "This survival of the fittest which I have here sought to express in mechanical terms, is that which Mr. Darwin has called 'natural selection, or the preservation of favored races in the struggle for life.'"[89] Spencer generously applied the principle to the evolutionary development of species' mental powers.[90] And in a short leap Spencer applied such formulations to humankind, focusing his lens on – extrapolating from and reifying – Darwin's central thought, placing it in realm of race, social caste and class.

Spencer then transferred and transmogrified this concept to the workings of unseen forces in emergent capitalist society, bringing his thesis up to date, then looking for verification in the past. "The poverty of the incapable, the distresses that come upon the imprudent, the starvation of the idle, and those shoulderings aside of the weak by the strong, which leaves so many [as Shakespeare remarks[91]] 'in shallows and in miseries,' are the decrees of a large, farseeing benevolence."[92]

Spencer offered corroboration: "We have unmistakable proof that throughout all past time, there has been a ceaseless devouring of the weak by the strong."[93] So that, he affirmed of the avaricious, "the first requisite to success in life is to be a good animal."

How, though, did this philosophical perspective dovetail with Carnegie, the steel industry, and the destiny of Homestead?

Spencer's long lifeline (to1903), overlapped with Carnegie's late 19th century eminence, when capitalism had reached the monopolist stage in both England and the United States. The great philosopher was alive to the forces at play: that capitalist production had transformed the simple homogeneous society of the past based on undifferentiated labor into a complex structure which integrated the *different* functional divisions of work and social processes. He abhorred maintaining centralized control over the system's dissimilar parts by state coercion, but also thought that, in the end, government would nonetheless inflate spending and undermine the ingenuity of each individual to maintain one's self.[94]

Carnegie leapt on, then modified Spencer's formulation. Why not let the forces of competition establish central control outside the realm of government? Why not let the captains of industry with ample resources establish market dominance over machinery and labor? Wrote Carnegie wistfully:

> While the law [of competition] may be sometimes hard for the individual, it is best for the race, because it insures that survival of the fittest in every department [of business and social affairs]. We accept and welcome, therefore, as conditions to which we must accommodate ourselves, great inequality of environment, the concentration of business, industrial and commercial, in the hands of a few, and the law of competition between these, as being not only beneficial, but essential for the progress of the race.[95]

For Carnegie, it was a short ideological jump to apply this theory of Social Darwinism to crushing his unwieldy, obstreperous workers at Homestead. In the struggle of "Survival of the Fittest," eviscerating disequilibrium of the forces of labor and capital, Carnegie would try to prove he was the "vital force" that grasped for and assumed overwhelming power, securing the "sacredness of property" resting at the core of civilization.

Was there no good side to this great captain of industry, then?

In his waning years Carnegie fell back on his 1889 insistence that: "Surplus wealth is a sacred trust which its possessor is bound to administer in his lifetime for the good of the community." Carnegie tried, lived to 1919, gave away millions, and sought to meet his own deadline for moral rehabilitation, that: "The man who dies . . . rich dies disgraced."[96] And yet, from the violent fugue of Homestead, he would never regain his full stature as a man of honor.

* * *

Though now there is no way to repair the past, there is a lesson here if only we can read between the lines. For the future choice for U.S. steelworkers remains clear: they must either give way or fight – and over what they will fight is still an open question given a scope and vision limited by their hope to secure steady work during an ongoing, technological revolution and market crisis eliminating jobs, while holding on to an influential place in a two-party political apparatus that straddles the lines of social class.

If union steelworkers persist in maintaining their position as a "monopoly of labor" operating within the confines of a "monopoly of mill capital" allied in political lobbying, as they did in Carnegie's day,[97] they will be forever dependent on the mill employers' competitive costs of production *vis-à-vis* the market price of steel.

In the end their job security will remain tenuous and their wage levels will turn on the employers' sliding scale and uncertain tariff protection – though steelworkers may "mutually pledge to each other" their "steadfast purpose as union men," their "honor" and their "very lives" – as they valiantly fight to the last mill worker in the nation!

And thus ends this grief-ridden, Homeric epic, each generation of steelworkers having had to face anew employers bent on their degradation and immizeration. Their struggle continues still, so much so that, even if today's remaining mill workers successfully join the big steel companies securing tariff and quota protection – momentary job security and livable wages – it might well be that the handful of steelworkers of tomorrow will be no closer to the resolution of their underlying difficulties than they were at Homestead in 1892.

## Notes

[1] See Jean Strouse, *Morgan: American Financier* (New York: Random House, 1999).

[2] John A. Fitch, *The Steel Workers* (New York: Russell Sage Foundation, 1911), 216-217.

[3] William Z. Foster, *Organizing Methods in the Steel Industry* (New York: Workers Library Publishers, 1936), 5.

[4] John Steuben, *Strike Strategy* (New York: Gaer Associastes, Inc., 1950), 194-195.
[5] Richard W. Stevenson, "Big Steel: An Invalid that Can Roar in Washington," *The New York Times*, December 17, 2001.
[6] Samuel P. Hays, *The Response to Industrialism: 1885-1914* (Chicago: University of Chicago Press, 1957), 8-9.
[7] Jesse S. Robinson, *The Amalgamated Association of Iron, Steel and Tin Workers* (Baltimore: John Hopkins Press, 1920); Hogan, Vol. I, 212-237, Vol. II, 444.
[8] Hamlin Garland, "Homestead and Its Perilous Trades," *McClure's Magazine*, June 1894; Steuben, *passim*; Fitch, 91, 213-220, *et passim*; Hogan, Vols. I, II, *passim*.
[9] David Brody, *Steelworkers in America* (Cambridge, Mass.: Harvard University Press, 1960), 52-63, 99-101; Foster, 5ff; Fitch, 91, *et passim*; Bridge 202, *et passim*.
[10] Foster, 5ff.; see André Gorz, *Strategy for Labor: A Radical Proposal* (Boston: Beacon Press, 1967), *passim*; "Organization of Labor," *Harvest Quarterly*, No. 1, 1975 (Santa Barbara: Harvest Publishers); Wolff, 41-44, 130.
[11] See Douglas A. Fisher, *The Epic of Stee*l (New York: Harper & Row, 1962); Paul H. Douglas, *Real Wages in the U.S., 1890-1926* (New York: Houghton Mifflin, 1930).
[12] Goldman, *Living My Life*, 87-88.
[13] Fitch, 215-217, *et passim*; M. Friedman, *The Pinkerton Labor Spy* (New York: Wilshire, 1907), *passim*; Foster, *passim*; Burgoyne, *passim*; Wolff, *passim*.
[14] Compare Chapter X, "The Militia and the People," with Steuben, 194.
[15] Emerson Jr., *A History of the Nineteenth Century*, Vol. III, 1746.
[16] Robert R. Brooks, *As Steel Goes . . .* , 28; John B. Commons, Assoc., *History of Labour in the United States*, Vol.II (New York: Macmillan Company, 1926), *passim*.
[17] Henry F. May, *Protestant Churches and Industrial America* (New York: Unpublished mss., 1949); Fitch, *passim*.
[18] Gunther, 618, fn. 3.
[19] See Matthew Josephson, *The Robber Barons* (New York: Harcourt, 1934), *passim*; see also *The Wall Street Journal*, 1901, *passim*.
[20] Jenifer Lee, "An Echo of Big Dreams of a World With Big Steel," *The New York Times*, December 5, 2001.
[21] See Strouse, *Morgan . . . passim*; Gustavus Myers, *History of the Great American Fortunes* (New York: Modern Library 1936), *passim*; Ferdinand Lundberg, *America's 60 Families* (New York: Vanguard Press, 1937), *passim*; Fisher, *The Epic of Steel*, *passim*.
[22] Brooks, 28-29; see Philip Taft, *Oganized Labor in American History* (New York: Harper & Row, 1970), 136-145.

[23] See Fitch, *passim*; Brooks; A. Austin, *The Labor Story* (New York: Coward-McCann, 1949); Charles M. Parker, *Steel in Action* (Landcaster, Pa: Cattell, 1943).

[24] This section is based on Foster, *Organizing Methods in the Steel Industry*; Fitch, *The Steel Workers*; Harry Braverman, *Labor and Monopoly Capital: The Degradation of Work in the Twentieth Century* (New York: Monthly Review Press, 1974).

[25] Burr W. Leyson, *Careers in the Steel Industry* (New York: E.P. Dutton, 1945); Robert A. Hadfield, *Work and Position of the Metallurgy Chemist* (Sheffield, England: University of Sheffield, 1921), Address; Fisher, *passim*; Brody, *passim*.

[26] Gunther, 617-620.

[27] Richard Krooth, *Empire: A Bicentennial Appraisal* (Santa Barbara: Harvest 1975), 12; A. Berglund, "U.S. Steel Corp.," *Columbia University Studies*, XXVII, 223-400.

[21a] For the theoretical framework, see Richard Edwards, et al., eds., *Labor Market Segmentation* (Lexington, Mass.: Harvard University Press, 1975).

[28] Foster, 3-4.

[22a] See Fisher, *passim* ; Foster, *passim;* Theodore A. Wertime, *The Coming of Age of Steel* (Netherlands: University of Chicago Press, 1962), *passim*.

[29] Gunther, 621.

[30] Foster, 5. For other views, see Brody, *passim*; Hogan, Vols. I, II, *passim*.

[31] Richard Krooth, *Arms & Empire: Imperial Patterns Before World War II* (Santa Barbara: Harvest Publishers, 1981), 67.

[32] *Ibid*., 78.

[33] Gunther, 621.

[34] See David M. Kennedy, *Freedom From Fear: The American People in Depression and War, 1929-1945* (New York: Oxford University Press, 1999); PBS, World War II Series, 1999; The History Channel, World War II Series 1999-2002..

[35] Gunther, 621-622.

[36] *Ibid*., 623.

[37] Amal Kumar Naj, "Steel Firms Strive for New Technology," *The Wall Street Journal*, February 21, 1989.

[38] Jonathan P. Hicks, "A Faster Path to Finished Steel," *The New York Times*, April 7, 1991.

[39] "Nucor Cuts Its Prices for Hot-Rolled Steel," *Bloomberg News* (Charlotte, N.C.) December 3, 1998. On December 3, 1998, the selling price of hot- and cold-rolled steel was cut by the second largest U.S. steelmaker, Nucor, to counter a wave of cheaper imports, as low as $200 a ton, from Japan, Russia and other nations. Nucor cut its price by 7.4 percent to $250 a ton to its largest customers for steel used in autos, appliances and machinery. The cut was not for everyone: smaller customers paid as much as $280 a ton; and steelworkers paid with wage reductions.

[40] See William T. Hogan, *Global Steel in the 1990s: Growth or Decline* (Lexington: Lexington Books, 1991); *Minimills and Integrated Mills: A*

*Comparison of Steelmaking in the United States* (Lexington: Lexington Books, 1987); *Steel in the United States: Restructuring to Compete* (Lexington: Lexington Books, 1984).

[41] See Clyde H. Farnsworth, "Pacts to Ban Subsidy of Steel Set," *The New York Times*, December 13, 1989.

[42] Alexander Cockburn and James Ridgeway, *The Village Voice* (New York, New York), March 8, 1983.

[43] See "Preface" by Morris Edelson.

[44] See Laurie Graham, *Singing the City: The Bonds of Home in an Industrial Landscape* (Pittsburgh: University of Pittsburgh Press, 1999).

[45] Joseph K. Stone, "Letter," *The New York Times*, December 18, 1989.

[46] Jonathan P. Hicks, "lnnovatiue Steel Mill Is Off To a Slow Start," *The New York Times*, December 2, 1989.

[47] "Talking Business With Uanous of Planecon," *The New York Times*, December 26, 1989.

[48] For other views of the dilemma facing steel workers, see Mark Reutter, *Sparrows Point: Making Steel – the Rise and Ruin of American Industrial Might* (New York: Summit Books, 1989).

[49] "Firm Insists Its Steel Was Made in America," *The San Francisco Examiner*, March 6, 1984; see John P. Hoerr, *And the Wolf Finally Came: The Decline of the American Steel Industry* (Pennsylvania: University of Pittsburgh Press, 1989).

[50] Michael Janofsky, "Pittsburgh at Progress Crossroads," *The New York Times*, September 21, 1997.

[51] Lenz v. Carnegie Bros. & Co., Limited, *Pittsburgh Legal Journal*, Number 42, May 18,1892, 383-385; see also Gordon J. Parr, *Man, Metals & Modern Magic* (Cleveland, Ohio: Iowa State College Press, 1958).

[52] U.S. Congress, *Pollution Prevention Act of 1990* [U.S.C.A. par. 13101 (b)].

[53] See John H. Cushman Jr., "Program to Clean Toxic Waste Sites is Left in Turmoil," *The New York Times*, January 15, 1996.

[54] See "Acid Rain's National Reach," Editorial, *The New York Times*, April 10, 1999.

[55] See William T. Hogan, *Global Steel in the 1990s: Growth or Decline*, (Lexington, Mass.: Lexington Books, 1991); *Steel in the 21st Century: Competition Forges a New World Order* (Lexington, Mass.: Lexington Books, 1994); *Steel in the United States: Restructuring to Compete* (Lexington, Mass.: Lexington Books, 1984).

[56] William J. De Lancy, Congressional Lobbying "*Memo*" (Washington, D.C.: American Iron and Steel Institute, June 1980).

[57] Robert J. Samuelson, "The Economics of Self-Pity," *Newsweek*, March 5, 1984, 78.

[58] Robert Ricklefs, "France Sends Workers Into Early Retirement, But Problems Persist," *The Wall Street Journal*, February 13, 1984; see also Earl Bohn, Associated Press, December 26, 1983.

[59] I.M. Drestler, "Protectionism and Election Year Politics: Why Reagan Is A Free-Trade Villain," *The New York Times*, March 18, 1984; see also "U.S. Claims Six Nations Dump Carbon Steel Goods," *The Wall Street Journal*, May 31, 1985.
[60] See American Iron and Steel Institute, *Report* (Washington, D.C.: June 1995).
[61] Quoted at Leslie Wayne, "Big Steel's Problems Are Home Grown," *The New York Times*, April 29, 1999.
[62] Quoted at Leslie Wayne, "A Divisive Steel-Quota Debate Moves to the Senate," *The New York Times*, March 21, 1999.
[63] William T. Hogan, *Global Steel in the 1990s . . . ; Minimills and Integrated Mills . . . ; Steel in the 21st Century . . . ; Steel in the United States . . ..*
[64] Hogan, *Global Steel in the 1990s . . . ; Minimills and Integrated Mills . . . ; Steel in the 21st Century . . . ; Steel in the United States . . . .*
[65] Leslie Wayne, "A Long-Shot House Vote to Limit Steel Imports," *The New York Times*, 19 March 1999.
[66] Richard W. Stevenson, "Big Steel. . ." *The New York Times*, December 17, 2001.
[67] *Ibid.*
[68] Wayne, *The New York Times,* March 19, 1999.
[69] Floyd Norris, "Starting a Trade war When Times Are Good," Editorial Observer, *The New York Times*, March 22 , 1999.
[70] Stevenson, *The New York Times*, December 17, 2001.
[71] Floyd Norris, "Protecting Steel: An Economic War That All Could Lose," *The New York Times*, December 21, 2001.
[72] Stevenson, *The New York Times*, December 17, 2001.
[73] Suzanne Kapner, "European Steel Makers Say a Merger in the U.S. is Overdue," *The New York Times*, December 6, 2001.
[74] Company Reports as of December 31, 200; *Bloomberg Financial Markets.* As charted by *The New York Times*, December 5, 2001.
[75] *Ibid.*
[76] Leslie Wayne, "Rivals and Others Criticize Steel Makers' Bid for U. S. Aid," *The New York Times*, December 6, 2001.
[77] Bethlehem Steel, citing International Iron and Steel Institute, 2001.
[78] Leslie Wayne, "Parched, Big Steel Goes to Its Washington Well," *The New York Times*, January 20, 2002.
[79] Included among the vast literature on the subject, the most outstanding is Christoper Tugendhat, *The Multinationals* (London, 1975); also see A. D. Smith, et al., *International Industrial Productivity: A Comparison of Britain, America and Germany* (Cambridge Univedrsity Press,1983).
[80] John Bernard, *Walter Reuther and the Rise of the Auto Workers* (Boston: Little Brown, 1983).
[81] See Richard W. Stevenson, "Greenspan Denounces Growing Protectionism," *The New York Times,* April 17, 1999.

[82] For historical background see Richard W. Rowan, *The Pinkertons*, (Boston: Little, Brown, 1931); see also U.S. Congress, House of Representatives *Report* 2447, 52nd Congress, 2nd Session, *Employment of Pinkerton Detectives.* For a contemporary view that Microsoft's Bill Gates conceivably could go beyond emulating Carnegie library grants to creating a private army, see Bill Eldridge, "Letter," *The New York Times*, 26 February 1999.
[83] See Barnaby J. Feder, "Securitas of Sweden to Acquire Pinkerton's in $384 Million Deal," *The New York Times*, February 25, 1999.
[84] Martha Frick Symington Sanger, *Henry Clay Frick: An Intimate Portrait* (New York: Abbeville Press, 1997).
[85] Robert L. Heilbroner, "Epitaph for the Steel Master," *American Heritage*, August, 1960; Robert L. Reynolds, "The Works are Not Worth One Drop of Human Blood," *American Heritage*, August, 1960; Carnegie, *Autobiography*; Roe Wellington, *Juggernaut* (New York: Lippincott and Co., 1948); PBS, *The American Experience: "The Richest Man in the World: Andrew Carnegie,"* January 20, 1997.
[86] Hays, 22; Carnegie, *Autobiography, passim.*
[87] Charles Darwin, *The Origin of Species* (New York: P. F. Collier & Son, 1909), 77.
[88] *Ibid.*, 318.
[89] Herbert Spencer, *Principles of Biology* (London: British Museum manuscript edition, 1864-1867), Part III, Chapter 12.
[90] *Ibid.*, 16.
[91] William Shakespeare, *Julius Caesar*, IV, iii, 217.
[92] Herbert Spencer, *Social Statics*, (London: British Museum manuscript edition, 1851), Part III, Chapter 25.
[93] Herbert Spencer, *First Principles*, (London: British Museum manuscript edition, 1861).
[94] Richard Krooth, *The Great Social Struggle*, Vol. III, (Santa Barbara: Harvest Publishers, 1980), 13.
[95] Andrew Carnegie, "The Gospel of Wealth," *The North American Review*, June 1889.
[96] Ibid.
[97] For historical parallels, see E. J. Smith, *The New Trades Combination Movement* (London: Rivingtons, 1899); Mike Holbrook-Jones, *Supremacy and Subordination of Labour: The Hierarchy of Work in the Early Labour Movement* (A Study of Work in Victorian Britain) (Exeter, New Hampshire: Heinemann Educational Books, 1983).

# Bibliography


"Acid Rain's National Reach." Editorial. *The New York Times* 10 April 1999.

Adamic, Louis. *Dynamite.* New York: Viking, 1934.

Adjutant-General of Pennsylvania. *Annual Report for the Year 1892.* Harrisburg, Pennsylvania.

"A Good Workman is Known by His Tools." *Popular Educator* Vol. VIII No10 (June, 1891): 321.

Alfonsi, Alice. "The Blue-Collar World." *Pittsburgh Post-Gazette* 3 February 1983.

Allen, E. C. "Advertisement." *Popular Educator* Vol. VIII No. 10 (June, 1891): 326.

Allen, Frederick L. *The Lords of Creation.* New York: Harper, 1935.

*Amalgamated Journal.* (w.) Thurs. Pittsburgh, Pennsylvania.: Amalgamated Association of Iron, Steel and Tin Workers of North America. 1899.

*American Farmer, The.* Cited at *The Local News* 29 October 1892: 2.

*American Federationist.* (m.) Washington, D. C. 1893.

*American Federation of Labor Weekly News Service.* Washington, D. C

American Iron and Steel Institute. "Report." Washington, D.C.: June 1995.

*American Manufacturer, The. 1* July 1892.

*American Review of Reviews.* 1892-1893.

Andrews, Benjamine E. *History of the United States.* Vol. 5. New York: Charles Scribner's Sons, 1927. 6 vols.

*Anthracite Labor News.* (w.) Saturdays. Shenandoah, Pennsylvania: M .E. Doyle, pub. and ed. 1904.

Ashley, Percy. *Modern Tariff History.* New York: E. P. Dutton, 1905.

*Associated Press Report.* Homestead, Pennsylvania. Postal Telegraph, 6 July 1892.

Austin, A. *The Labor Story*. New York: Coward-McCann, 1949.
Baker, Ray Stannard. *The New Industrial Unrest: Reasons and Remedies* Garden City, New York: Doubleday, Page and Company, 1920.
Barnard, John. *Walter Reuther and the Rise of the Auto Workers*. Boston: Little Brown & Co., 1983.
Barr, Amelia E. "The Servant Girl's Point of View." *The North American Review* June, 1892.
Barry, Dan. "Preserving a Buried Legacy." *The New York Times* 6 March 1999: A10.
Beard, Charles A. and Mary R. *The Rise of American Civilization*. New York: Macmillan, 1934.
*Beaver Valley Labor News* (w.) Thurs. Beaver, Pennsylvania: Official organ Central Labor Council. 1915.
Beer, Thomas. *The Mauve-Decade*. New York: Garden City, 1926.
Bemis, Edward. "The Homestead Strike." *Journal of Political Economy* 1893-1894.
Bent, Quincy (vice president of Bethlehem Steel Company). "*75 Years of Steel.*" *A Newcomen Address*. Princeton: Princeton University Press, 1939.
Berglund, A. "The U.S. Steel Corp." *Columbia University Studies* Vol. XXVII: 223-400.
Berkman, Alexander. *Prison Memoirs of an Anarchist*. New York: Mother Earth Press, 19l2.
Bernard, John. *Walter Reuther and the Rise of the Auto Workers*. Boston: Little Brown, 1983.
Bimba, Anthony. *History of the American Working Class*. New York: International Publishers, 1929.
*Bloomberg News*. Charlotte, N.C. 3 December 1998.
______. *Bloomberg Financial Markets*. Charted by *The New York Times* 5 December 2001.
Bohn, Earl. *Associated Press* 26 December 1983.
Boorstein, Edward. *Allende's Chile: An Inside View*. New York: International Publishers, 1977.
Boyer, Richard O., and Herbert M. Morais,. *Labor's Untold Story*. New York: Cameron & Kahn, 1955.
Braverman, Harry. *Labor and Monopoly Capital: The Degradation of Work in the Twentieth Century*. New York: Monthly Review Press, 1974.
Brecher, Jeremy. *Strike!* San Francisco: Straight Arrow Books, 1972.
Brechin, Gray. *Imperial San Francisco: Urban Power, Earthly Ruin*. Berkeley: University of Califoirnia Press, 1999.
Bridge, James H. *The Inside Story of the Carnegie Steel Company*. New York: Aldine Press,1903.
Brody,David. *Steelworkers in America*. Cambridge, Mass.: Harvard University Press, 1960.
Brooks, Robert R. *As Steel Goes: Unionism in Basic Industry*. New Haven: Yale University Press, 1940.
Burgoyne, A. G. *A Complete History of the Struggle of July, 1892, Between the*

*Carnegie Steel Company, Limited, and the Amalgamated Association of Iron and Steel Workers*. Pittsburgh: Self-Published, 1893.

Byington, Margaret F. *Homestead: The Households of a Mill Town*. New York: Charities Publication Committee, The Pittsburgh Survey, Russell Sage Foundation, 1910.

Carnegie, Andrew. *Autobiography*. New York: Houghton Mifflin, 1920.

______. "An Employer's View of the Labor Question." *Foru.m* Vol. I: 114-125.

______. "Cable to H. C. Frick." Sunningdale, Scotland, 8 July 1892.

______. "The Gospel of Wealth." *North American Review* June 1889.

______. "Results of the Labor Struggle." *Forum* Vol. I: 538-551.

*Chicago Tribune, The*. Chicago, Illinois. 1892-1893.

Childs v. Pennsylvania. *Pittsburgh Legal Journal* No. 50 (July 13, 1892): 466-468.

*Chronical Telegraph, The*. Pittsburgh, Pennsylvania. 1891-1893.

*Cleveland Leader, The*. Cleveland, Ohio. 1892-1893.

*Cleveland Leader, The*. Cleveland, Pennsylvania. 1891-1893.

*Cleveland Press, The*. Cleveland, Ohio. 1892-1893.

Cockburn, Alexander and James Ridgeway. *The Village Voice*. New York, N.Y. 8 March 1983.: 1, 3.

Commanger, Henry Steele. *Documents of American History*. 7th ed. New York: Meredeth Publishing Company, 1963.

*Commercial Gazette*. Pittsburgh, Pennsylvania. 1892-1893.

Commons, J. R., and Associates. *History of Labor in the United States*. Vol. II. New York: Macmillan, 1926. 17+ vols.

*Courier-Herald*. (w.) Thurs.. Wilkes-Barre, Pennsylvania. 1899. H. W. and C.F. Yetter, pub. and ed. Western Newspaper Union.

Cronon, William. *Nature's Metropolis: Chicago and the Great West*. New York: W. W. Norton, 1991.

Cushman, John H. "Program to Clean Toxic Waste Sites is Left in Turmoil." *The New York Times* 15 January 1996.

*Daily Dispatch, The*. Pittsburgh, Pennsylvania.1891-1893.

*Daily Gazette, The*. Pittsburgh, Pennsylvania. 1892-1893.

*Daily Record, The*. Baltimore, Maryland. 1892.

David, Henry. "Upheaval at Homestead." *America in Crisis*. Ed. Daniel Aaron. New York: Knopf, 1952.

Darwin, Charles. *The Origin of Species*. New York: P. F. Collier & Son, 1909.

De Lancey, William J. *Congressional Lobbying "Memo."* Washington, D.C.: American Iron and Steel Institute, June 1980.

Del Rio, Daniel. *Simon Bolivar* . Clinton, Mass.: The Colonial Press Inc., 1965.

*Democrat, The*. Niagra, New York. 1892.

*Despatch*. Pittsburgh, Pennsylvania. 1892.

Diamond, Jared. *Guns, Germs, and Steel*. New York: W. W. Norton Co., 1999.

*Dictionary of American Biography*. 1936.

Douglas, Paul H. *Real Wages in the U.S., 1890-1926*. New York: Houghton Mifflin, 1930.

Drestler, I. M. "Protectionism and Election Year Politics: Why Reagan Is A Free-Trade Villain." *The New York Times* 18 March 1984.

Drinnon, Richard. *Rebel in Paradise*. Chicago: University of Chicago Press, 196l.

Durant, John and Alice. *Pictoral History of American Presidents*. New York: A. S. Barnes and Company, 1955.

Edwards, Richard, Michael Reich and David M. Gordon, eds. *Labor Market Segmentation*. Lexington, Mass.: D. C. Heath, 1975. Based on papers presented at the Conference on Labor Market Segmentation, Harvard University, 1973.

______. Contested Terrain: The Transformation of the Workplace in the Twentieth Century. New York: Basic Books, 1979.

Eldridge, Bill. Letter. The New York Times 26 February 1999.

Emerson. Edwin, Jr. *A History of the Nineteenth Century*. 3 Vols. New York: P. F. Collier & Son, 1900.

Engels, Frederick. *Socialism: Utopian and Scientific*. New York: International Publishers, 1945.

Erwin, W.W. Speech of November 22, 1892. *A Complete History of the Struggle of July, 1892, Between the Carnegie Steel Company, Limited, and the Amalgamated Association of Iron and Steel Workers*. By A. G. Buurgoyne. Pittsburgh, Pa.: Self-Published, 1893. iii-iv.

Farnsworth. Clyde H. "Pacts to Ban Subsidy of Steel Set." *The New York Times* 13 December 1989.

Feder, Barnaby J. "Securitas of Sweden to Acquire Pinkerton's in $384 Millon Deal." *The New York Times* 23 February 1999.

"Firm Insists Its Steel Was Made in America." *San Francisco Examiner* 6 March 1984.

Fisher, Douglas A. *The Epic of Steel*. New York: Harper Row, 1963.

Fitch, J. A. *The Steel Workers*. New York: Charities Publication Committee, The Pittsburgh Survey, Russell Sage Foundation, 1911.

Foner, Philip S. *History of the Labor Movement in the U.S.* New York: International Publishers, 1955.

*Forum* . September 1892.

Foster, William Z. *Organizing Methods in the Steel Industry*. New York: Workers Library Publishers, 1936.

Franks, Lucinda. "Where Carnegies Rusticated." *The New York Times* 1 May 1983: Sec. XX, 27.

*Freiheits Freund*. Pittsburgh, Pennsylvania. 1892.

Friedman, M. *The Pinkerton-Labor Spy*. New York: Wilshire, 1907.

Friedman, Thomas L. *The Lexus and the Olive Tree*. New York: Farrar, Straus & Giroux, 1999.

Gabriel, Ralph. *Course of American Democratic Thought*. New York: Ronald, 1940.

Garland, Hamlin. "Homestead and Its Perilous Trades." *McClure's Magazine* June 1894.

Garland, M. M. *Letter of February 17, 1893 to Samuel Gompers*. National and International Union Correspondence: AA File, US, MSS, 117A, IIA, Box 44. Madison, Wisconsin: State Historical Society of Wisconsin.

Gibbon, Edward. *Rise and Fall of the Roman Empire*. New York: Wise & Co., 1943.

Ginger, Ray. *The Bending Cross*. New Brunswick, N..J..: Rutgers University Press, 1949.

*Globe, The*. Toronto, Canada. 1892-1893.

Goldberg, Harvey, ed.. *American Radicals*, New York: Monthly Review, 1957.

Goldman, Emma. *Living My Life*. New York: Knopf, 1931; Salt Lake City, Utah: Peregrine Smith, 1982.

Goldman, Eric. *Rendezvous with Destiny*. New York: Knopf. 1958.

Gompers File. "General Correspondence." Copy of July 15, 1892, *Letter to Gompers from H. D. Lloyd*: US. MSS. 117A, 11A, Box 1. Madison, Wisconsin: State Historical Society of Wisconsin.

Gompers, Samuel. *Seventy Years of Life and Labor: An Autobiography*. New York: E.P. Dutton, 1925, 1953.

Goodale, Stephen J., comp., Speer, J. Ramsey, ed. *Chronology of Steel*. Pittsburgh, Pa.: Pittsburgh Iron and Steel Foundaries Co., 1920.

Gorz, André, *Strategy for Labor: A Radical Proposal* . Boston: Beacon Press, 1967.

Gunther, John. *Inside U.S.A.* New York: Harper & Brothers, 1947.

Graham, Laurie. *Singing the City: The Bonds of Home in an Industrial Landscape*. Pittsburgh: University of Pittsburgh Press, 1999.

Hacker, Louis M., and Kendrick, Benjamine B. *The United States Since 1865*. New York: F. S. Crofts, 1932.

Hadfield, Robert A. *The Work and Position of the Metallurgical Chemist, Also References to Sheffield and Its Metallurgy*. An Address delivered before the Sheffield Association of Metallurgists and Metallurgist Chemists. Sheffield: The University, Mappin Hall, October 24, 1921.

______. *Metalurgy and Its Influences on Modern Progress With a Survey of Education and Research*. London: Chapman & Hall, 1925.

Harbison-Walker Refractories Co. *Catalogue*. Pittsburgh, Pa.: Harbison-Walker Refractories Co., 1908.

*Harper's Weekly*. 1891-1893.

*Harvest Quarterly* . Santa Barbara, CA. No. 1 (1975) ; No. 2 (June-September 1976).

Harvey, George. *Henry Clay Frick*. New York: Scribner's, 1928.

Hays, Samuel P. *The Response to Industrialism: 1885-1914*. Chicago: University of Chicago Press, 1957.

Heilbroner, Robert L. "Epitaph for the Steel Master." *American Heritage* August 1960.

Hendrick, Burton J. *Life of Andrew Carnegie*. New York: Doubleday Doran, 1932.

*Herald, The*. Baltimore, Maryland. 1892.

*Herald, The*. New York, N.Y. 1892.

Hicks, Jonathan P. "A Faster Path to Finished Steel." *The New York Times* 7 April 1991.

______. "Innovative Steel Mill Is Off To a Slow Start." *The New York Times* 2 December 1989.

Hoerr, John P. *And the Wolf Finally Came: The Decline of the American Steel Industry*. Pennsyluania: University of Pittsburgh Press, 1989.

Hogan, William T. *Economic History of the Iron and Steel Industry in the United States*. 5 Vols. Lexington, Mass.: D, C, Heath, 1971.

______. *Global Steel in the 1990s: Growth or Decline*. Lexington, Mass.: Lexington Books, 1991.

______. *Minimills and Integrated Mills: A Comparison of Steelmaking in the United States*. Lexington, Mass.: Lexington Books, 1987.

______. *Steel in the 21st Century: Competition Forges a New World Order*. New York: Lexington Books, 1994.

______. *Steel in the United States: Restructuring to Compete*. Lexington, Mass.: Lexington Books, 1984.

Holbrook-Jones, Mike. *Supremacy and Subordination of Labor: The Hierarchies of Work in the Early Labor Movement*. (A Study of Work in Victorian Britain). Exeter, N. H.: Heinemann Educational Books, 1983.

Holbrook, Stewart H. *The Age of the Moguls*. New York: Doubleday, 1953.

*Homestead Daily Messenger, The*. Homestead, Pennsylvania. Irreg. 1892-1893.

Hunter, Robert. *Violence in the Labor Movement*. New York: Macmillan, 1914.

*Independent, The*. (w.) Harrisburg, Pennsylvania: Federated Press. 1922. Endorsed by unions of vicinity.

*Industrial Advocate*. (w.) Pittston, Pennsylvania. Central Labor Union.

*Industrial World, The*. Cardiff Wales. 1892.

*International Labor News Service*. (irreg.) Washington, D. C.: International Labor Press of America.

*Iron City Trades Council*. Pittsburgh, Pennsylvania.

"Iron Industry, The." Editorial. *The San Francisco Chronicle* 1 July 1892: 6.

Janofsky, Michael. "Pittsburgh at Progress Crossroads." *The New York Times* 21 September 1997.

Jeans, J. S., ed. *American Industrial Conditions and Competition*. London: Report of the British Iron Trade Commission, 1902.

Josephson, Matthew. *The Robber Barons*. New York: Harcourt, 1934.

*Journal of the Knights of Labor*. 1892-1893.

Kapner, Suzanne. "European Steel Makers Say a Merger in the U.S. is Overdue," *The New York Times* 6 December 2001.

Kelland, Clarence Budington. *Dynasty*. New York: Harper & Bros., 1929.

Kennedy, David M. *Freedom From Fear: The American People in Depression and War, 1929-1945*. New York: Oxford University Press, 1999.

Krooth, Richard. *Arms & Empire: Imperial Patterns Before World War II*. Santa Barbara: Harvest Publishers, 1981.

______. *Empire: A Bicentennial Appraisal*. Santa Barbara: Harvest Publishers, 1975.

______. "The 48'ers." *The Wisconsin Patriot*. Madison, Wisconsin. Vol. I No. 2 (January 1972): 3-4.

______. Au. ed. *The Great Social Struggle.*. 3 vols. Santa Barbara: Harvest Publishers, 1978, 1979, 1980.

*Labor Age.* (m.) Harrisburg, Pennsylvania: Labor Publication Society, Inc. 1921. Cooperatively owned by a group of national, state and local unions.

*Labor Free Press.* (w.) Thurs. Pittsburgh, Pennsylvania. 1922. Endorsed by State Federation of Labor, Pittsburgh Central Labor Union, etc. International Labor News.

*Labor Journal* . (w.) Fridays. Monongahela, Pennsylvania. 1921. Endorsed by trades councils of district.

*Labor Leader*. Lancaster, Pennsylvania.

*Labor News.* (w.). Fridays. Wilkes Barre, Pennsylvania: Labor News Publishing Co. 1924. Official organ, Central Labor Union. King Features, International Labor News, Federated Press.

*Labor World, The.* (w.) Thurs. Pittsburgh, Pennsylvania.:The Journal Publishing Co. 1892.

Lee, Jenifer. "An Echo of Big Dreams of a World With Big Steel," *The New York Times* 5 December 2001.

Le Garde, Ellen. "Concerning Women." *Popular Educator* Vol. VII, No. 10 (June, 1891): 324.

Lenz v. Carnegie Bros. & Co., Limited. *Pittsburgh Legal Journal* Number 42 (May 18,1892): 383-385.

Leyson, Burr W. *Careers in the Steel Industry*. New York: E. P. Dutton & Co., 1945.

Livesay, Harold C. *Andrew Carnegie and the Rise of Big Business*. Ed. Oscar Handlin. Boston: Little Brown & Co., 1975.

*Local News, The.* Homestead, Pennsylvania. 1891-1893.

Lohr, Steve. "Gates to Aid Libraries, in Footsteps of Carnegie." *The New York Times,* 24 June 1997.

*Loyalty Press*. Lancaster, Pa.: Published by R.F.D. 2.

Lundberg, Ferdinand. *America's 60 Families.* New York: Vanguard Press, 1937.

Luxemburg, Rosa. *The Accumulation of Capital.* Trans. Agnes Schwarzschild. New York: Monthly Review Press, 1964.

Madison, Charles A. *Critics and Crusaders.* New York: Henry Holt, 1947.

*Mail and Express, The.* New York, N.Y. 1892.

Manchester, William. *The Arms of Krupp, 1587-1968.* New York: Bantam Book, 1968.

Mario, L. "The Winner in the End." *C. S. Green's Monthly Musical Journal.* January, 1895: 3-5. .Fall River, Mass.

May, Henry F. *Protestant Churches and Industrial America.* New York: Self-Published, 1949.

McCandless v. Allegheney Bessemer Steel Company. Court of Common Pleas No. 1. *Pittsburgh Legal Journal* No. 32 (March 9, 1892): 299-302.

McCandless v. Allegheney Bessemer Steel Company. Supreme Court of Pennsylvania. *Pittsburgh Legal Journal,* N. S. Vol XXIII, O. S. Vol. XI

No. 25 (January 18, 1893): 223-228.
*Metal Trades Department Bulletin.* (m.) Washington, D. C.: American Federation of Labor. 1919.
Metcalf, R.C. "A Few Problems in Education." *Popular Educator* Vol. VII No. 10 (June, 1891): 309.
*Milwaukee Daily News. The.* Milwaukee, Wisconsin. 1891-1893.
Morison, Samuel E., and Henry S. Commanger, *The Growth of the American Republic.* Volume II. New York: Oxford University Press, 1950. 2 + vols.
Myers, Gustavus. *History of the Great American Fortunes.* New York: Modern Library, I936.
*Nation.* July 14, July 21, July 28, August 11, 1892.
*National Labor Journal* (w.) Pittsburgh, Pennsylvania: Union Labor Temple.
*National Labor Tribune* (w.) Thursdays. Pittsburgh, Pennsylvania: National Labor Tribune, Inc.
Naj, Amal Kumar. "Steel Firms Strive for New Technology." *The Wall Street Journal* 21 February 1989.
Nearing, Scott and Joseph. Freeman, *Dollar Diplomacy.* New York: Monthly Review Press, 1966.
*New York Times*, The. New York, N.Y. 1891-1893, 1994-2002.
*New York Tribune*, The. New York, N.Y. 1891-1893.
Norris, Floyd. "Starting a Trade War When Times Are Good." Editorial Observer. *The New York Times* 22 March 1999.
______. "Protecting Steel: An Economic War That All Could Lose." *The New York Times* 21 December 2001.
Norris, James D. *Frontier Iron: The Maramec Iron Works, 1826-1876.* Madison, Wisconsin: The State Historical Society of Wisconsin, 1964.
*North American Review.* September 1892.
O'Connor, Harvey. *Steel-Dictator.* New York: The John Day Company, 1935.
Orth, S. P. *The Armies of Labor.* New Haven, Conn.: Yale University Press, 1921.
Ostrogorsky, M. *Politics in Nineteenth Century America.* 2 vols. Santa Barbara: University of California, Santa Barbara: Special Collections, bound, undated manuscript.
*Panther Creek News.* (d.) Coaldale, Pennsylvania: District 7, United Mine Workers of America. 1920. Associated Press.
Parker, Charles M. *Steel in Action.* Lancaster, Pa.: The Jaques Cattell Press, 1943.
Parr, J.Gordon. *Man, Metals and Modern Magic.* Cleveland, Ohio: American Society for Metals; and Ames, Iowa: Iowa State College Press, 1958.
PBS. The American Experience. "The Richest Man in the World: Andrew Carnegie." 20 January 1997.
PBS. "World War II Series." 1999-2002.
*Pennsylvania Labor Herald.* (w.) Saturdays. Allentown, Pennsylvania: Labor Herald Publishing Co. 1913. Official organ: Central Trade and Labor Council. Owned by organized labor movement of Lehigh Valley.
Perlman, Selig. *History of Trade Unionism in the U.S.* New York: Augustus

McKelley, 1950.
*Philadephia Press, The*. Philadephia, Pennsylvania. 1891-1893.
*Pittsburgh Revealed: Photographs Since 1850.* Carnegie Museum of Art, Pittsburgh, Pennsylania. October 1998-January 1999.
*Pittsburgh Times, The*. Pittsburgh, Pennsylvania. 1891-1893.
*Post, The.* Pittsburgh, Pennsylvania. 1891-1893.
Powderly, T. V. *Thirty Years of Labor, 1859 to 1889.* Columbus, Ohio: Excelsior Publishing House, 1889.
*Press, The.* Cleveland, Ohio. 1892.
*Press, The*. Pittsburgh, Pennsylvania. 1891-1893.
"Primary Manual Work." *Popular Educator* Vol. VIII No. 10 (June, 1891): 297.
*Progressive Labor World.* (w.) Philadelphia, Pennsylvania.
*Public Opinion*. Lehigh Valley.
Reutter, Mark. Sparrows Point: Making Steel - the Rise and Ruin of American Industrial Might. New York: Summit Books, 1989.
Reynolds, Robert L. "The Works Are Not Worth One Drop of Human Blood." *American Heritage* August 1960.
Rickard, T. A. *Iron in Antiquity*. Reprinted from *The Journal of the Iron and Steel Institute* Vol. CXX No. 2 (1929). London: Iron and Steel Institute, 1930.
Ricklefs, Robert. "France Sends Workers Into Early Retirement, But Problems Persist." *The Wall Street Journal* 13 February 1984.
Robb v. Carnegie Bros. & Co., Limited. *Pittsburgh Legal Journal* Number 13 (October 28, 1891): 108-109.
Robinson, Jesse S. *The Amalgamated Association of Iron, Steel and Tin Workers*. Baltimore: Johns Hopkins Press,1920.
Roe, Wellington. *Juggernaut.* New York: Lippincott, 1948.
Rowan, Richard W. *The Pinkertons*. Boston: Little, Brown, 1931.
Safire, William. "Essay: 'We Never Sleep'." Op-Ed. *The New York Times* 31 May 1999.
Samuelson, Robert J. "The Economics of Self-Pity." *Newsweek* 5 March 1984: 78.
*San Francisco Chronicle, The.* San Francisco, California. 1892-1893, 1995-2002.
Sanger, Martha Frick Symington. *Henry Clay Frick: An Intimate Portrait.* New York: Abbeville Press. 1997.
Shakespeare, William. *Julius Caesar* IV, iii,: 217.
Sims, Calvin. "G. A. O. Says Contract Unfairly Favored I.B.M." *The New York Times* 23 June 1989.
Sinclair, Upton. *The Goose Step*. Pasadena, California: Self-Published, 1923.
*Slovensky Obcan* (*Slovak Citizen*). (w.) Fridays. Hazelton, Pennsylvania: Citizen Publishing Co. 1912. Endorsed by Slovak working men's societies.
Smith, Adam. *An Inquiry into the Nature and Causes of the Wealth of Nations.* Book IV. New York: Modern Library, 1937.
Smith, A. D., D.M.W.N., Hitchens, and S.W. Davies. *International Industrial Productivity: A Comparison of Britain, America and Germany.* Cambridge

University Press, 1983.
Smith, E. J. *The New Trades Combination Movement*. London: Rivingtons, 1899.
Smith, J. Russell. *The Story of Iron and Steel*. New York: D. Appleton & Co., 1908.
Social Economic Union of Cuba, The. *Commercial Relations Between Cuba and the United States of America*. Havana, Cuba:1936.
Spencer, Herbert. *First Principles*. London: British Museum manuscript edition, 1861.
______. *Principles of Biology*. London: British Museum manuscript edition, 1864-1867.
______. *Social Statics*. London: British Museum anuscript edition, 1851.
Stanwood, Edward. *American Tariff Controversies in theNineteenth Century*. Vol. II. Boston: Houghton, Mifflin & Co., 1903. 2 vols.
Steinmetz, Charles P. *America and the New Epoch*. New York and London: Harper & Brothers, 1916. The author founded General Electric.
Steuben, John. *Strike Strategy* . New York: Gaer Associastes, Inc., 1950.
Stevenson, Richard W. "Greenspan Denounces Growing Protectionism." *The New York Times* 17 April 1999.
______."Big Steel: An Invalid that Can Roar in Washington." *The New York Times* 17 December 2001.
Stone, Joseph K. Letter. *The New York Times* 18 December 1989.
Stowell, M. R. *Fort Frick*. Pittsburgh: Self-Published, 1893.
Strouse, Jean. *Morgan: American Financier*. New York: Random House, 1999.
Sullivan, Mark. *Our Times*. Vol. I. New York: Scribner's,1926. 2 vols.
*Sun, The*. Baltimore, Maryland. 1892-1893.
*Sun, The*. New York, New York. 1892-1893.
Taft, Philip. *Oganized Labor in American History*. New York: Harper & Row, 1970.
"Talking Business With Uanous of Planecon." The New York Times 26 December 1989.
Taussig, F. W. *Tariff History of the United States*. 4th ed. New York: Putnam, 1888, 1898.
Temin, Peter. *Iron and Steel in 19th Century America: An Economic Inquiry*. Cambridge, Mass.: M.I.T. Press, 1964.
*Times, The*. Philadelphia, Pennsylvania. 1892-1893.
*Times, The*. Pittsburgh, Pennsylvania. 1892-1893.
*Trades Union News*. (w.) Thurs. Philadelphia, Pennsylvania: Trades Union News Publishing Co. 1902. Endorsed by local unions.
*Transit Journal, The*. Pittsburgh, Pennsylvania. 1892.
Tugendhat, Christopher. *The Multinationals*. New York: Random House, 1972.
Tunney v. Carnegie Bros. & Co., Limited. *Pittsburgh Legal Journal* No. 24 (13 January 1892): 219-221.
*Union News*. Allentown, Pa.
"U.S. Claims Six Nations Dump Carbon Steel Goods." *The Wall Street Journal* 31 May 1985.

U.S. Congress, House of Representatives, 52nd Congress, 2d Session. *Report 2447: Employment of Pinkerton Detectives.* Washington, D.C.: Government Printing Office.

______. *Pollution Prevention Act of 1990* . U.S.C.A. par. 13101 (b).

______. Senate, 52nd Congress, 2d Session. *Report 1280: Investigation of Labor Troubles.* Washington, D.C.: Government Printing Office.

______. Senate, 61st Congress, 2d. Session. National Monetary Commission. *Document No. 578 : Statistics for Great Britain, Germany and France, 1867-1910.* Washington, D.C.: Government Printing Office, 1910.

Veblen, Thornstein. *Absentee Ownership and Business Enterprise in Recent Times; the Case of America.* New York: B. W. Huebsch, 1923.

Waitzkin, Howard. "A Marxist View of Medical Care." *Annals of Internal Medicine* Vol. 89 No. 2 (August, 1978).

*Wall Street Journal, The.* New York, N.Y. 1892-2002.

*War Supplement* to *Compton's Pictured Encyclopedia.* Chicago: F..E. Compton & Co., 1939.

Wayne, Leslie. "A Divisive Steel-Quota Debate Moves to the Senate." *The New York Times* 21 March 1999.

______. "A Long-Shot House Vote to Limit Steel Imports." *The New York Times* 19 March 1999.

______. "Big Steel's Problems Are Home Grown." *The New York Times* 29 April 1999.

______. "Parched, Big Steel Goes to Its Washington Well." *The New York Times* 20 January 2002.

"Rivals and Others Criticize Steel Makers' Bid for U. S. Aid." *The New York Times* 6 December 2001.

Wertime, Theodore A. *The Coming of the Age of Steel.* Netherlands: The University of Chicago Press, 1962.

Williams. William Appleman. *The Tradedy of American Diplomacy.* New York: The World Publishing Company, 1959.

Wilson. Edmund. "American Heroes: Fremont and Frick." *The Shores of Light.* New York: Farrar, Straus & Young, 1952.

Winkler, J. K. *Incredible Carnegie.* New York: Vanguard Press, 1931.

Winston, Diane. *Red-Hot and Righteous. The Urban Religion of the Salvation Army.* Cambridge, Mass.: Harvard University Press, 1999.

*Wisconsin Banner und Volksfreund.* Milwaukee, Wisconsin. 1892.

*Wisconsin History Calendar.* Madison, Wisconsin. 1974. Graphic, May 1886.

Wolff, Leon. *Lockout: The Story of the Homestead Strike of 1892: A Study of Violence, Unionism, and the Carnegie Steel Empire.* New York: Harper & Row, 1965.

*Working World* . (w.) Uniontown, Pa.

*World, The.* Baltimore, Maryland. 1892.

*World, The.* New York, New York. 1891-1893.

Wright, Carroll D."The Amalgamated Association of Iron and Steel Workers." *Quarterly Journal of Economics* July 1893.

Yellen, Samuel. *American Labor Struggles.* New York: Harcourt, Brace, 1936.

*York Labor Advocate.* (w) York, Pennsylvania: Federation Trades Union. 1917.

## INDEX

**D.**

**H.**

**O.**

**R.**

**Y.**

# About the Author

Richard Krooth was educated at De Paul University in business and economics and went on to practice public accounting, receive a Juris Doctor degree at the University of Wisconsin and practice law.

He was a graduate student in economics at Atlanta University, and later entered the Sociology Department of the University of Wisconsin as a special student of Hans H. Gerth. He earned his doctorate in sociology at the University of California, Santa Barbara.

He has taught criminology, industrial and labor studies, and community and welfare sociology at the University of California in Santa Barbara and Riverside. His credits include teaching in the Criminal Justice Department at California State University at Sonoma; as well as lecturing on economics, political economy and public services at the Madison Area Technical College, California State University at San Francisco, Golden Gate University, and the California Institute of Management (affiliate of the *Institut supérieur de gestion*, Paris, France).

Periodically over the last decade he has also been a Visiting Scholar in Sociology at the University of California, Berkeley. Retired from a teaching career, he is currently a political, economic, and jury consultant.

Mr. Krooth has been the U.S. correspondent for the *Economic and Political Weekly* (Bombay), former editor of *Harvest Quarterly*, and editor of *The Great Social Struggle*, 3 vols. (1978-80). He is author of

*Empire: A Bicentennial Appraisal* (1975); *Arms & Empire: Imperial Patterns Before World War II* (1981); *Mexico, NAFTA and the Hardships of Progress* (1995); *Fate of the Planet: Ecosystems in Distress* (2003); and many professional journal articles. He is a contributor to the Los Angeles Riots edited by Mark Baldassare (1994). And he is co-author of *Common Destiny: Japan and U.S. in Global Age* (1990); *Quest For Freedom: The Transformation of Eastern Europe in the 1990s* (1993); *Race and the Jury: Inequality and Justice* (1993); *The Middle East: A Geopolitical Study* (1995); *Anatomy of the McMartin Child Molestation Case* (2001); and *The Racialized Jury Box: Affirmative Action in Jury Selection and Racially Mixed Juries* (2002).

His oral history with Ann Baxandall of political activists from Madison, Wisconsin, covering the period from the 1930s to the 1970s, is forthcoming. And he is working on a multi-volume study of global ecology.

With his co-authors he received the 1994 Outstanding Book Award for Human Rights in North America from the Gustavus Myers Center for the Study of Bigotry and Human Rights in North America. Listings of his bibliographical background can be found *at Who's Who in America–West*, *Who's Who in the World*, *Outstanding Writers of the 20th Century*, *Intellectuals of the 20$^{th}$ Century*, *Outstanding People of the 20$^{th}$ Century*, *Dictionary of International Biography*, the Gustavus Myers Center, and *Contemporary Authors.*

www.ingramcontent.com/pod-product-compliance
Ingram Content Group UK Ltd.
Pitfield, Milton Keynes, MK11 3LW, UK
UKHW041102150926
475561UK00001BA/9

9 780761 824152